# of Maryland, Delaware, and Washington, D.C.

John Means

Illustrated by
Matthew Moran and Suzannah Moran

2010
MOUNTAIN PRESS PUBLISHING COMPANY
Missoula, Montana

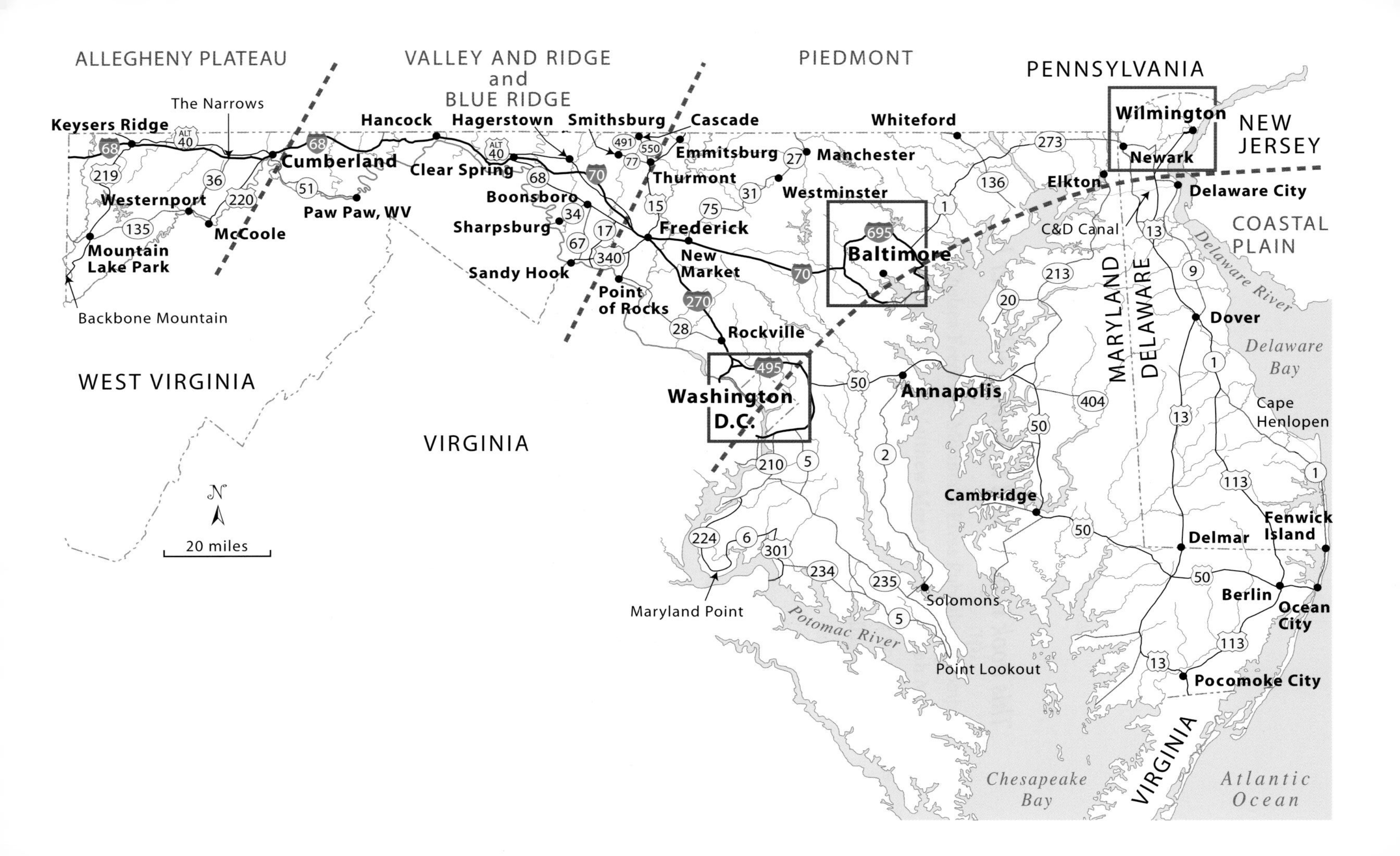
ALLEGHENY PLATEAU
VALLEY AND RIDGE and BLUE RIDGE
PIEDMONT
COASTAL PLAIN
PENNSYLVANIA
NEW JERSEY
MARYLAND
DELAWARE
WEST VIRGINIA
VIRGINIA
Keysers Ridge
The Narrows
Cumberland
Hancock
Hagerstown
Smithsburg
Cascade
Emmitsburg
Thurmont
Manchester
Westminster
Whiteford
Wilmington
Newark
Elkton
Delaware City
C&D Canal
Delaware River
Dover
Delaware Bay
Cape Henlopen
Westernport
McCoole
Mountain Lake Park
Backbone Mountain
Paw Paw, WV
Clear Spring
Boonsboro
Sharpsburg
Sandy Hook
Frederick
New Market
Point of Rocks
Rockville
Baltimore
Washington D.C.
Annapolis
Cambridge
Fenwick Island
Delmar
Berlin
Ocean City
Pocomoke City
Solomons
Maryland Point
Potomac River
Point Lookout
Chesapeake Bay
Atlantic Ocean
20 miles

• — Contents — •

Acknowledgments

My companions on this roadside journey have been Matt and Zana Moran. They spent many hours doing the painstaking and meticulous work involved in designing the many maps into a computerized format. I cannot thank them enough. Matt also drew some great illustrations. The three of us have worked together on this project from the beginning, seven years ago, and they have been a joy to work with.

If you live in Maryland, Delaware, or the District, and you want to learn about the geology around you, you have a wealth of resources from which to draw. Most of the information in this book comes from the Maryland Geological Survey (MGS), Delaware Geological Survey (DGS), and United States Geological Survey (USGS)—their reports, their maps, and personal responses from their knowledgeable geologists.

I want to express particular gratitude and thanks to the geologists who reviewed drafts of different sections of my manuscript and who helped me greatly with geological interpretations through their detailed comments and their answers to my e-mail questions: Scott Southworth of the USGS, Reston; Jim Reger, now retired from the MGS, Baltimore; and A. Scott Andres and William "Sandy" Schenck of the DGS, Newark.

During my years of teaching geology and during the years of work on this book, I have relied repeatedly on one indispensable book: *Maryland's Geology*, by Martin F. Schmidt, Jr. I have never met Mr. Schmidt, but I want to thank him for his book and recommend it highly to anyone interested in learning more about geology. I have used my copy so much that it is literally falling apart.

I want to thank another person whom I have never met: Jenn Carey, my editor at Mountain Press. Before I was a geology teacher, I was an English teacher. These qualifications I cite to give weight to my evaluation of her as a superb editor. Besides helping me at literally hundreds of stages with hundreds of answers to hundreds of questions, she has continuously encouraged me. Otherwise, this project might have weathered away and eroded into the sea.

Also at Mountain Press is Chelsea Feeney. I thank her for making additional figures, manipulating and matching up the computer maps, and implementing the new geologic color scheme.

Finally, I go back in time a couple of decades to express, once again, my heartfelt admiration, gratitude, and respect for my guide and mentor, Professor William R. Shirk, now retired, from Shippensburg University, Pennsylvania. His *A Guide to the Geology of Southcentral Pennsylvania*, his teaching, his field trips, and his personal guidance and help on my first geological guide book were the beginning of an exciting new life for me. To me Professor Shirk will always be "The Rock Man."

I want to thank Bill and Richard for never failing to ask how things were going with the book. I thank Michael for convincing me, a couple of decades ago, not to "burn my bridges." I thank my wife for toughing it out for better or for worse, usually the latter.

My dearest wish is that this book will fall into the hands of teachers and students.

•— Introduction —•

Maryland has been called Little America. In the morning you can surf in the ocean and in the afternoon snowboard in the mountains. Maryland, Delaware, and the District of Columbia as a region contain geologic formations as diverse as Pleistocene Ice Age dunes and billion-year-old rock that was part of the ancient North American tectonic plate, or craton. The region shows the effects of at least four mountain building episodes and each subsequent mountain erosion and basin deposition. The last such cycle—the assembly and rifting of the supercontinent Pangea—left many marks on the landscape and is primarily responsible for the creation of the five geological provinces you can see today. Each has its own distinctive, recognizable terrain: Allegheny Plateau, Valley and Ridge, Blue Ridge (here included with Valley and Ridge), Piedmont, and Coastal Plain. In order to understand the regional geology more fully, we need to understand the history of the earth.

UNDERSTANDING THE EARTH

Our solar system is thought to have formed—or been formed—by the collapse of interstellar material into a rotating disk, with about 90 percent of the material concentrated into an early, central sun. Within this rotating disk were eddies, smaller pockets of rotation where gases, liquids, and solids condensed and coalesced into planets. About 4.6 billion years ago, earth and the other planets formed. In its early days, earth was probably of uniform density and composition throughout. Gradually, earth temperature was elevated by meteorite impacts, gravitational compaction, and radioactive decay. Layers of different composition and density began to form. This stratification was probably the most important process in earth history because it created the thin, hard outer crust that supports life.

Plate Tectonics

Today the earth consists of three general layers: the innermost core, 16 percent of earth volume; the mantle, 83 percent; and the outermost crust, only 1 percent—the hard rock upon which we live. Core temperature is about 8,500 degrees Fahrenheit, a result of heat energy that remains from the creation of the earth. For hundreds of millions of years this heat energy has driven the major crustal interactions that create earthquakes, volcanoes, and mountain ranges.

Age	Period	mya	Geologic Events in Maryland, Delaware, and D.C.
CENOZOIC	Holocene Epoch	.01	Sea rises to near modern levels by 3,000 years ago.
	Pleistocene Epoch Quaternary	1.8	Wisconsinan ice sheet develops about 80,000 years ago; deposits terminal moraine at Long Island about 22,000 years ago then begins receding, dumping meltwater into the Delaware and Susquehanna Rivers. Interglacial periods, with higher sea levels, occur 320,000 and 120,000 years ago.
	Tertiary: Pliocene, Miocene, Oligocene, Eocene, Paleocene	65	Tertiary Upland Gravel, Beaverdam, and other formations deposited. Chesapeake Group deposited 20 to 10 million years ago on growing coastal plain wedge. Meteorite hits Chesapeake Bay 35 million years ago.
MESOZOIC	Cretaceous	145	Potomac and Matawan Group sediments eroded from Appalachians and deposited on edge of continent.
	Jurassic	208	Pangea splits apart and Atlantic Ocean opens about 200 million years ago. Rift basins, including the Gettysburg and Culpeper Basins, open along the margin of the North American continent.
	Triassic	248	
PALEOZOIC	Permian	286	Alleghanian mountain building event, between 320 and 250 million years ago, finalizes assembly of the Pangean supercontinent when Africa collides with North America.
	Pennsylvanian	320	Ancestral Appalachian Mountains form. Swampy inland basins accumulate organic material—future coal.
	Mississippian	355	
	Devonian	417	Acadian mountain building event occurs 420 to 375 million years ago when Avalon microcontinent collides with New England. Catskill Wedge forms west of mountains.
	Silurian	443	
	Ordovician	490	Taconic mountain building event occurs 480 to 440 million years ago as volcanic arc collides with proto–North America. Queenston Wedge deposited to west of eroding mountains.
	Cambrian	545	Setters, Cockeysville, and Loch Raven Formations deposited on Baltimore Gneiss blocks on ocean floor. Limestones of Frederick and Hagerstown Valleys begin forming and continue into early Ordovician time. Volcanic island arc forms in Iapetus Ocean.
PRECAMBRIAN	Late Proterozoic	900	Sediments of Westminster Terrane deposited in ocean. Rifting splits Grenville gneisses; Iapetus Ocean fills gap 640 to 580 million years ago. Catoctin basalts and rhyolites flow onto surface.
	Middle Proterozoic	1,600	Grenvillian mountain building event forms Baltimore Gneiss about 1.3 to 1.0 billion years ago.
	Early Proterozoic	2,500	
	Archean Eon		Earth forms 4.5 billion years ago.

mya=millions of years ago

Geologic time scale with geologic events that shaped the rocks and landscape in Maryland, Delaware, and D.C.

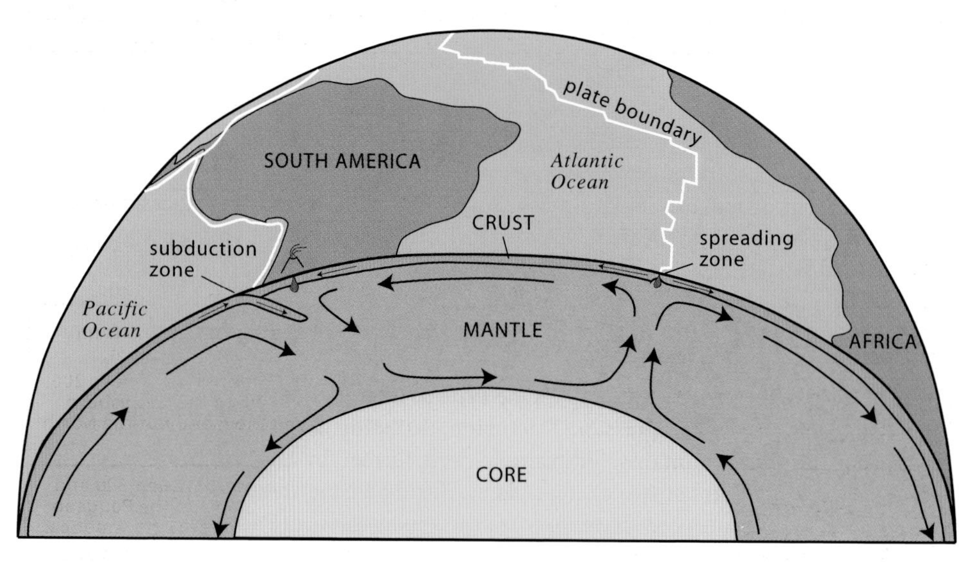

Currents in inner-earth convection cells, driven by heat energy from the core, cause pieces of the crust—the plates—to move. —Modified from Kious and Tilling, 1996

Surrounding the core is the substance that occupies the predominant volume of the earth, the mantle. While the lower mantle is solid, the middle of the mantle is plastic—capable of being bent or molded. Its claylike consistency can support a slow current or flow. This middle section is called the asthenosphere, from Greek *asthenes*, meaning "weak." Because the core is hotter than the layers above it and because heat energy moves from a region of higher temperature to one of lower temperature, slow currents rise from the core through the plastic asthenosphere, but only at the rate of inches per year. These convection currents, as they are called, can be observed in other fluids: the column of heated, rising air above a campfire, candle, or radiator; the swirling currents in a pan of heated water or soup.

On top of the asthenosphere floats the solid lithosphere, about 10 to 120 miles in thickness and composed of the rigid upper mantle and the crust. The lithosphere is broken into several pieces, as the shell of a hard-boiled egg might be. As a crude analogy, the white of a hard-boiled egg could be considered a plastic material similar to the mantle, while the thin, brittle, and broken but still-intact shell resembles the irregularly shaped pieces of the relatively thin lithosphere.

These pieces are known as plates, and they have been moved around for millions of years at rates of inches per year by the underlying, slowly moving, heat-driven convection currents in the plastic asthenosphere upon which they float and ride. This process, known as plate tectonics, while being understood and accepted for only a few decades, helps to explain many major phenomena of geology.

In very early earth history, before plates existed, the planet was without continents. Magma erupted from the hot, underlying mantle onto the surface and solidified into volcanic islands. Over hundreds of millions of years, these early volcanic landmasses were moved around by the heat currents from below, and as they collided and stuck together, they formed small plates that later converged and sutured together to form small continents. By about 2.5 billion years ago (2,500 million years), a substantial amount of continental crust existed, but only about 30 to 40 percent of the present landmass. Today these ancient crustal pieces, known as cratons or as basement rock, usually form the interiors of continents and are generally buried beneath layers of younger rock. In Maryland and Delaware, portions of this very old basement rock have been thrust upward and are exposed.

Interactions at plate boundaries, which exist both on land and on seafloors, have caused major changes throughout earth history in topography and in geographical locations of continents and oceans. One type of interaction, the slow convergence of two plates, causes compression, fracturing, thrusting, and folding of rock layers. The Appalachian Mountains, for example, were formed by a slow inching together over millions of years of two ancient plates, during a time about 300 million years ago when many plates converged to form the supercontinent Pangea.

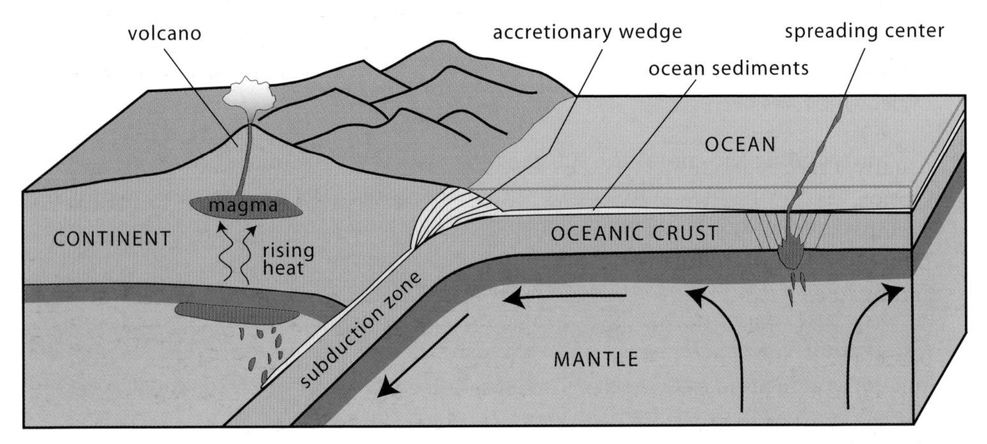

New ocean crust forms at spreading centers and older ocean crust is subducted beneath continental crust at subduction zones. —Artwork by Mountain Press

A second type of interaction, called divergence or rifting, causes rock to fracture and sink downward. The topographic results can be low-elevation valleys on land or massive, broken chunks of crust subsiding below sea level. Divergence over hundreds of miles for millions of years creates an ocean. In Maryland's Frederick Valley, crust fractured and slid downward thousands of feet during the divergence of tectonic plates in the breakup of Pangea, creating an inland basin. Fractures in earth crust created by divergence have also been avenues for magma—molten rock—to ooze upward into subterranean cracks or pour onto the surface as lava flows.

A third type of plate-to-plate interaction is a transform boundary, where one plate slides laterally, or sideways, relative to another. The San Andreas Fault of California is an active example of this earthquake-causing movement. Maryland contains older, now inactive examples.

Rocks

All rocks on earth fall into one of three types: igneous, sedimentary, and metamorphic. Igneous rocks form by the cooling and solidification of hot, viscous, liquidy magma. When magma reaches the surface, it is known as lava, which then cools rapidly and solidifies into rock, such as basalt or rhyolite. Magma can also cool more slowly beneath the surface into igneous rock, such as granite or gabbro.

Sedimentary rocks form when the products of weathering—pebbles, sand, silt, and clay—are removed or eroded from existing rock and transported downslope by water, wind, and ice. These small pieces are called clasts, and over time they accumulate in the lower areas of the earth's crust, most of which are underwater. Most of the eroded sediments are carried by rivers and deposited in thickly layered wedges under the ocean in offshore regions, such as the present-day Atlantic continental shelf, or in inland seas. Hundreds or thousands of feet of sediments accumulate and compact over millions of years. Eventually, eroded minerals that are dissolved in the water cement the sediments together into sedimentary rock. Sandstone is cemented sand particles, and shale is cemented clay. Thus, new rock is made from the eroded pieces of older rock—part of a process known as the rock cycle, which has been going on during most of earth history.

Clastic erosion and deposition are dependent on the existence of a sourceland, usually a mountain range. As mountains wear down after millions of years of erosion, clastic deposition slows, and if the environment of deposition is located in a temperate or tropical climate, carbonate deposition can replace clastic deposition. Limestone, the most common carbonate, forms from the precipitation of calcium carbonate from seawater or from the fragmented remains of marine organisms.

Metamorphic rocks form when existing igneous and sedimentary rock, called parent rocks, are changed by heat and pressure. New minerals may form and may be aligned in a foliation such as the platy mica layers in a schist. Slate, phyllite, schist, and gneiss are metamorphic rocks that can be produced from shale. The parent limestone can be metamorphosed into marble, and granite into gneiss. Convergence of tectonic plates and mountain building episodes often supply the pressure and heat of deep burial necessary for metamorphism to occur.

Geologic Time

The amount of time involved in the formation of the earth is hard to fathom. If Julius Caesar or Jesus lived a few seconds ago, then humans have existed for a few minutes, and the earth has been around for a full year.

Until the post-nuclear twentieth century, geologists had no techniques for establishing the absolute ages of different rocks. With sedimentary rock layers, geologists could determine only relative ages with the law of superposition: in a sequence of undisturbed sedimentary formations, the age progresses sequentially from oldest on the bottom to youngest on the top.

Within stacks of sedimentary layers around the world, geologists found fossilized remains of thousands of extinct plants and animals. According to their locations in the sedimentary continuum, the ages of particular fossils were established relative to one another through detailed study, classification, and comparison. Of all species that have ever existed, 99.9 percent are now extinct, and many life forms existed only during a time when a particular sedimentary layer was being deposited. A geographically widespread fossil that survived for a relatively short period is known as an index or guide fossil, and it can be used to date the formation in which it is found and correlate it with other layers located elsewhere.

The fossil record is analogous to a sequential list of the American presidents, with a fossil's survival interval comparable to a president's term of office. For many decades, however, the fossil record was like the presidential list with no historical dates assigned. Geologists knew oldest to youngest but not how old or young.

As nuclear physics emerged during and after World War II, scientists developed a technique for establishing the absolute age of an igneous rock. Certain isotopes—variant forms—of some elements undergo radioactive decay, whereby the "parent" isotope decays or transforms into the "daughter" isotope. Once rates of decay were determined, a nuclear laboratory could find the ratio of parent-to-daughter and plot how long the decay had been in progress. Because the decay clock starts when magma solidifies into rock, the ratio plot could yield the age of a given igneous rock.

Because most sedimentary sequences contain igneous intrusions, lava flows, or volcanic ashfalls, geologists could now bracket particular sedimentary layers and fossils within the detailed paleontological record with absolute dates. Using the presidential analogy, if we had determined that Lincoln's term included 1863 and Wilson's 1917, we could then establish with certainty that Teddy Roosevelt's term existed between those two dates, but closer to 1917. Before the absolute dates were determined, historians would not have known if Teddy was in the twentieth, nineteenth, or even eighteenth century. That is to say, geologists would not have known if the Catoctin metabasalt was 10,000, 100,000, 100 million, or 1 billion years old (it is about 600 million years old). Detailed cross-referencing and comparisons of fossils and rock layers across different regions enabled geologists, using the absolute dates assigned to igneous rocks, to establish educated estimates of absolute ages for most sedimentary and metamorphic rocks.

GEOLOGIC HISTORY OF MARYLAND, DELAWARE, AND WASHINGTON, D.C.

Ancient Cratons, Grenville Mountains, Rifting, and Offshore Deposition

Over a billion years ago, relatively small cratons increased their sizes as separate landmasses were tectonically transported and accreted to their edges. These convergences resulted in episodes of mountain building that included volcanic activity, thickening and compression of the crust, burial and metamorphism, and folding, fracturing, and thrusting of rock layers. In eastern North America, the ancient Grenville Mountains formed 1.3 to 1.0 billion years ago in an elongate band just east of where the Appalachian Mountains stand today. By 700 million years ago, the mountains had been mostly eroded, and these leveled roots of the Grenvilles later became the underlying basement rock of the eastern United States and the Appalachians. In the Baltimore area, this 1.1-billion-year-old rock is exposed in several domelike structures.

Weathering and erosion of the ancient Grenville Mountains must have been rapid for at least two reasons. First, glaciers are known to cause extensive weathering and erosion, and from about 900 to 600 million years ago, there were four episodes of worldwide glaciation, perhaps the most widespread that has ever occurred. Second, land plants did not yet exist on the planet, leaving the rock totally exposed to weathering agents. The once lofty Grenville Mountains were reduced to low hills.

During very late Proterozoic time, about 640 to 580 million years ago, the eroded rock of the Grenvilles underwent a tectonic pulling apart, or rifting. This extension of the crust caused bedrock to fracture in many places. Magma flowed upward through the cracks from the hot, underlying mantle and poured onto the surface as layer upon layer of lava, reaching thicknesses of up to 2,000 feet. Today these lava flows, the Catoctin rhyolites and basalts, form part of the broad anticline that is the Blue Ridge Province, and they are exposed in Catoctin Mountain in Frederick County, isotopically dated at 600 to 560 million years old. Below them lies the gneiss of the Grenville basement rock.

As separation of the two ancient plates continued, huge blocks or massifs of the Grenville rock, some covered with the lava and some not, slid downward along nearly vertical fractures in a process known as normal faulting. As these blocks were losing elevation, some even being transported away from the mainland, the ocean level was rising. Eventually, somewhere around 560 million years ago, some of the blocks became submerged offshore in the ancient Iapetus Ocean.

As the crustal divergence continued in early Paleozoic time, the Iapetus Ocean widened. Offshore, the Grenville massifs or large chunks lay at different ocean depths and different distances from the continent, separated by ocean-floor basins or troughs.

Offshore along the eastern edge of the continent, a thick wedge of clastic deposits was laid down on top of the submerged, lava-covered Grenville basement rock. One of the first deposits was the hard beach sands that eventually became the highly resistant, ridge-forming quartzites of the Cambrian Weverton Formation, which today tops the crests of South Mountain and Catoctin Mountain of the Blue Ridge Province. Then, as water depths and climates changed, siltstones and sandstones were formed in the offshore wedge.

As the highlands wore down and clastic deposition slowed, carbonate limestones and dolomites formed on this offshore wedge from the chemical precipitation of minerals dissolved in seawater and from the fragmented remains of tropical marine organisms. In the warm, tropical seas to which tectonic movement had taken the continent about 500 to 450 million years ago, both carbonate precipitation and marine life were abundant. The offshore depositional wedge, including the carbonates, stretched from present-day Newfoundland to Alabama and was about 1 mile thick. Today the limestones comprise most of the floor of the Hagerstown

1. LATE PROTEROZOIC TIME (about 650 to 580 million years ago)
As the continent rifted apart in more than one place, blocks of gneiss dropped down along faults. Sediments were deposited on submerged blocks and lava poured out on land.

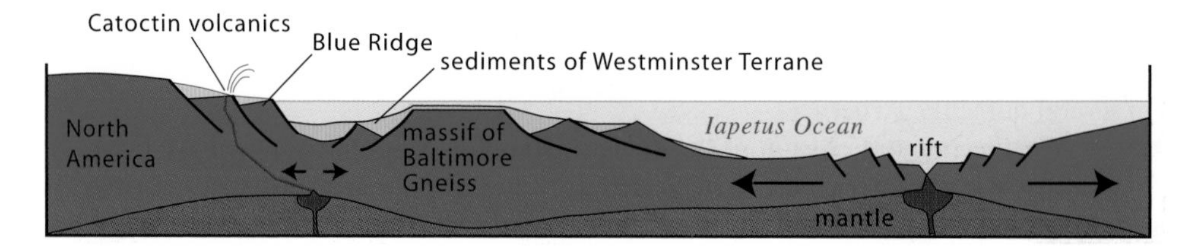

2. MIDDLE CAMBRIAN TO ORDOVICIAN TIME (about 500 to 450 million years ago)
An island arc forms offshore. Sediments are deposited undersea and carbonate banks form in shallow waters.

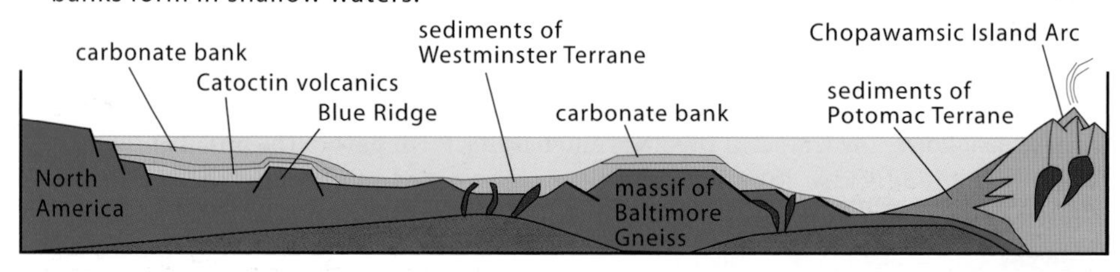

3. MIDDLE ORDOVICIAN AND SILURIAN TIME (about 480 to 420 million years ago)
The Chopawamsic Island Arc collides with North America in the Taconic mountain building event. The huge mountains erode with sediments accumulating to the west in the Queenston Wedge.

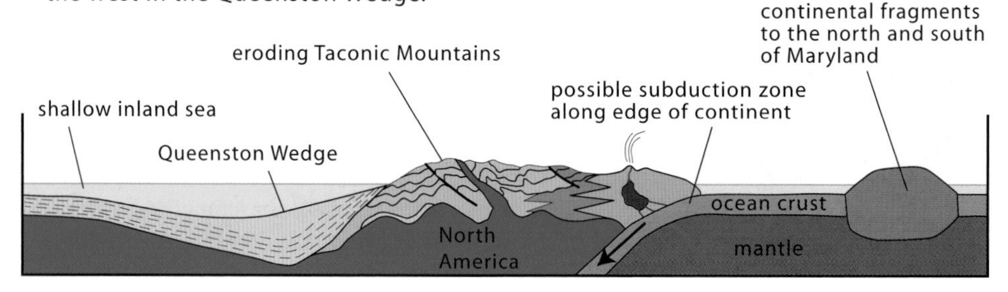

4. DEVONIAN TO MISSISSIPPIAN TIME (about 410 to 350 millian years ago)
Microcontinents collide with North America north and south of Maryland and Delaware in the Acadian mountain building event. Sediments eroded from the mountains collect in a basin to the west, creating the Catskill Wedge.

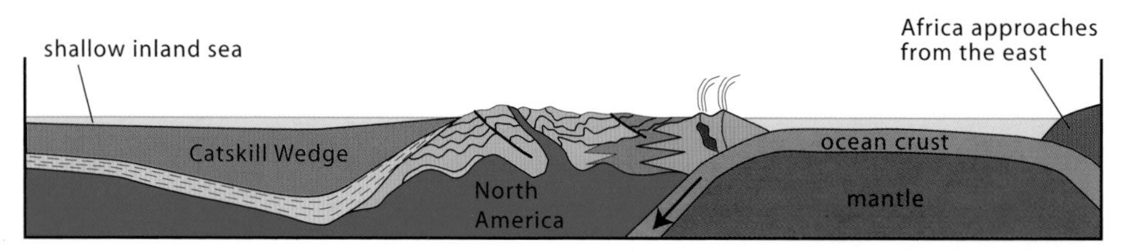

5. PENNSYLVANIAN AND PERMIAN TIME (about 320 to 250 million years ago)
Africa collides with North America in the Alleghanian mountain building event, forming the supercontinent Pangea.

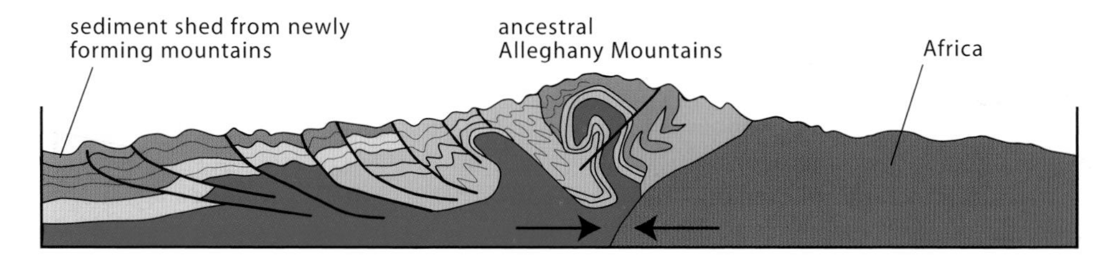

6. LATE TRIASSIC TIME (about 200 million years ago)
Pangea rifts apart, and the Atlantic Ocean fills the growing gap between North America and Africa.

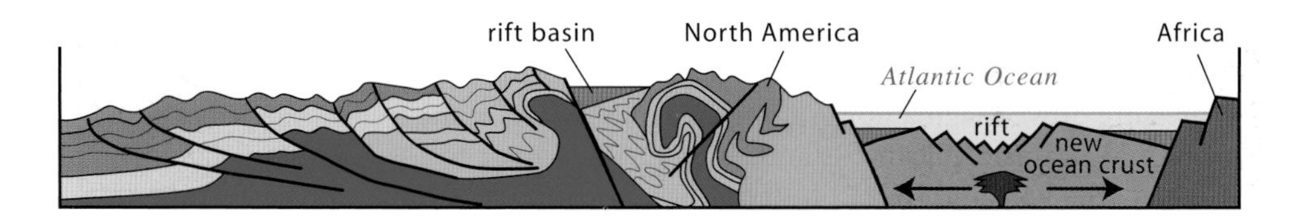

7. CRETACEOUS TO MODERN TIME (about 145 million years ago to present)
The mountains erode and huge amounts of sediment accumulate on the continental shelf, forming the modern coastal plain.

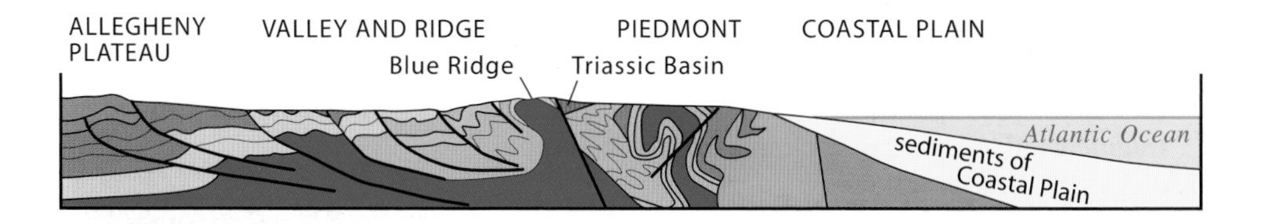

These seven sections show the major events in the geologic history of Maryland, Delaware, and D.C. The drawings are intended to present an overview of the big picture but do not in any way represent the exact positions or shapes of the rocks in times past. —Modified from Gates, Muller, and Valentino, 1991; Kunk and others, 2004

Valley in Maryland, part of the Great Valley that runs from Pennsylvania to Tennessee. Major north-south routes, US 11 and I-81, have been conveniently located in this valley.

Farther out in the stillness of the deep rift basins, fine clay and silt particles fell out of suspension to form thick, muddy sediments. These formed the shales that were later metamorphosed into the schists and phyllites of the Westminster Terrane of the central Piedmont.

Still farther out to sea, large, rifted blocks lay on the ocean floor. On their tops were deposited muds, sands, and carbonates similar in age and composition to those of the offshore wedge. These blocks or massifs of Grenville-aged gneiss and their overlying deposits, the Glenarm Group, are today known as the Baltimore Terrane, and they are found in a band of domes, or nappes, west of I-95 from D.C. to Philadelphia.

Out in the ocean past these massifs, about 550 million years ago, a tectonic plate began converging from the east. Oceanic crust was diving under or subducting beneath this plate, forming a deep ocean trench. From beneath the trench, tectonic pressure drove magma upward to generate, seaward of the trench, the James Run or Chopawamsic volcanic island arc, a chain similar to the Aleutian Islands or Japan. The volcanic rocks produced here became the Chopawamsic Terrane in Maryland, today found just west of the Coastal Plain, and the Wilmington Complex in Delaware.

Another package of rocks formed in the deep trench adjacent to the volcanic island chain as subduction and consumption of oceanic crust brought it closer to collision with the Baltimore Terrane. Fine sediments eroded from the islands and slid into the steep-sided trench, where they mixed chaotically with rock shearing off from subducting ocean floor—creating an accretionary wedge known as a mélange. This mix of sedimentary rock and mafic igneous rock—some of it from perhaps as deep as the mantle—is the Potomac Terrane, exposed in several parts of the eastern Piedmont, and especially well at Potomac Gorge and Great Falls, where you can see fragments of one type of rock in a matrix of another.

Taconic Mountain Building Event and the Queenston Wedge

Continued convergence progressively squeezed together island arc, accretionary wedge, Grenville-massif platforms, basin deposits, and offshore wedge. In Ordovician time, about 480 to 440 million years ago, the Taconic mountain building event occurred as the island arc collided with North America. The intervening terranes were metamorphosed, folded, and thrust westward over the craton. Their previous, primarily horizontal widths were mashed into mountainous heights. The offshore clastic and carbonate wedge of today's Blue Ridge Province was thrust onto the continent, the Westminster Terrane onto eastern Blue Ridge rocks along the Martic Fault, the Potomac Terrane onto the Westminster Terrane along the Pleasant Grove Fault, and the Chopawamsic onto the easternmost margins. Massifs of the Baltimore Terrane, caught between the Chopawamsic and Potomac Terranes, were folded into overturned anticlines called nappes, or domes.

The Taconic Mountains, built by uplift of inches per year over millions of years, stood where the Piedmont is today. Much of eastern North America was involved in the complex folding and faulting. Heat and pressure generated by the compression and by increased rock depth metamorphosed much of the existing sedimentary rock: shale to schist and phyllite, sandstone to quartzite, limestone to marble, granite to gneiss, and basalt to metabasalt.

The hard, metamorphic rocks of the easterly Chopawamsic and Potomac Terranes are responsible for what became known to white colonists as the Fall Line, or Fall Zone, the boundary between the older Cambrian Piedmont and the much younger Cretaceous deposits of the Coastal Plain. Where major rivers descend this resistant rock, rapids occur. Below the rapids are the deep, tidal, and formerly navigable portions of the rivers. At these upper limits of Coastal Plain navigation, early settlers established ports: Washington, Baltimore, Wilmington, Philadelphia.

Westward of the Taconic Mountains, the tectonic collision had flexed the land downward into an inland basin holding water. Just before the Taconic uplift, during early Ordovician time, carbonates dominated in the basin, but as a low sourceland emerged, clay eroded and mixed with carbonates to form the Chambersburg Limestone. As the mountains pushed higher, erosion accelerated. Clays and dirty sands eroded, deposited, and cemented into shales and sandstones in the subsiding basin—later becoming the half-mile-thick Martinsburg Shale of today's Great Valley. With the basin filling and seawater receding, alluvial deposition predominated. Exposure of the deposits to air resulted in oxidation of iron minerals present in the sediments, imparting a distinctive red coloration to the sandstones and shales of the Juniata Formation. Then, in early Silurian time a new inland sea developed, with beaches of white quartz sands. These became the resistant Tuscarora Sandstone, later elevated to become a ridge-forming rock in western Maryland. As water levels fluctuated, more shales and red beds were laid down, and finally, as the sourceland eroded to the point that it was contributing little clastic sediment, the Tonoloway and Helderberg limestones formed at the end of Silurian time.

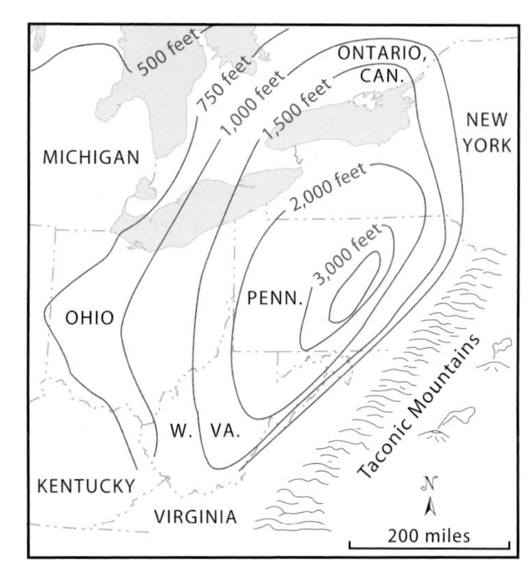

The Queenston Wedge, a thick pile of sediments, was deposited in an inland basin to the west of the eroding Taconic Mountains. The thickness of the sediments is shown in feet.
—Modified from Van Diver, 1990

Deposits were thickest, up to 4,000 feet, in the east closest to the mountains, and gradually thinned out for over 200 miles to the west, running from New England southwest to Virginia. These remains of the once lofty Taconic Mountains came to be known as the Queenston Wedge or Queenston Delta. Today you can see most of these rocks in roadcuts, folded by later mountain building events.

Acadian Mountain Building Event and the Catskill Wedge

During Devonian time, about 410 to 360 million years ago, the Acadian Mountains were built. The event began as continental fragments, or microplates, converged and sutured onto the American northeast and southeast. Most geologists agree that in eastern Maryland the main effect was not mountain building but just some metamorphism in the Piedmont. If you look at a map of the United States and ignore Delmarva, which was deposited much later, you can see that the continent edge seems to have a gap or missing piece in the Baltimore-Washington area. No microplates were added here.

Western Maryland did, however, receive many sediments from the emerging Acadian Mountains. Generally on top of the Queenston Wedge, from New England to Ohio to Tennessee, up to 9,000 feet of sediments were deposited in the Catskill Wedge—about three times as much as in the Queenston. The thickest part of the wedge was in the east. The Acadian Mountains were much higher than the Taconic Mountains.

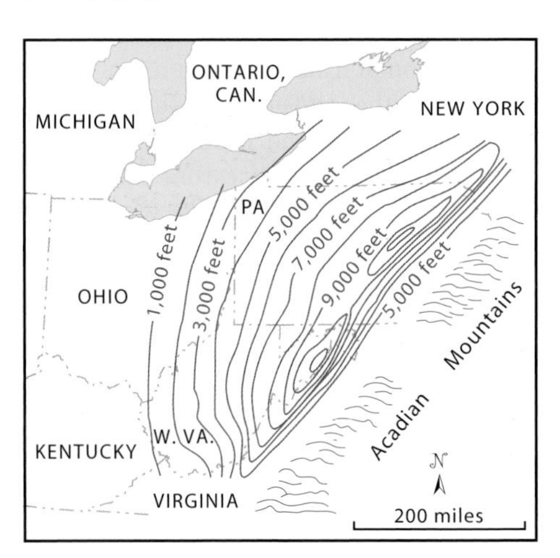

The Catskill Wedge, a thick pile of sediments, was deposited in an inland basin to the west of the eroding Acadian Mountains. The thickness of the sediments is shown in feet, with the thickest section near the base of the mountains. —Modified from Van Diver, 1990

The depositional pattern was similar. First, the convergent pressure created a downwarped inland basin. Here low-oxygen waters received thousands of feet of material that became shales and sandstones of several formations: Needmore, Marcellus, Mahantango, Brallier, and Foreknobs. Then, as with the Queenston Wedge, the basin filled and red alluvial deposits dominated: the Hampshire Formation. Next were stream deposits of the Pocono Group, mixed with shaly coals generated from low-lying swampy, vegetated areas. Finally, shallow seas returned and the appearance of the limestone of the Greenbrier Formation signaled that clastic sediments had slowed and that, in Mississippian time, about 330 million years ago, the Acadian Mountains had eroded as the Taconics before them had.

You can see these formations and others of the Catskill Wedge in western Maryland along I-70 and I-68 between Clear Spring and Cumberland, and in intervals on the Allegheny Plateau west of Frostburg. The formations repeat themselves in roadcuts because they were folded by a later mountain building event. The most readily identifiable formation along I-68 is the red Hampshire, which you can see alternately dipping east and west for about 70 miles.

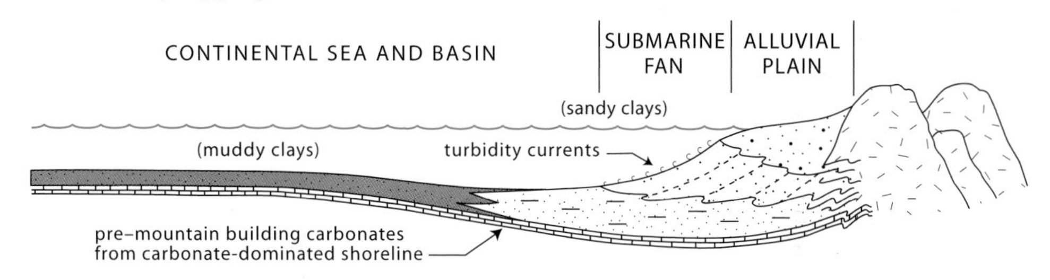

Sediment eroded from mountains is deposited in alluvial fans and river bottoms at the base of the mountains, in beaches along shores, in nearshore submarine fans, and in deep basins out from land. —Modified from Fichter and Poche, 1979

Alleghanian Mountain Building Event

On top of the Greenbrier Formation, deposition of the red sand, silt, and clay of the Mauch Chunk Formation from the east meant that yet another mountain range was slowly forming. During the Alleghanian mountain building event, the African Plate, approaching from the east, would slowly converge into the North American Plate over the course of some 70 million years, about 320 to 250 million years ago. This convergence was part of the assemblage of Pangea, a supercontinent made of all the world's continents. As Africa collided, land was uplifted in the Piedmont region again. The Pangea Appalachians, a high, long range toward the interior of the supercontinent, were much like the Rockies or the Alps of today. The wide and several-thousand-foot-thick Queenston and Catskill Wedges were compressed into less than half of their precollision width by the planetary-scale forces of the tectonic plate collision.

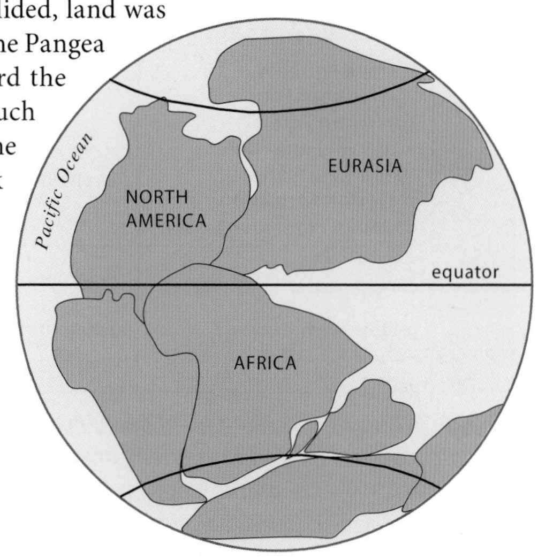

All continents were once joined in Pangea, a supercontinent, in Permian time. —Modified from Kious and Tilling, 1996

TIME	FORMATION		DEPOSITIONAL SETTING	EVENT
Permian	Dunkard Group			
Penn.	Monongahela Group		Alternating shallow marine	
Penn.	Conemaugh Group		and alluvial plain with swamp and	
Penn.	Allegheny Group		marsh vegetation compacting to coal	ALLEGHANIAN MOUNTAIN BUILDING
Penn.	Pottsville Group			
Miss.	Mauch Chunk Formation		Alluvial plain and stream deposits	
Miss.	Greenbrier Formation	CATSKILL WEDGE	Acadian Mountains nearly eroded; carbonate deposition with few clastic rocks,	
Miss.	Purslane Sandstone	CATSKILL WEDGE	River channel and alluvial plain deposits	
Dev./Miss.	Rockwell Formation	CATSKILL WEDGE		
Devonian	Hampshire Formation	CATSKILL WEDGE	Alluvial fans, stream and river deposits	
Devonian	Foreknobs Formation	CATSKILL WEDGE	Inland basin, deeper in west, shallower in east	
Devonian	Scherr Formation	CATSKILL WEDGE		
Devonian	Brallier Formation	CATSKILL WEDGE		
Devonian	Harrell Shale	CATSKILL WEDGE		
Devonian	Mahantango Formation	CATSKILL WEDGE	Clays and silts deposited as shales in deep inland basin	
Devonian	Marcellus Shale	CATSKILL WEDGE		
Devonian	Needmore Shale	CATSKILL WEDGE		
Devonian	Oriskany Sandstone	CATSKILL WEDGE	Resistant quartz sands in beach areas	ACADIAN MOUNTAIN BUILDING
Devonian	Shriver Chert	CATSKILL WEDGE		
Sil./Dev.	Helderberg Group	QUEENSTON WEDGE	Carbonate deposition with few clastic rocks	
Sil./Dev.	Keyser Limestone	QUEENSTON WEDGE		
Silurian	Tonoloway Limestone	QUEENSTON WEDGE	Taconic Mountains nearly eroded; carbonate deposition with few clastic rocks	
Silurian	Wills Creek Formation	QUEENSTON WEDGE	Shallow waters in inland basin	
Silurian	Bloomsburg Formation	QUEENSTON WEDGE	Alluvial deposits	
Silurian	McKenzie Formation	QUEENSTON WEDGE	Inland basin, deeper in west, shallower in east	
Silurian	Rose Hill Formation	QUEENSTON WEDGE		
Silurian	Tuscarora Sandstone	QUEENSTON WEDGE	Resistant quartz sands in beach areas	
Ordovician	Juniata Formation	QUEENSTON WEDGE	Alluvial fans, stream and river deposits	TACONIC MOUNTAIN BUILDING
Ordovician	Martinsburg Shale	QUEENSTON WEDGE	Clays and shales deposited in inland basin	
Ordovician	Chambersburg Limestone			
Ordovician	St. Paul Group			
Ordovician	Beekmantown Group		Before Taconic Mountain building begins; carbonate deposition along continent edge	
Ordovician	Conococheague Formation			
Cambrian	Elbrook Formation			
Cambrian	Waynesboro Formation			
Cambrian	Tomstown Formation			
Cambrian	Antietam Formation		Clastic deposits collect offshore as Grenville Mountains erode	
Cambrian	Harpers Formation			
Cambrian	Weverton Formation			
Cambrian	Loudoun Formation			
Proterozoic	Catoctin Formation		Lava flows	
Proterozoic	Middletown Gneiss		Basement of ancient craton	GRENVILLE MOUNTAIN BUILDING

Depositional history of the Blue Ridge, Valley and Ridge, and Allegheny Plateau. Mountain building events, sediments eroded from them, and depositional environments are shown. The oldest rock formations are at the bottom and the youngest are at the top. Note that the limestone formations (shown in red) were deposited after old mountains were eroded and prior to new mountain building.
—Modified from Schmidt, 1993; Cleaves, Edwards, and Glaser, 1968; Glaser, 1994a, 1994b

Areas in western Maryland, Pennsylvania, West Virginia, and Kentucky were again downwarped into a shallow inland sea. Sea levels rose and fell, generating several cycles of marine and nonmarine deposition. In addition, this area was tectonically located in the tropics, and one type of sedimentary environment that recurred periodically was the coastal swamp. Large masses of lush tropical vegetation collected in swamp bottoms and then, as conditions alternated between continental and marine, the decaying organic material was buried and compacted under sand, clay, and carbonate deposits. The partial decay and compaction of this tropical vegetation formed the Appalachian coal field, which extends from northern Pennsylvania to central Alabama on the Allegheny Plateau. In western Maryland coal is found in five synclinal basins, where it is interbedded with sandstone, shale, and limestone layers. By about 285 million years ago, the inland sea was gone from western Maryland and the swampy environments had largely disappeared.

As the African Plate inched relentlessly into the North American Plate—as today the Indian Plate is shoving into the Eurasian Plate and continuing to elevate the Himalayas—this Alleghanian mountain building event pushed farther westward than had the Taconic or the Acadian. In the Piedmont and Blue Ridge, slices of ancient Grenville massifs and thick sections of covering rock were metamorphosed, folded, shoved over, and thrust many miles inland along deep planes into the sedimentary rocks of the clastic wedges to the west.

The folding and thrusting rock layers of the Valley and Ridge and the Allegheny Plateau rode on the same basal thrust fault, a nearly horizontal sliding of tens of miles through weak beds of Cambrian Elbrook and Waynesboro shales and laminated limestones. Shoved westward at depths of about 30,000 feet (about 6 miles), this sole thrust broke upward out of Cambrian beds under the Allegheny Front and leveled off in the even weaker Silurian Wills Creek Formation at less than 10,000 feet deep under the plateau.

Because the Queenston and Catskill Wedges were thickest in the Valley and Ridge Province, they offered the greatest resistance there, resulting in layers

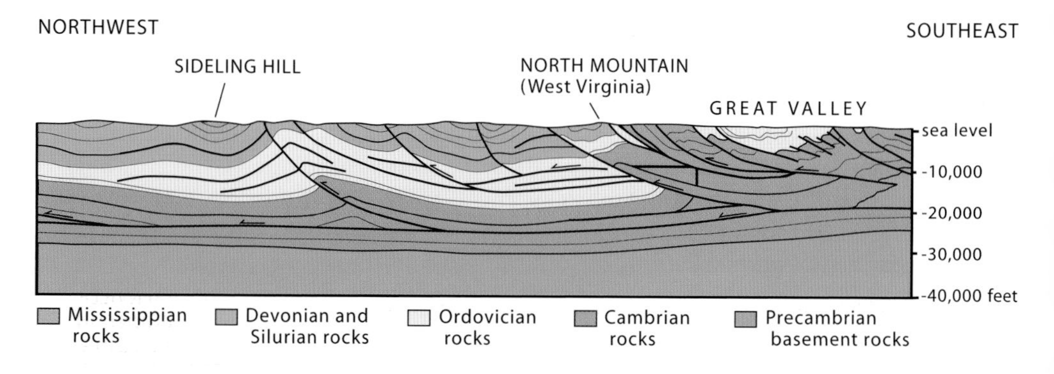

This cross-section of the Blue Ridge Province and Allegheny Plateau shows how deep layers were sheared and thrust westward and upward by the strong continent-continent collision that produced the Alleghanian mountain building event. Notice how the folding and upward thrusting resulted in multiple surface exposures, east-to-west, of the same formations. —Modified from Cardwell, Erwin, and Woodward, 1968

progressively slicing from the sole thrust and ascending at steep angles. Compressed movement along these steep thrust faults—still at great depth and under intense heat and pressure—caused the tight folding of the Valley and Ridge.

Farther west, where the Queenston and Catskill beds thinned out, the layers offered less resistance and folded only slightly into the open anticlines and synclines of the Allegheny Plateau. In addition, these layers were subjected to less pressure because they were located farther away from the tectonic collision zone. If you drive I-70 and I-68 between Baltimore and the West Virginia line, a distance of 180 miles, you will see only tilted layers in the roadcuts—the evidence of three mountain building events across 200 million years, the Alleghanian having to date exerted the final and most powerful compression.

The Rifting of Pangea and Triassic Basins

About 200 millions years ago, the plates that had converged to form Pangea began to diverge, or rift. Large blocks of lithospheric crust broke and slid downward along large-scale fractures in what is known as normal faulting. Because lithosphere floats on softer, plastic mantle, it can sink when it is pulled apart, somewhat as a large ship sinks if its hull is fractured. As fractures opened, magma oozed up into some of these faults, somewhat as water rushes into a sinking hull. Nearing the surface, the magma cooled and solidified into hard, tabular structures called dikes. Some of these hard structures survive as ridges in the western Piedmont of Maryland.

The Triassic Border Fault, a continental-scale downslippage, produced an intermittent trough or valley running from New Jersey to North Carolina, roughly parallel to the coastline and 50 to 100 miles inland. In Maryland, land to the east of the fault slid downward thousands of feet. Lying just east of the still lofty Alleghanian Mountains in what is known today as Frederick Valley, this deep trough received large amounts of eroded sediments, the red alluvial deposits of the Newark Group, which contain dinosaur footprints. These Triassic rocks were laid down on the eroded surface of Piedmont rocks of Ordovician age, representing an unconformity, or gap in the rock record, of about 300 million years.

Erosion of the Appalachians and Building of the Coastal Plain

During Cretaceous time, sediments continued to erode from the Appalachians and were deposited on the continental shelf. In late Cretaceous time, during an extended episode of global warming, the earth was without glaciers, and a worldwide rise in sea level flooded many coasts (as well as the American Midwest). The growing offshore continental margin subsided and tilted seaward, possibly due to the weight of the offshore sediments continuing to pile up. This tilting and regional reactivation of normal faulting created the Salisbury Embayment, which covered Virginia, Maryland, Delaware, and southern New Jersey up to the Fall Line during late Cretaceous and most of Tertiary time. Because water levels fluctuated in this bay, shorelines shifted, and continued tectonic movements in basement rock made the margin anything but passive. In this large bay, much of Coastal Plain Maryland and Delaware was built by millions of years of marine deposition.

What was happening in the sediment-contributing highlands to the west during these times? By about 100 to 50 million years ago, the agents of erosion would have had time to level the Appalachian range, but eroded material continued to flow toward the sea. How could the mountains have rejuvenated without plate convergence? One explanation is that the mountains uplifted more than once as a result of isostatic rebound, a kind of buoyant floating up of the crust that occurs as cumulatively larger weights of sediment are removed from the top by erosion. Earth crust rides or floats on the plastic, viscous mantle, similar to the way a cargo ship floats on the sea. After a mountain building event, the thick load of compressed rock sinks into the mantle—just as a ship full of cargo rides low in the water. As erosion removes the load, the crust begins to buoy up or float higher, and the mountains rise again.

When the Appalachians wore down during Cretaceous time, streams and rivers established meandering courses across the nearly level landscape. Then, as the slow uplift occurred—a movement that ultimately created today's topography—the rivers continued their downcutting action as the land rose. The meanders incised themselves into solid rock, a pattern you can see today in the many limestone cliffs along the Potomac River or along the meanders of Conococheague Creek, incised in shale. Also, as bands of highly resistant rock slowly rose to form elongate ridgetops, the Potomac River had time to cut through the rock and establish passages or water gaps through the emerging ridges. Today many such gaps exist in elongate ridges, such as South Mountain and Sideling Hill, where the Potomac flows through them.

One period of intense uplift about 20 to 10 million years ago caused an increase in sediments eroded into the Salisbury Embayment. These deposits, the Chesapeake Group, contain many fossils and make up the distinctive Calvert Cliffs on the western shore of Chesapeake Bay.

Pleistocene Ice Ages

During the Pleistocene Epoch, from about 1.6 million years ago to about 10,000 years ago, there were four glacial stages in North America. During these times, when ice covered large areas of the United States south of the Canadian border, sea levels dropped dramatically—more than 200 feet below present sea level about 18,000 years ago. During three interglacial stages, global climate warmed, glaciers

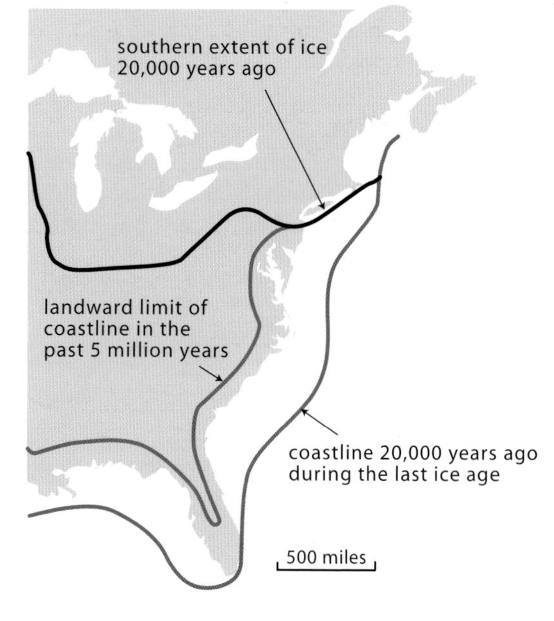

As we worry today about rising sea level, we should know that within the geologically brief span of the last 5 million years the coastline has moved in and out within a 300-mile range. —Modified from Williams, Dodd, and Gohn, 1997

retreated, and sea levels rose to over 40 feet above present sea level about 320,000 years ago and to about 20 feet about 120,000 years ago.

While the glaciers did not cover Maryland or Delaware, they did occupy the headwaters of the Susquehanna and Delaware River drainage basins to the north. The melting glaciers contributed large volumes of water that carried much sediment, which had been eroded from the land by the glaciers. These sands and gravels (and some ice-rafted boulders) washed down and fell out of suspension on the ever-thickening Coastal Plain wedge.

In addition to the mounting evidence that current global warming is partly human induced, some geologists think we may also be in just another of the many interglacial, natural warming periods documented throughout earth history. During worldwide glacier retreat and sea level rise, deposits are laid down in encroaching sea and estuarine environments. During colder periods of sea level drop, rivers cut deep channels into Coastal Plain terrains. During the last Ice Age, the ancestral Susquehanna, Potomac, and Delaware Rivers incised such valleys as they ran down to a sea that was about 50 miles east of where it stands today and scores of feet lower. Then, as climate warmed over the last 10,000 years, the sea flooded into the lower river reaches and created Chesapeake and Delaware Bays and their many wide, tidal tributary rivers.

Before they melted, the glaciers left their mark on the mountains of western Maryland. Even though the continental glacier did not advance as far south as Maryland—its terminus lying near Wilkes-Barre, Pennsylvania—the colder highlands of Maryland were subjected to about one hundred days per year of freeze-and-thaw cycles. During the days, temperatures would climb above freezing and melt part of the abundant snow cover. Water would trickle into porous rock or into cracks in rock. At night, freezing temperatures would turn the water to ice, causing an expansion in volume of about 10 percent. Rock would be wedged apart, broken from cliffs, and further disintegrated on the ground. Today in the Maryland mountains, you can find these legacies of the Ice Age in stone streams, boulder fields, and talus fields, some visible from many miles away.

Geologic history is not finished, of course. Humans live for all too brief a span to notice or record very many dramatic changes in the landscape. Photographs from the nineteenth century reveal that landscape features have changed very little in a century and a half. However, the geologic record is being preserved all around us. It is written not just in stone but in unconsolidated sediments carried by modern rivers and deposited in floodplains, deltas, estuaries, and ocean basins.

•— Allegheny Plateau —•

La Vale, French for "the valley," lies at the western edge of the Valley and Ridge Province, a region of compressed, tightly folded rock. If you drive 10 miles west from La Vale, you will recognize immediately that you have entered another geologic province. You will climb from the valley elevation of about 1,000 feet past Dans Mountain to Big Savage Mountain at over 2,700 feet—up what is known as the Allegheny Front and onto the Allegheny Plateau.

Early pioneers who tried to construct transportation routes up this escarpment barrier would marvel at the modern roads that cross it today. Going around the Allegheny Front was not an option because it extends from northern Alabama northeast through Tennessee, Kentucky, Virginia, West Virginia, Maryland,

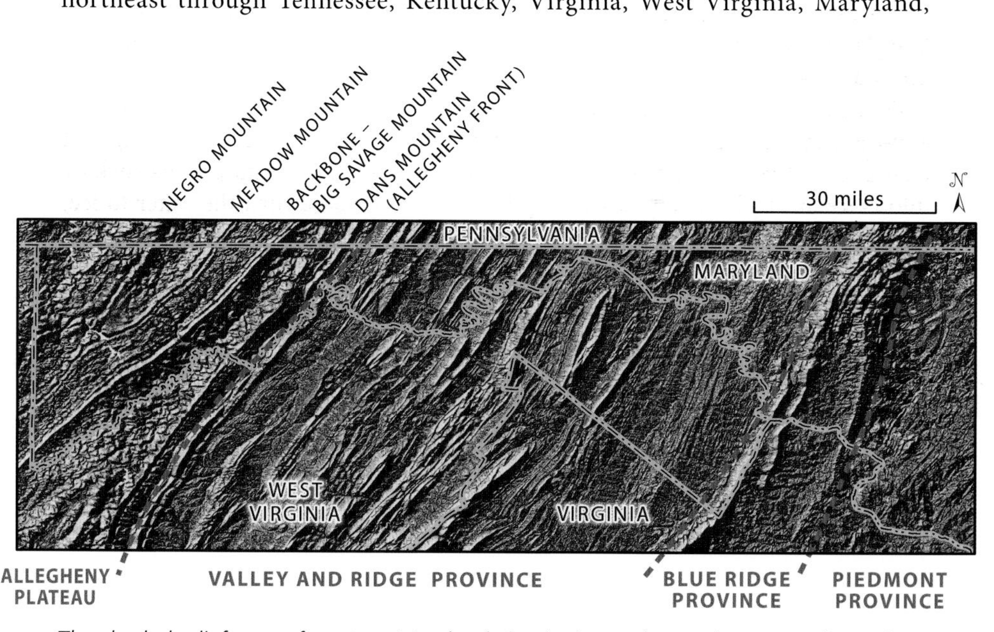

The shaded relief map of western Maryland clearly shows the northeast-trending ridges and valleys of the Allegheny Plateau. —Compiled at Mountain Press from U.S. Geological Survey, National Elevation Database

Pennsylvania, and into New York. Elevations on the Allegheny Plateau run over a thousand feet higher than those of the Valley and Ridge.

The rock of the plateau is sedimentary and gently warped into open folds—a series of low-amplitude, wavelike anticlines and synclines, the tops of which have been eroded. The folds in the plateau are not as tight as those in the Valley and Ridge because the rock layers were not as thick—offering less resistance to the tectonic compression of the Alleghanian mountain building event—and they were located farther from the zones of tectonic collision during the assembly of Pangea. In addition, the convergent forces caused Valley and Ridge rock layers to detach along deeply buried Cambrian layers and thrust westward until they finally ramped upward at the present Allegheny Front, releasing lateral pressure and reducing folding to the west. As you drive across the Allegheny Plateau, you will cross the eroded limbs of these gentle folds and see the same rock formations—dipping at about the same angle but in the opposite direction—on each limb.

Why is the region called a plateau? A plateau is defined as a relatively high area of comparatively flat land, but as you drive across it, you will go up and down long slopes, much as you did through the Valley and Ridge and the western Piedmont. The land may have once been like a plain, but over the 200 million years since the assembly of Pangea and the Alleghanian mountain building event, streams have dissected the landscape, and different kinds of rocks have weathered at different rates. Tilted sandstone limbs of anticlines and synclines resisted weathering and stand as high ridges, while shales and siltstones eroded into intervening valleys. In the Valley and Ridge Province, the northeast-trending ridges are all at elevations around 1,500 to 1,800 feet. On the Allegheny Plateau, the northeast-trending ridges are at elevations above 2,500 feet.

horizontal layers of Paleozoic sedimentary rock
prior to Alleghanian mountain building event

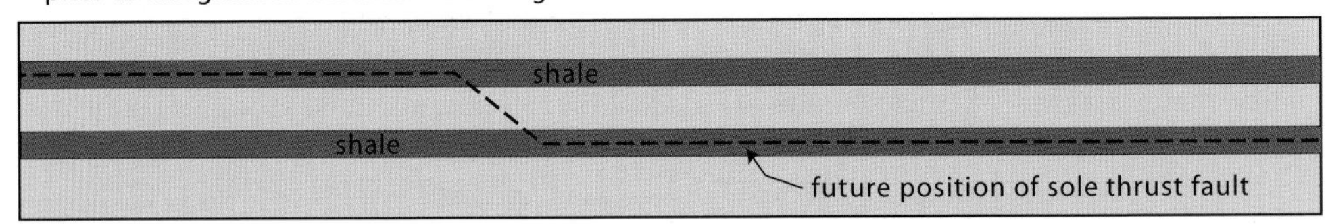

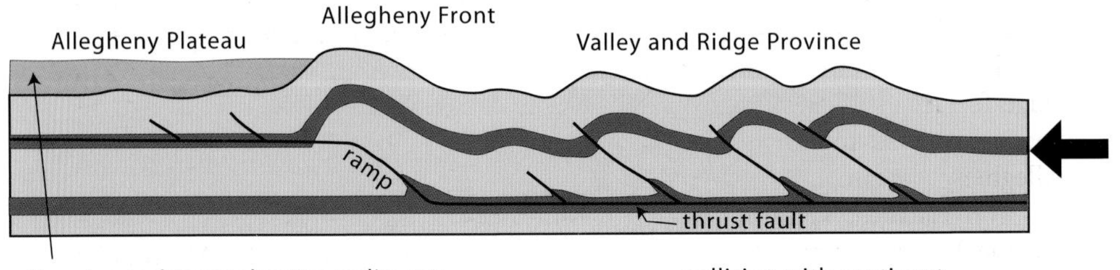

A thrust fault moved upward and west during the Alleghanian mountain building event, forming the Allegheny Front. —From Van Diver, 1990

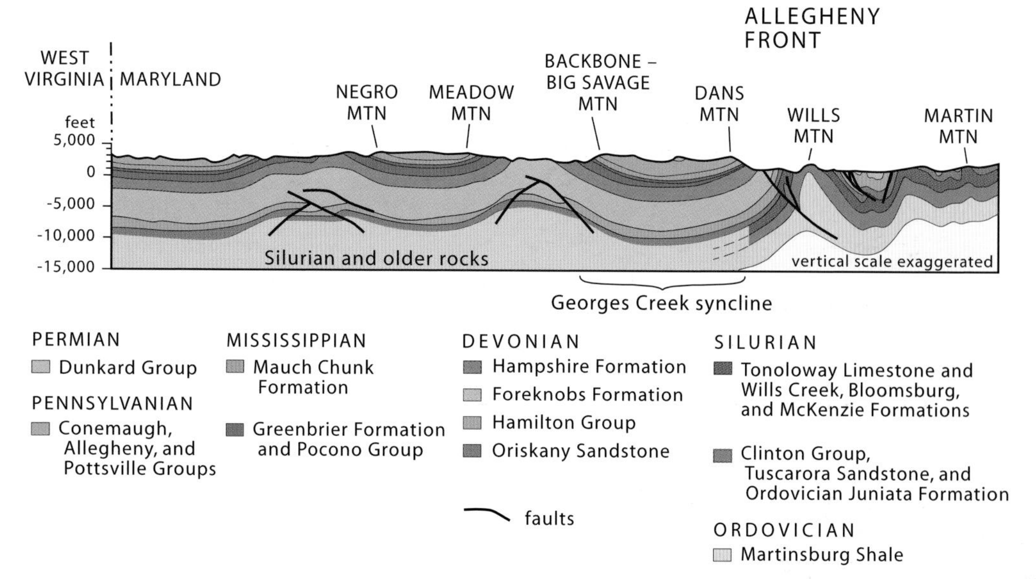

A cross section of the Allegheny Plateau reveals less compression and faulting than in the adjacent Valley and Ridge Province to the east, which is closer to where the tectonic collision occurred. —Modified from Cleaves, Edwards, and Glaser, 1968

On the plateau, the eroded anticline crests (upward-arching folds) today form valleys, while the resistant synclinal limbs (downward-arching folds) form ridges. This pattern may seem the opposite of what might be expected, but there are several reasons why anticline crests might erode more quickly than synclinal troughs: (1) the folded tops of anticlines are fractured and thus more exposed to penetration and subsequent weathering by precipitation, (2) anticline crests have been shoved up to altitudes higher than the synclinal downfolds, and weathering at the colder, higher altitudes proceeds more quickly, and (3) the steeper slopes of the anticline limbs erode more quickly because of mass wasting and rapidly downcutting streams.

When a resistant rock layer in the anticline, such as sandstone or quartzite, becomes exposed to weathering, the anticline will erode from the center outward, and less resistant rock underneath will become exposed. This weaker rock, perhaps shale or limestone, will then erode more quickly than the surviving outer limbs of harder rock. The weak rock will become valley, and the harder rock, ridge. If the anticline (or syncline) is plunging—or tilted from the horizontal—the resistant, ridge-forming rock will create a pocket or bowl-shaped landscape where water can become trapped and form swamps or bogs. There are several such locations on the Allegheny Plateau.

Why is the plateau so much higher than the adjacent Valley and Ridge? During Mesozoic time, the tighter folds and higher elevations of the Alleghanian Mountains—positioned at that time above the present-day Valley and Ridge—would have been subjected to much more rapid erosion and lowering than the land to

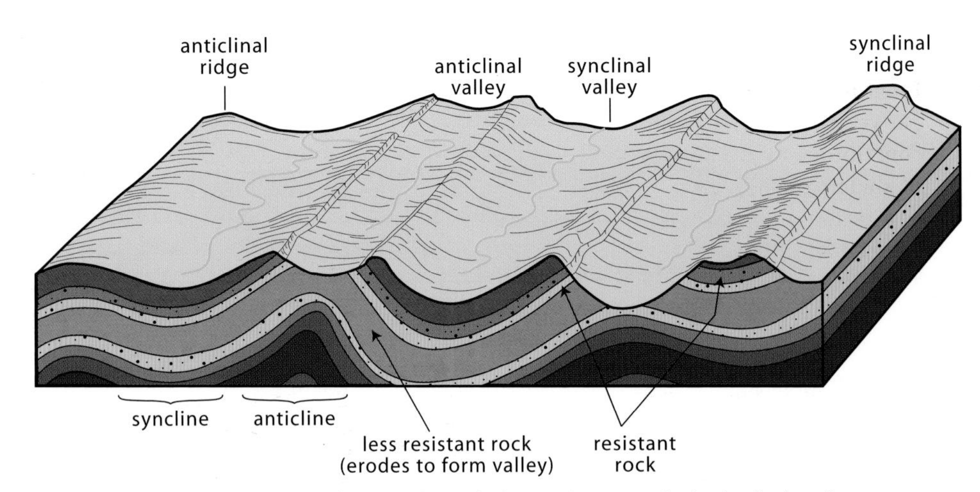

Folded rock layers often erode such that resistant rocks in the limbs of synclines are left as ridgetops.

the west where the Allegheny Plateau is located today. Higher-elevation erosion is generally more rapid than lower-elevation erosion due to extreme climate and steep slopes. Also, the mountain sediments eroded west onto the plateau, so deposition rather than erosion was occurring there. Ultimately, the Valley and Ridge Province wore down to elevations about a thousand feet lower than the plateau.

The highest ridge of the Allegheny Plateau in Maryland runs from Backbone Mountain on the West Virginia border to Big Savage Mountain northwest of La Vale. Except for Savage River cutting through it from west to east, this high ridge forms what is called the Eastern Continental Divide. The Casselman and Youghiogheny Rivers flow northwest to the Monongahela River, which flows into the Ohio and Mississippi Rivers and on south to the Gulf of Mexico. The Savage and other streams cut down the Allegheny Front and flow east into the Potomac, which enters Chesapeake Bay and the Atlantic Ocean.

Also flowing out of this province for well over one hundred years have been natural gas and coal. Most of the natural gas fields have been associated with the upward-arching, gentle anticlines, while the limbs of the synclines have been the locations of sedimentary coal formations.

The high plateau has an effect on precipitation. When humid air is cooled, it cannot hold as much moisture as it can at a higher temperature. A common demonstration of this phenomenon is the condensation of dew on the ground during a cool night. A humid air mass moving across the plateau on prevailing westerly winds is cooled as it is moved to higher elevations and loses its moisture as precipitation. Rain or snow is dropped in the highlands, and the air that descends into the Valley and Ridge is much drier than that which entered the plateau from the west. This rain shadow effect is well documented on the downwind or lee side of all mountain ranges.

Parts of Garrett County in extreme western Maryland receive as much as 49 inches of precipitation per year, while parts of Allegany County east of the Allegheny

Front receive as little as 36 inches per year. This lower precipitation contributes to the development of shale barrens, areas of sparse vegetation that resemble some desert environments.

TIME	FORMATION
Permian	Dunkard Group
Pennsylvanian	Monongahela Group with Waynesburg and Pittsburgh coals Conemaugh Group with Barton coal Allegheny Group with Upper Freeport and Brookville coals and Mount Savage clay Pottsville Group
Miss.	Mauch Chunk Formation Greenbrier Formation Pocono Group Purslane Sandstone
Dev./Miss.	Rockwell Formation
Devonian	Hampshire Formation Foreknobs Formation Scherr Formation Brallier Formation Harrell Shale

Sedimentary rocks of the Allegheny Plateau in Maryland. —Compiled from Abbe and others, 1900, 1902; Cleaves, Edwards, and Glaser, 1968; Glaser, 1994b

Road Guides to the Allegheny Plateau 

Interstate 68 and US 40 Alternate
La Vale—West Virginia Border
43 MILES

Between La Vale and Keysers Ridge, I-68 and US 40A are roughly parallel. East of Big Savage Mountain, I-68 may be better for viewing the terrain because US 40A goes through populated areas, but US 40A is generally better and safer for roadside stops than the interstate.

West from the I-68 La Vale exit, I-68 climbs the long, steep Dans Mountain, the high ridge that in Maryland is the Allegheny Front. Between Vocke Road at exit 40 and a point on I-68 about 2 miles west, the road crosses nearly all of the Catskill Wedge, rock formed from the eroded sediments of the Acadian Mountains. The formations dip gently to the west, indicating that you are on the eastern limb of a large syncline known as the Georges Creek syncline. Just west of exit 39, you can see the red shales and siltstones of the Devonian-age Hampshire Formation dipping west. Ten miles west on Big Savage Mountain near the Finzel exit (exit 29), you will see this red formation emerge as the east-dipping limb of the Georges Creek syncline. Where I-68 crosses the Hampshire Formation farther east at the

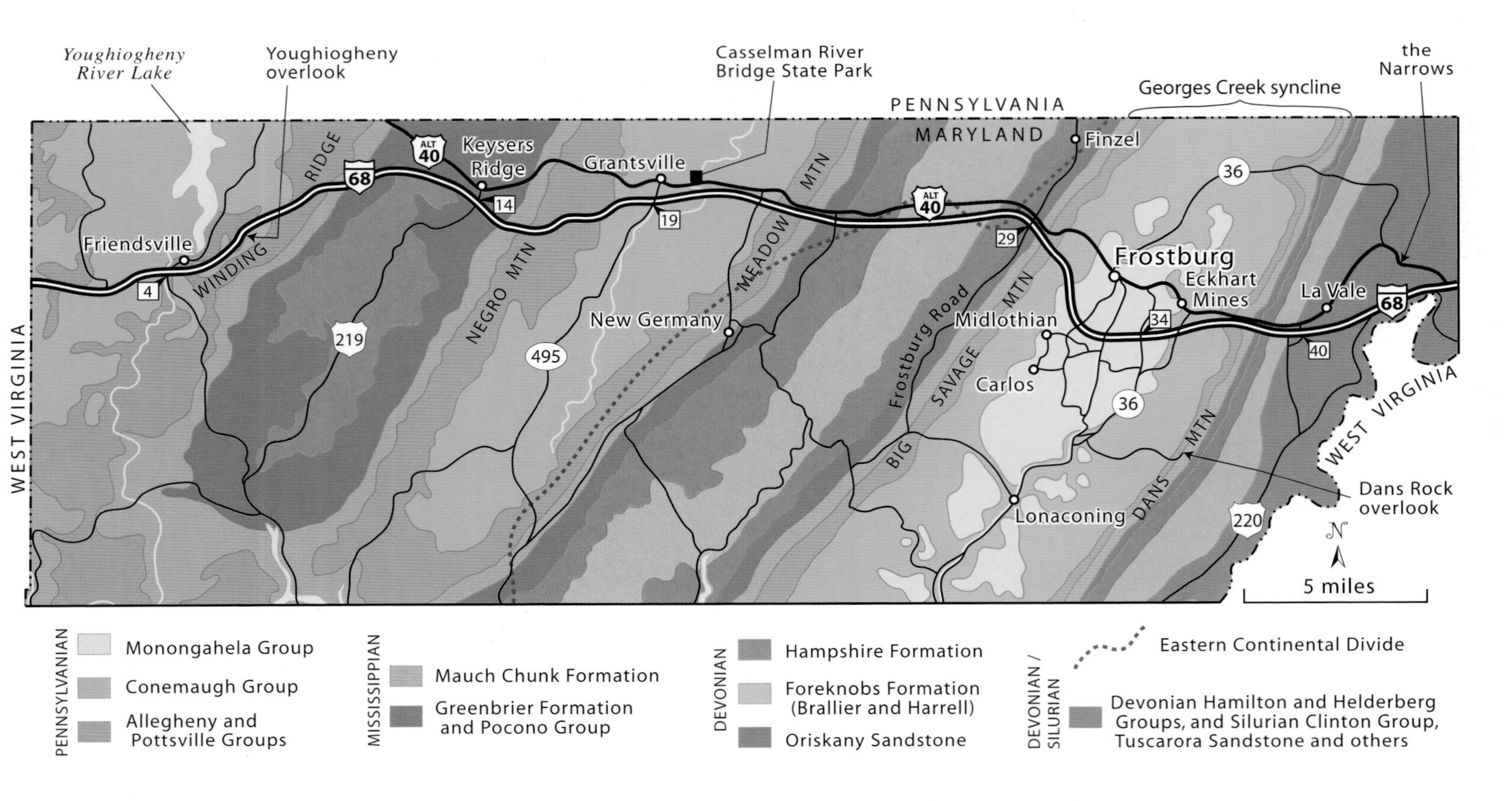

Geology along I-68 and US 40A between La Vale and the West Virginia border. —Modified from Cleaves, Edwards, and Glaser, 1968; Darling and Slaughter, 1962; Glaser, 1994b

Visible from I-68 near exit 33 is a long-worked surface mine in the western limb of the Georges Creek syncline.

Sideling Hill syncline in the Valley and Ridge Province, the distance between east- and west-dipping limbs is only about 3 miles. This 10-to-3 mile difference reflects the general ratio of the intensities involved in the tectonic compressions of the Alleghanian mountain building event. The folds become much more tight—and the thrust faults more frequent—toward the east, where the tectonic collision with the African Plate occurred.

Erosion of the Georges Creek syncline by its main stream, Georges Creek, has exposed coal beds on the slopes above the creek to east and west (see **MD 36: The Narrows—Westernport (Coal Country)**). On I-68 just east of exit 34 for MD 36, you can see coal beds in the roadcut in the east limb of the syncline. West of exit 33 for Midlothian, you can see extensive coal strip-mining operations to the south on the east slope of Big Savage Mountain, above Frostburg, the west limb of the syncline.

Just east of Frostburg and exit 34 at Eckhart Mines, between Old National Pike and US 40A, you can explore on foot up the valley and find the remains of Consolidation Coal Company Mine No. 4 in the Pennsylvanian-age Big Vein. At Frostburg you can get a great view of the Allegheny Front to the east from the old train station (or the Thrasher Carriage Museum).

Resistant sandstones of the Pottsville Group of Pennsylvanian age form the Dans and Big Savage ridges that enclose the Georges Creek Basin. Look for the sandstone, along with underlying black shale and coal, near the top of Big Savage, just east of exit 29 for Finzel. Also, above Frostburg just east of the top of the mountain in the westbound lane, you can pull off and look back for a great view of Dans Mountain and the gap cut by Braddock Run. More rock is visible along US 40A, and going up Big Savage you get a better sense for the steepness and length of the climb. At the top of Big Savage you can stop for a great view of the terrain to the east and west, erosion-resistant sandstones topping all of the ridges you see.

From above Frostburg on Big Savage Mountain—the west limb of the Georges Creek syncline—you can pull over in the westbound lane and look east to see where Braddock Run (and parallel I-68) has cut through Dans Mountain, the east limb of the syncline.

At the westbound exit 29 ramp, don't miss the east-dipping red Hampshire rock of the syncline limb. Then, just west of the sign for the Eastern Continental Divide or about 1 mile east of exit 24 (about 4 miles west of exit 29) you will see a west-dipping limb of the Hampshire Formation. These two exposures are the limbs of a gentle anticline, and they are not far apart because the erosion of the top of the anticline has only "recently" reached the Hampshire Formation. You may have figured out by now that in a succession of synclines and anticlines, the limb of a syncline is also the limb of the adjacent anticline, and so forth.

Topography and stream drainage on the Allegheny Plateau are determined mostly but not entirely by the presence of anticlines and synclines. Relative resistance to weathering of the rock layers presently exposed and the presence of water gaps are also factors. The recurring Pottsville Group forms synclinal ridges of sandstone that stick up from intervening anticlinal valleys. Each valley contains a major stream. The Eastern Continental Divide separates streams that drain into the Mississippi River and the Gulf of Mexico (Casselman and Youghiogheny Rivers) from those that drain into the Atlantic Ocean via the Potomac River and Chesapeake Bay (Jennings Run, Braddock Run, Georges Creek, and Savage River).

As you travel through valley bottoms, look for swampy wetlands. Deposition of clay and poor drainage are common in the Allegheny Plateau, and many glades or bogs occupy low areas or pockets. See **US 219: Keysers Ridge—Backbone Mountain Trailhead, West Virginia** for a detailed treatment of one of these, Cranesville Swamp.

Meadow Mountain and Negro Mountain occupy a syncline that is very similar in size, cross section, and geologic formations to the Georges Creek syncline. Just east of exit 22, near the top of Meadow Mountain, you can see the same brownish white Pottsville sandstone and shaly coal that topped Big Savage Mountain. From the top of Meadow Mountain you can see the synclinal limb, Negro Mountain, to the west. Near exit 19 in the syncline, you can see coal in a roadcut. In the valley west of Grantsville, look north for the diggings of an abandoned coal mine. On Negro Mountain, at the Amish Road overpass, you can see the sandstone and coal beds that form the west limb of the syncline, and at the top of Negro Mountain, more rimrock sandstone. On lower slopes, you'll see less resistant, crumbly shales.

The pattern continues west. Look for this same sandstone atop Keysers Ridge at the exit 14A ramp, atop Winding Ridge above Friendsville and the eastbound Maryland rest stop, and atop Evans Hill, just over the West Virginia border. If you are eastbound, you can stop at the rest stop and look below to Youghiogheny River Lake, a reservoir along the Youghiogheny River, which flows northwestward to the Ohio River.

At Casselman River Bridge State Park near Grantsville, you can walk over this beautiful stone bridge and look down at water that will flow down the Mississippi to the Gulf of Mexico. This stone arch, at 80 feet of span, was the largest in the United States when it was completed in 1813.

Side Trip along the Top of Maryland

At exit 29, signed for MD 546 and Finzel, go south on Beall School Road for an unpopulated drive on secondary roads along the slopes of the highest ridge in Maryland: Big Savage Mountain. Follow Beall School Road to its end and take Frostburg Road to the south (not Old Frostburg Road to the west). Here you will have good views of meadows and fields to the west—the anticlinal valley between Big Savage and Meadow Mountain that is downcut and drained by the Savage River, the westernmost river in Maryland that flows east into the Potomac.

At the stop sign, take Avilton-Lonaconing Road to the east (left) to the top of Fourmile Ridge, where you will see a stone quarry in the Pocono Group sandstone. Here there are good views of Meadow and Negro Mountains to the west, and the top of Big Savage to the east. All of these are capped by the resistant Pottsville Group sandstone.

After you pass over the top of Big Savage, you descend east into the Georges Creek Basin, where the younger layers of the Conemaugh Group exposed within the eroding syncline are veins of coal that have been mined for over one hundred years. At Lonaconing the road becomes Douglas Avenue, and at the stoplight is the junction with MD 36.

US 219
Keysers Ridge—Backbone Mountain Trailhead, West Virginia
40 MILES

US 219 leads to some unique geologic features of the high Allegheny Plateau: Arctic bogs, waterfalls, a plunging syncline, and Maryland's highest point.

About 1.5 miles south of I-68 (exit 14) on US 219 is an overlook into the anticlinal valley between the sandstone ridges of Negro Mountain to the east and Winding Ridge to the west. This area was far enough removed from the tectonic collision zone of the Alleghanian mountain building event that the rock is only gently folded. At this overlook you can get a good sense of the stream-dissected landscape of a high plateau and a good view of the Cove, one of the finest farming districts in the area. From the overlook, the Cove looks like a huge bowl surrounded on all sides by mountain ranges.

At the junction with Bear Creek Road, there is a good exposure of the red shale and sandstone of the Hampshire Formation, Maryland's westernmost outcropping of this folded layer that intersects the surface in nine different locations on I-68 between here and a few miles east of Hancock. Its iron-red coloration clearly marks the ascents and descents of the limbs of each anticline and syncline.

At McHenry you can look across an arm of Deep Creek Lake and see a recreational business that depends for its very existence on differential weathering—downhill skiing and snowboarding. Here the calcareous shale and clayey limestone of the Greenbrier Formation of Mississippian age has weathered more rapidly than sandstone of the Pottsville Group, creating a hill sloping down to a valley. East of US 219 and opposite the ski area is the southern end of Negro Mountain.

Deep Creek Lake State Park

At Thayerville you can drive to the opposite side of the lake on Glendale Road to Deep Creek Lake State Park. There, the strenuous, steep, rocky, half-mile-long Fire

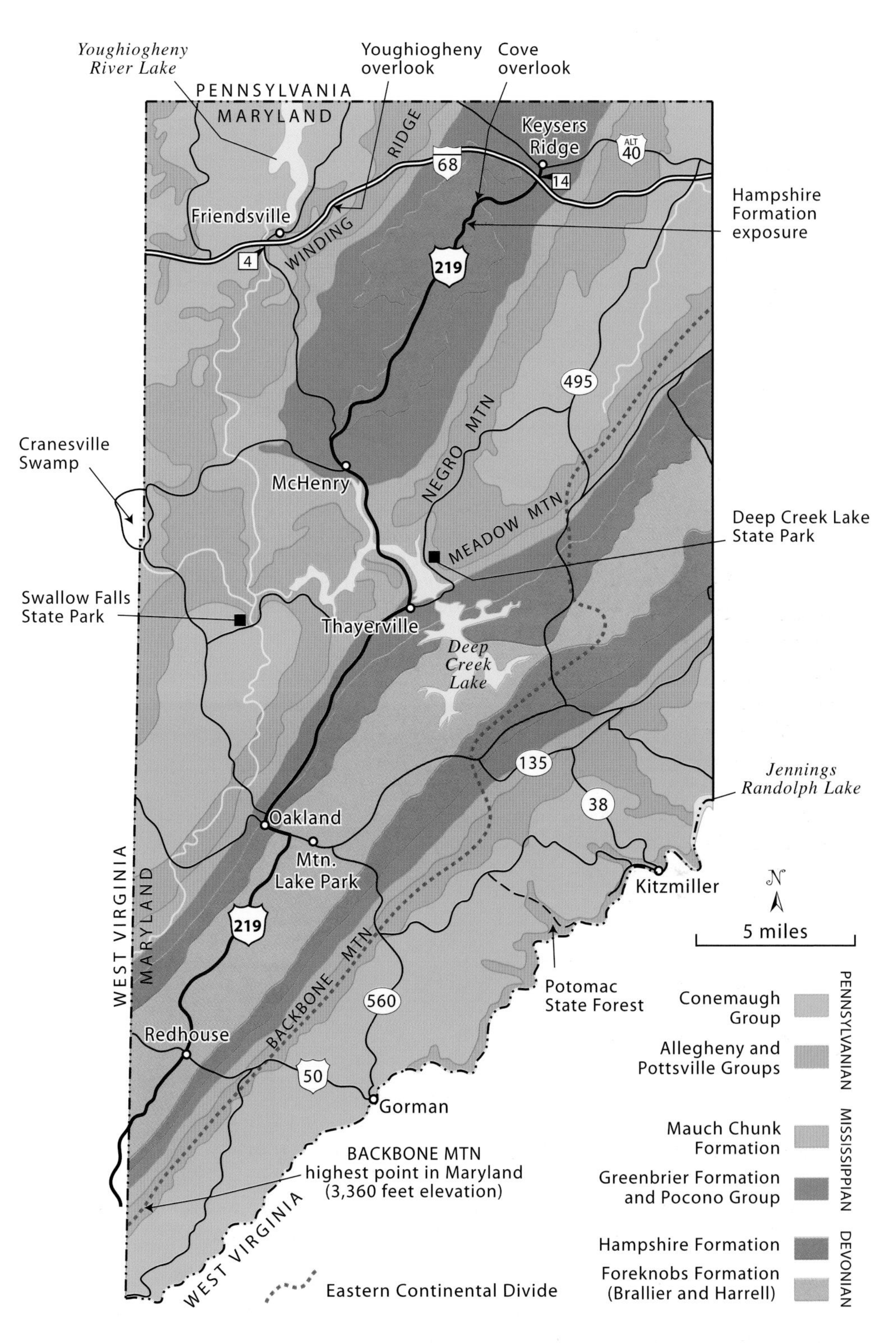

Geology along US 219 between Keysers Ridge and the Backbone Mountain Trailhead in West Virginia. —Modified from Cleaves, Edwards, and Glaser, 1968; Glaser, 1994b

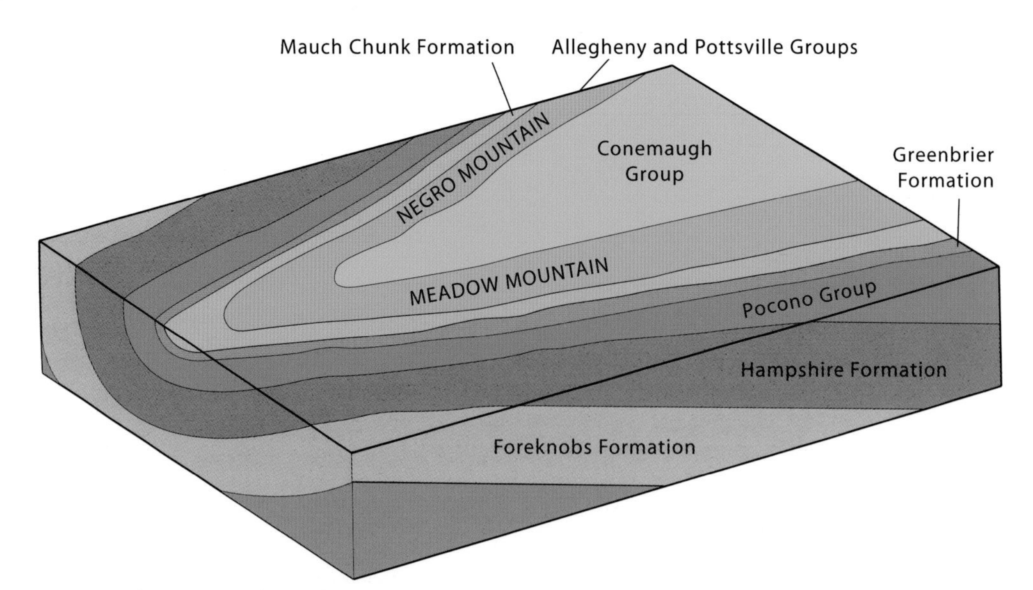

The merger of Meadow Mountain and Negro Mountain, each topped with Pottsville sandstone, occurs because the general U shape of the syncline is tilting or plunging into the earth toward the northeast. On the landscape the two mountain ranges are more parallel than depicted here. —Modified from Cleaves, Edwards, and Glaser, 1968

The entrance to a coal mine along Brant Mine Trail in Deep Creek Lake State Park. Underground coal mining is claustrophobic, dangerous work.

Tower Trail climbs the south end of Meadow Mountain. At the top you will see the resistant sandstone of the Pottsville Group supporting communications towers. This sandstone forms the long ridge of Meadow Mountain, which to the north on I-68 is separated by about 7 miles from the sandstone ridge of Negro Mountain. At Deep Creek their tops essentially join, as they are only about 1 mile apart and the Pottsville sandstone is continuous between them. This meeting of ridges results from the northeast plunge of the syncline lying between the two. On a geologic map or aerial photo, the sandstone is exposed in a C-shaped parabola pointing northeast, with each ridge occupying half of the C. Another such structure lies southwest of Deep Creek Lake. Plunging folds remind us that the processes of plate tectonics occur on a three-dimensional earth, not on a planar surface.

While you are at the park, you can hike the half-mile looping Brant Mine Trail and see the opening of a coal mine that was worked by two men in the early 1920s. They sold the coal for home heating. It is said that the two men who operated this mine worked themselves to death.

Swallow Falls State Park and Cranesville Swamp

To reach Swallow Falls State Park, turn west off US 219 onto Mayhew Inn Road about 1 mile south of Thayerville. To the south you can see another of the typical high-plateau bogs. Follow the road about 4.5 miles to its end. Turn left onto Oakland-Sang Run Road, and after about a quarter of a mile turn onto Swallow Falls Road, which leads to the state park.

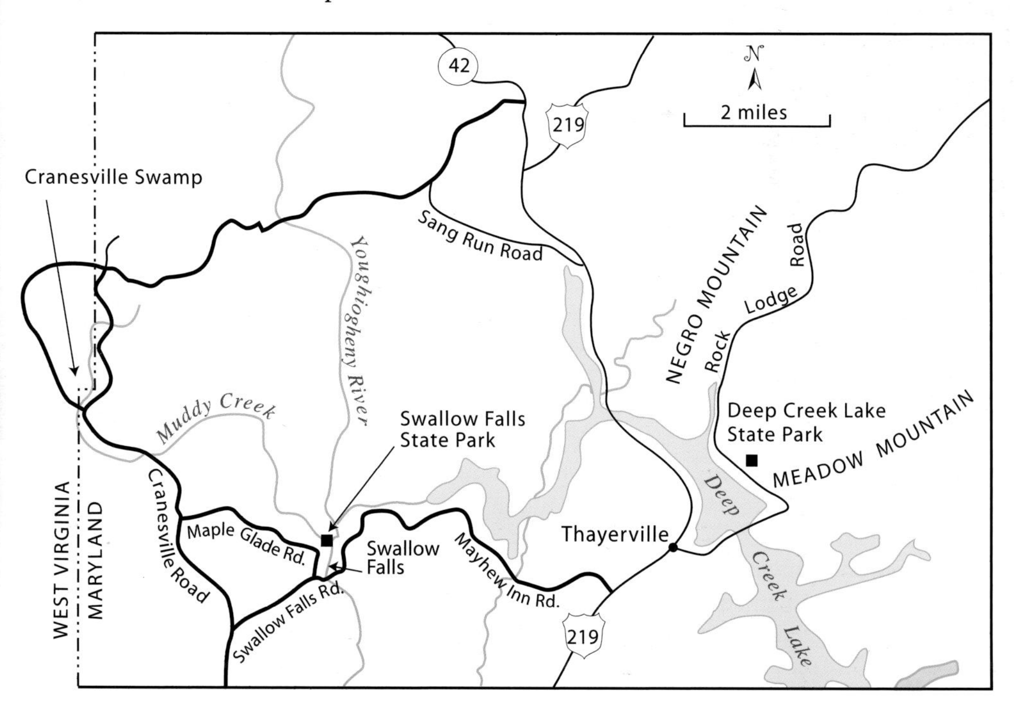

Road map to Swallow Falls and Cranesville Swamp. —Modified from Cleaves, Edwards, and Glaser, 1968

These crossbeds in Pottsville sandstone in Swallow Falls State Park reflect differing directions of incoming water currents carrying sediment during deposition.

At Swallow Falls State Park, you can take a beautiful circuit hike through virgin hemlocks to the different waterfalls. Muddy Creek emerges from the high plateau bog of Cranesville Swamp and flows down here, plunging about 60 feet over interbedded shales, siltstones, and sandstones of the Pottsville Group. The sandstone presently forms a resistant top ledge to Muddy Creek Falls, Maryland's highest straight-drop waterfall, listed by various sources between 51 and 63 feet. Over time, the falls retreat upstream, with the weaker shale eroding below the stronger sandstone cap until the unsupported cap breaks off.

On the path between where Muddy Creek enters the Youghiogheny River and Swallow Falls on the Youghiogheny, you can see cliff exposures, and in one place right beside the trail, excellent examples of crossbedding in sandstone. These crossbeds formed when sand was dropped from a moving current—not the Youghiogheny, but some water current that flowed 300 million years ago. The downward angle of a bed indicates the direction of the current, and cross-angled beds indicate deposits from currents flowing in different directions. On the path below Upper Swallow Falls on the Youghiogheny River you can see a tower of interbedded shales and sandstones eroded by water.

Upstream from Swallow Falls State Park, Muddy Creek passes through the 600-acre Cranesville Swamp, which straddles the Maryland–West Virgina line. To get there from the park, turn right (west) as you exit the park and follow Swallow Falls Road to a right (north) turn onto Cranesville Road. Right after Cranesville Road leaves Garrett State Forest and crosses Muddy Creek, you will begin to see swampy areas to your left, and at Lake Ford you will see a sign for Cranesville Swamp indicating a left turn. Turn and go into West Virginia about a half mile and follow signs to Wildlife Viewing Area and Nature Conservancy Parking Area. The White Trail, which heads out from where the power line crosses the road just before the parking area, is the shortest passage to the loop walkway over the swamp. Out on the swamp, if you turn 360 degrees, you will see hills all around.

This high mountain bog formed during the Ice Ages and retains northern flora and fauna far south of their normal habitats because of several geologically determined features. First, the high, cool elevation of over 2,500 feet receives abundant rain and snow. Second, the swamp lies in the middle of the Briery Mountain anticline, and like other anticlines of the Allegheny Plateau, it has eroded into a valley, but because this is a plunging anticline, erosion has created a bowl-shaped terrain instead of two parallel ridges. Third, the rim of the surrounding bowl is over 300 feet above the swamp, and at night the heavier cold air descends to form what is called a frost pocket. In this temperature inversion, conditions can be below freezing in the swamp while higher, surrounding slopes are above freezing. Sphagnum moss serves as an insulating layer to keep swamp water cold. Fourth, the bog retains much of its water because of poor drainage caused by underlying, impermeable clay layers deposited in the bottom of the bowl. Thus, many plants

Exotic vegetation of many different shapes and hues exists in Cranesville Swamp.

and animals that migrated here ahead of the advancing glaciers thousands of years ago during the Ice Ages have managed to survive in this southern niche. You can get a good view of the bowl and surrounding mountains if you continue north on Cranesville Road.

Backbone Mountain

On US 219 south of Oakland, you will see to the east the long, high ridge of Backbone Mountain, which stretches northeast to become Big Savage Mountain above Frostburg. At Redhouse you can drive east on US 50 over the top of Backbone Mountain, which is 3,095 feet here. South of US 50, look for the huge propellers of wind-powered electrical generators on Backbone Mountain.

If you want to climb to Maryland's highest point, at elevation 3,360 feet on Backbone Mountain, you will need to go a few miles into West Virginia. Follow US 219 to Silver Lake and then, about 1 mile south of there, look carefully for a small sign by the road for Backbone Mountain. If you come to the sign for Monongahela National Forest, you have gone too far. You can park by the road and hike up about 1.5 miles on a forest road as you return to Maryland. The view is mostly to the east and not as spectacular as the views from Big Savage Mountain and Dans Rock.

Source of the Potomac River

Continue a few miles farther south on US 219 into West Virginia and look for signs that will lead you to the Fairfax Stone, which marks the spring (often dry) that is the traditional source of the Potomac River. If you follow a river upstream from

The man shaft and company town of Kempton, circa 1920. —Photo courtesy of the Maryland Geological Survey

its mouth and at forks always follow the one discharging the most water (which can vary, of course), you will eventually reach the traditional source of the river. Locating the source of the Potomac was important in determining state boundaries in the nineteenth century. The real source of the North Branch of the Potomac lies about 1 mile to the west.

Just over the border in Maryland is Kempton, a coal-mining ghost town. An interpretive plaque at the former man shaft site tells about the 1 million tons of coal mined annually from 1915 to 1921.

US 220
McCoole—Cumberland
21 MILES

US 220 between McCoole and Cumberland generally follows the North Branch of the Potomac River. Between Dawson and Rawlings, the road swings away from the river and travels through a small water gap, a trough cut by streams flowing off the eastern slope of Dans Mountain. The small ridge to the east of the road is Fort Hill, one of the westernmost exposures of the tightly folded Valley and Ridge Province. The erosion-resistant Oriskany Sandstone, here dipping almost vertically, formed a barrier to the streams flowing east from Dans Mountain and forced them to turn north or south to cut a gap along the strike of the dip before entering the North Branch.

US 220 travels through a trough, or water gap, in the Oriskany Sandstone. Cliffs tower above each side of the road.

North of Rawlings, US 220 travels mostly on the broad floodplain of the North Branch of the Potomac, very close and parallel to the thrust fault that marks the border between the Allegheny Plateau to the west and the Ridge and Valley Province to the east. The route provides good views of the towering Allegheny Front, here the long ridge known as Dans Mountain. The Allegheny Front runs from Alabama to New York.

Between Rawlings and Cresaptown, you can get particularly good views of Dans Rock, which, at almost 2,900 feet, is about 2,200 feet above the valley floor.

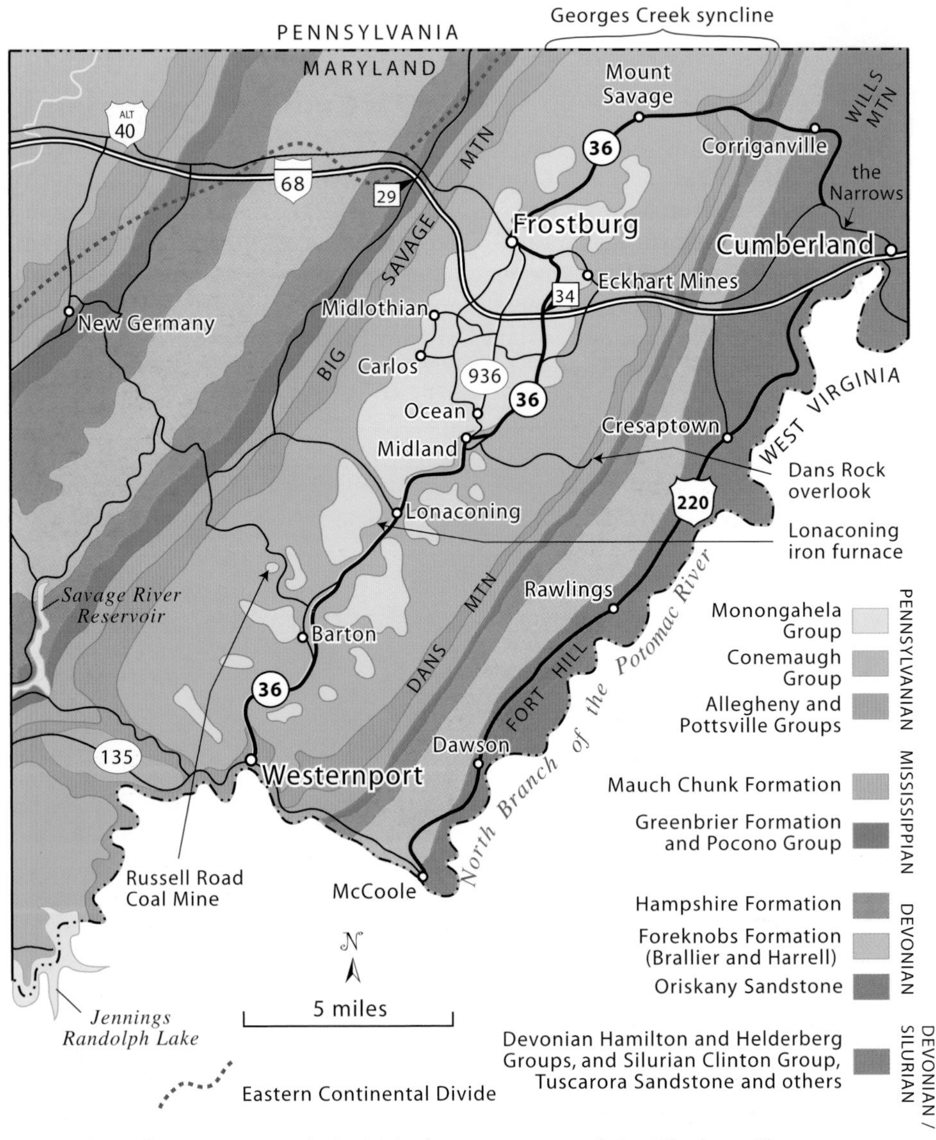

Geology along US 220 and MD 36 in the eastern part of the Allegheny Plateau.
—Modified from Cleaves, Edwards, and Glaser, 1968; Darling and Slaughter, 1962; Glaser, 1994b

Dans Mountain, part of the Allegheny Front, looms above the western edge of the Valley and Ridge Province and the broad floodplain of the North Branch of the Potomac River.

You can easily identify it by the many communications towers located there. The Pottsville Group sandstone, very resistant to weathering, forms the ridge that is Dans Mountain, the eastern limb of the broad Georges Creek syncline.

Draining streams from the high Allegheny Plateau, the North Branch of the Potomac River has established a broad floodplain just east of the Allegheny Front. Look for parts of some fairly long meanders characteristic of streams that cut into highly erodable shale formations of the Helderberg and Hamilton Groups.

Maryland 36
The Narrows—Westernport (Coal Country)
30 MILES

From just west of the Haystack Mountain–Wills Mountain water gap known as the Narrows, MD 36 leaves the Valley and Ridge Province, climbs the Allegheny Front, and then travels south along Georges Creek, which flows south along the axis of the Georges Creek syncline to the Potomac River at Westernport. The highway passes through iron and coal country, and you can easily follow the road because it is signed as Maryland Scenic Byways: Coal Heritage Tour.

The first 2 miles of MD 36 parallel Wills Creek, and on the western slope of Wills Mountain you can see a huge quarrying operation in the Tuscarora Sandstone of Silurian age. MD 36 is a more gradual climb to Frostburg than either I-68 or US 40A. From Corriganville it ascends along the floodplain of Jennings Run, which descends from the high eastern slope of Big Savage Mountain above Frostburg.

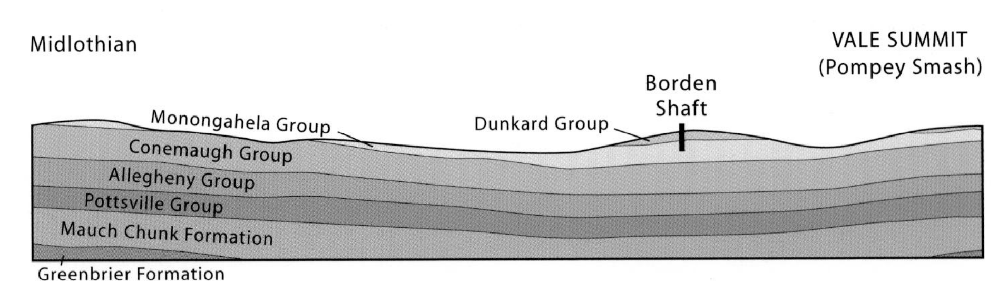

A cross section of Georges Creek syncline through Vale Summit, Borden Shaft, and Midlothian shows the major coal-bearing groups—the Monongahela, Conemaugh, and Allegheny. The coal veins, exposed by weathering and erosion, intersect the surface of the slopes of Dans and Big Savage Mountains. This geometry reveals why you can often see strip mines at comparable elevations on either side of Georges Creek valley and why a vertical shaft was dug at Borden Shaft in the middle of the syncline. The 14-foot Big Vein, or Pittsburgh coal, is (or mostly, was) located at the bottom of the Monongahela Group. —Modified from Abbe and others, 1900

Just east of Barrelville at the junction with MD 47, you can see massive sandstone blocks of the Pottsville Group that have tumbled downslope as the weaker shales of the underlying Mauch Chunk Formation have eroded during the downcutting of Jennings Run.

At Mount Savage, iron furnaces were completed in 1840 on a site that today is marked by a sign on New School Road. The furnaces utilized coke produced from nearby coal and carbonate ores. Output was about 9,000 tons per year. The rolling mill here produced the first solid railroad tracks in the United States. Firebrick was produced here in 1840 for the iron furnaces, with the fireclay mined above the town on Big Savage Mountain. The iron works ceased in 1865, as attention was shifted to more profitable coal mining, but the production of firebrick outlasted the iron production. At Zihlman on MD 36 is a plant that still produces firebrick, which is used for oven linings subjected to high temperatures, but because most of the Maryland fireclay is exhausted, much of this plant's clay comes from Pennsylvania.

After winding your way up to Frostburg, you can visit the old railroad station (follow signs for the Thrasher Carriage Museum) for a great view of the Allegheny Front–Valley and Ridge border to the east.

Coal in the Georges Creek Basin

MD 36 south of Frostburg heads into the Georges Creek Basin, in which several veins or beds of high-quality, bituminous coal were naturally exposed by erosion on the east and west limbs of the Georges Creek syncline. The Jennings Run area is often grouped with the Georges Creek Basin in coal-mining references, giving this area in Maryland dimensions of about 5 miles in width and over 20 miles in length, from Pennsylvania to Westernport. This synclinal fold extends, with its coal seams, into Pennsylvania to the north and West Virginia to the south.

How did organically based coal become layers within clastic sedimentary deposits? About 300 million years ago, during Pennsylvanian time, this area was

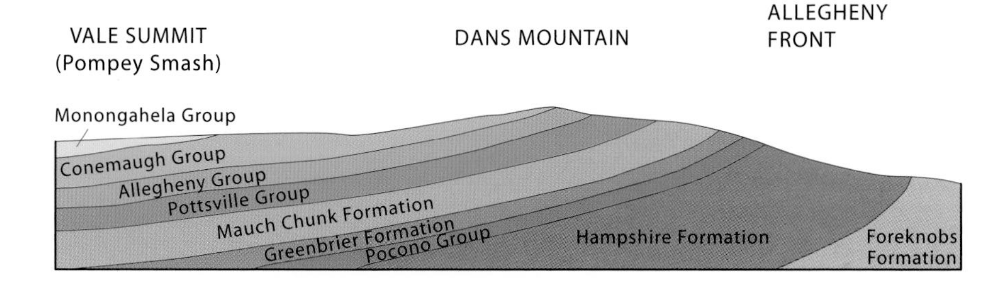

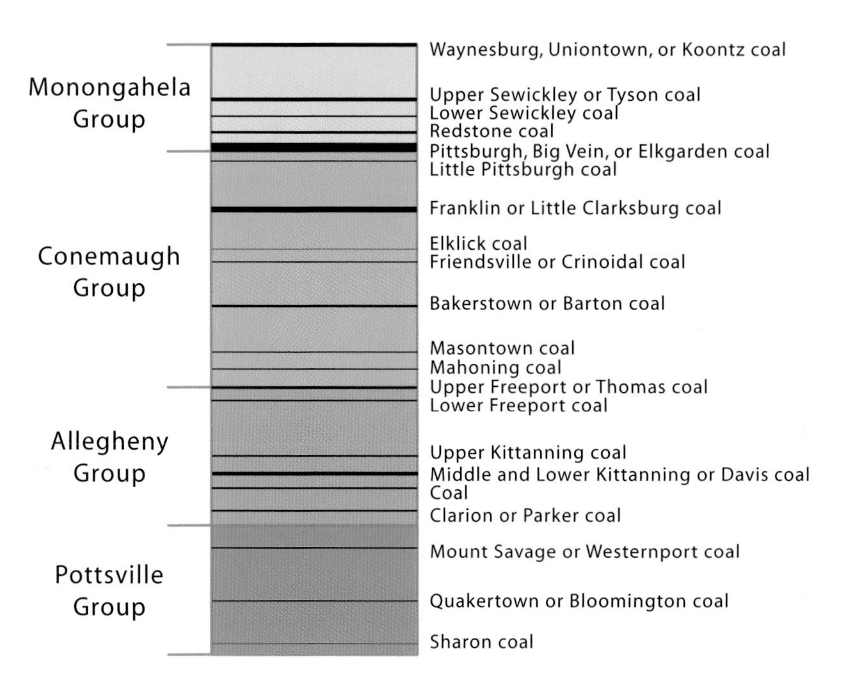

The many coal veins in the groups of the Georges Creek syncline. —Modified from Abbe and others, 1902

tectonically located in the tropics and covered by a shallow inland sea. Bordering the sea, shallow coastal swamps produced prolific vegetation, resembling the Everglades in Florida. Falling leaves, branches, and trunks collected and settled on the bottoms of the swamps, while other growth propagated from the accumulating plant debris. Stagnant, low-oxygen, shallow water prevented the plant decay that would have released its stored energy into the atmosphere. This partially decomposed, compressed mass of plant material became peat.

Inland seawater levels rose and fell. When levels rose, the swamps became submerged in deeper water, and the peat was buried under thick layers of sand and mud, protecting it from decay. Such burial also generated the heat and pressure necessary to drive off moisture and gases and to compact the peat into coal. The deeper the burial, the lower the oxygen and hydrogen content, and the higher the carbon content—and the higher the heat value of the mined coal. Maryland's coal

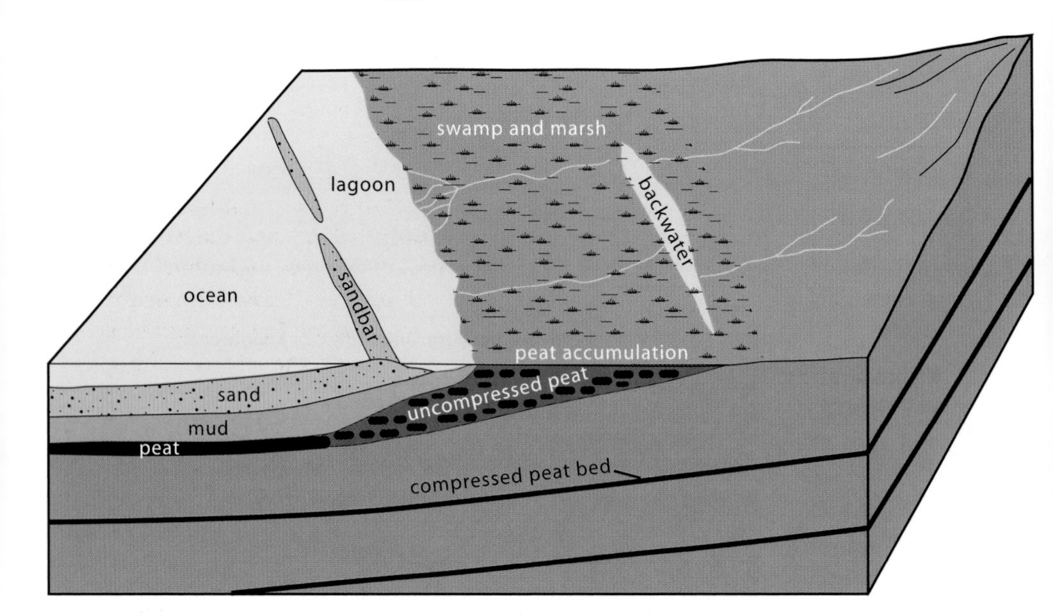

Ancient swamp vegetation becomes coal when it is buried and compacted under many layers of sediment. —Modified from Edmunds and Koppe, 1968

has always ranked high. In 1836 a geologist familiar with European coals concluded that none compared to Georges Creek coal.

Falling sea levels caused the return of swamps on top of the thick sand and mud deposits, and later rising waters repeated the burial and compaction process. Numerous rises and falls resulted in cycles of coal formation so that today beds of coal are found interbedded with sandstones, shales, siltstones, and even limestones. These rock layers formed the footwalls and hanging walls of underground mines, and as the coal was removed, they had to be braced and timbered, or they would collapse. Today the stone layers are removed by strip mining on the land surface.

In Maryland the sedimentary beds containing coal total about 2,000 feet in thickness, and they have been compared to a thick book with every twentieth page made of black paper, representing coal. In the Georges Creek Basin, the book is more like a series of stacked bowls such that the coal is buried at the center of the synclinal basin but crops out at intervals between the center and the bowl rims—the west slope of Dans Mountain and the east slope of Big Savage Mountain.

Tens of feet of vegetation were compacted to form 1 foot of coal, and the heat energy transferred from the sun of 300 million years ago into plant growth is today being transferred back into heat energy when combined with oxygen during combustion—thus the term *fossil fuel*. Half of the nation's electricity is produced by burning coal to generate high-pressure steam, which rotates turbines connected by shafts to electrical generators. Ultimately, we are returning ancient heat to today's global energy system, and we have generated accelerated global warming in the process.

To the east and west on the slopes above Georges Creek and MD 36 between Frostburg and Westernport, the three major layers that compose the Monongahela

Group coal beds are exposed. From Ocean south to Westernport, Georges Creek and its steep tributaries have eroded the land more deeply than to the north of Frostburg, and the coal veins are thus exposed on the slopes above the creek. Many were unearthed by massive floods in 1810. Being in the middle of a very gently folded syncline, the beds are level or nearly so—a condition that enabled nineteenth-century mining companies to start at a hillside exposure and pick and tunnel into the almost horizontal seam, a method known as drift mining. Because the seams were not buried and did not require the vertical shafts of deep mining, groundwater drainage could be achieved with minimal pumping. Because the miners did not have to work far below the surface, mining in Maryland has been relatively safe—if mining can ever be called safe.

The Monongahela Group of Pennsylvanian age has (or in the case of mined areas, had) at its base the 14-foot-thick Pittsburgh coal, better known as the Big Vein. About 130 and about 240 feet above that are two other coal veins, the Sewickley and the Uniontown. Much of the Big Vein has been mined out.

The coal boom began in Maryland during the decades preceding the Civil War. Coal was shipped over the C&O Canal for many decades, but in the 1890s cheaper and faster rail shipping replaced it. The coal was high quality, and during the era of steam power, it was used for locomotives, steamboats, factories, and American and British naval ships. The year of greatest production was 1907, with over 5 million tons, but 1920 was the last year in which totals exceeded 3 million tons. After World War II, deep mining and drift mining decreased, and strip (or

This old mine entrance is on the east side of MD 936 south of Frostburg about one-quarter mile north of Ocean. You can see "1908" and "Mine No. 10" on the concrete portal. We owe much to those who passed through it.

surface) mining increased and continues today. Gravity water drainage from old mines and from old and current surface pits has caused overly acidic conditions in streams—western Maryland's most serious water pollution problem. Coal mining will, however, continue for a while here because western Maryland still has several hundred million tons of recoverable reserves.

You can see evidence of past and present mining in many places. From MD 936, which forms a loop with MD 36 south of Frostburg, you can see strip mining on the eastern slope of Big Savage Mountain. If you want a closer look you can turn onto Shaft Road Southwest at Borden Shaft, follow it to Midlothian, and turn onto Carlos Road. A rough, dirt road called Fairview Farm Road takes you up Big Savage Mountain to the Fairview Surface Mine. You do not need to drive all the way in to get a good look at it. If you continue south on Carlos Road to Carlos, you will see the Carlos Surface Mine. Both operations cover large areas.

Dans Rock Overlook

At Paradise Avenue, less than a half mile north of Midland and the junction of MD 36 and MD 936, you can drive east above the coal beds to the top of the Georges Creek syncline. Look for a sign on MD 36 for Paradise Avenue/Dans Rock Overlook. Turn onto Paradise Avenue and then turn east onto Dans Rock Road and go up the mountain. From Dans Rock, at an elevation of almost 2,900 feet, you can look east and far below to the Valley and Ridge Province, where the ridgetops are in the 1,500 to 1,700 foot range. This is the eastern edge of the Allegheny Plateau. Don't forget to look down beneath your feet at sandstone of the Pottsville Group.

Lonaconing Iron Furnace

Just north of Lonaconing on MD 36, you can see spoil piles above Georges Creek. In Lonaconing you can stop at a park commemorating the old Lonaconing iron furnace as well as Lonaconing native Lefty Grove, American League MVP in

The Lonaconing iron furnace was built against a hillside so that horse-drawn wagons could dump iron ore, coke, and flux into the top.

1931. Because of the isolation of Lonaconing, local geologic materials were used in building the furnace: clays near here were good for firebrick, local limestone was good for mortar (but not for furnace flux), and quarries on the slopes above the town supplied sandstone for the furnace walls. Unfortunately, poor-quality iron ores forced the end of production in 1856. It was, however, thought to have been the first successful coke-fired blast furnace in the United States, reaching its highest output in 1839. The search for fuel to fire the furnace resulted in the discovery of the finest bituminous coal to date, and coal mining eventually supplanted iron smelting as the profit of choice.

Coal Mining between Barton and Westernport

In Barton turn west on Bartlett Street, drive up about 1 mile to a Y, and take the gravel Russell Road to the right—it is a mining road, but it is also a public road. Within the next mile you will see that the slope has been strip-mined extensively, and you may see work going on. Look across the valley, and you can see evidence of mining at about the same elevation on the opposite slope—indicating a nearly symmetrical synclinal fold.

Between Barton and Westernport you can see further alteration of terrain from mining to both the east and west. At Westernport, Georges Creek enters the North Branch of the Potomac.

Extensive surface mining continues above Barton off Russell Road in one of the coal veins of the Georges Creek Basin.

Maryland 135
Mountain Lake Park—McCoole
25 MILES

MD 135 travels from the Allegheny Plateau to the Valley and Ridge. It drops down and roughly parallels the 116-feet-per-mile Seventeen Mile Grade of the railroad over Backbone Mountain. Then the highway passes through the water gap cut through Dans Mountain (the Allegheny Front) by the North Branch of the Potomac from Westernport to McCoole.

East of Deer Park, just west of the MD 495 junction, is an exposure of the red Hampshire Formation, here a limb of the Deer Park anticline. Then, to the east, MD 135 climbs Backbone Mountain, Maryland's highest—a continuous Pottsville sandstone ridge that is called Big Savage Mountain to the northwest. A roadside overlook provides a vista of the anticlinal valley of Deep Creek to the west.

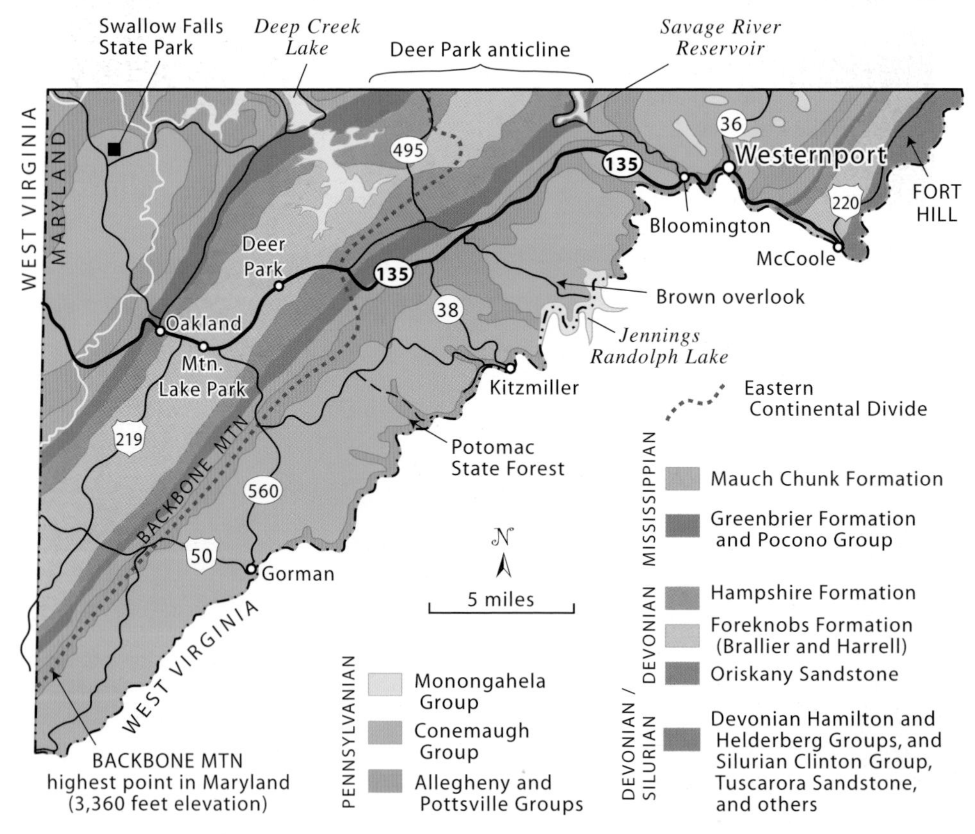

Geology along MD 135 between the West Virginia border and McCoole. —Modified from Cleaves, Edwards, and Glaser, 1968; Darling and Slaughter, 1962; Glaser, 1994b

Kitzmiller Side Trip

East of the top of Backbone Mountain is an interesting side trip down MD 38 to Kitzmiller on the North Branch of the Potomac River. Stop near the bridge and take a look at the flood wall built by the U.S. Army Corp of Engineers. Most of the time the North Branch does not carry much water here, but the tributaries upstream descend steeply from high plateau gulleys, and the water can rise quickly. From Kitzmiller you can drive upstream to the old coal mining company town of Shallmar.

The North Branch of the Potomac, a small creek here in Kitzmiller, can become a torrent, and the flood wall is designed to protect the town.

On the road between Kitzmiller and Shallmar, you can see beds of coal in between more resistant, protruding sandstone layers of the Allegheny and Pottsville Groups (mapped together here).

Brown Overlook

On MD 135 look for the sign for Randolph Lake Boat Launch and follow it to Brown overlook for a view of the North Branch valley and coal mining operations on the slopes above Shallmar. You might also see the stacks of the coal-fired electrical generating plant at Mount Storm, West Virginia. You can drive on down for a look at the reservoir and dam at Jennings Randolph Lake, a recreational and flood-control facility.

Savage River

At Bloomington, you can drive up Savage River Road, which parallels the Savage River to the Savage River Reservoir. On the floodplain you will see a few dwellings that would be in harm's way during heavy rains or sudden snowmelt were it not for the dam. The river gradient is steep here: Olympic whitewater canoeing and kayaking trials and many other events have been held here. Water is released from the dam for these competitions.

An outcrop of the Pottsville Group along Savage River Road. When this small anticline of weak shale erodes, the sandstone will lose support and collapse.

Savage River has cut a gorge through the interbedded sandstones and shales of the Pottsville Group. It may not be wise to stop when you see a "Falling Rock" sign, but it is usually a good place to observe differential weathering, where less resistant shale and coal layers collapse and cause overlying sandstone to lose support, fracture, and tumble down. During spring freeze-and-thaw season, you may see stone on the road.

At the dam, you used to be able to walk across to the spillway to observe the large, exposed cut of dipping, interbedded limestones and red, calcareous shales of the Greenbrier Formation. You may be able to see them from the fence near the road.

Westernport Area

Between Bloomington and Westernport you may see railroad cars filled with coal, and if you look carefully along the way, you can see coal in one of the roadcuts. In the eighteenth century, Westernport acquired its name because it was the limit of navigation on the North Branch of the Potomac. During the nineteenth century, coal shipments were floated down during high water. Just east of Westernport look for a west-dipping seam of coal—a Georges Creek Basin coal—that you can see in cross section.

About 2 miles east of Westernport, along MD 135 on the outside of a river bend, you can find recently deposited stone. Stony Run enters the North Branch here, and you will see that the stones are rounded, indicating transport of some distance in high, fast-moving water that rounded the corners and edges of the stones. Determining whether they were deposited by Stony Run or by the North Branch could be done by correlating or matching the stones with formations existing within each drainage basin.

Between Westernport and McCoole you can see several good rock exposures. Here the North Branch has cut down through Dans Mountain of the Allegheny Front, exposing progressively older rock as you travel east through the limb of the broad syncline. As you drive the road, notice that sandstone cliffs just above the road were cut by human machines, whereas cliffs higher up were cut by a river ancestral to the North Branch, flowing at a higher level. About midway between Westernport and McCoole, look for exposures of crossbedded sandstones of the ubiquitous, red Hampshire Formation. As you go under US 220 but before entering it, look at the vertical-dipping shale along the road, the carbonaceous Marcellus Shale of the Hamilton Group. If you look west from McCoole, you can see high Backbone Mountain in the distance.

In the Hampshire Formation on the north side of MD 135 between Westernport and McCoole, resistant sandstone layers form sheer faces, while shale erodes into soil where vegetation takes hold.

•— Valley and Ridge —• and Blue Ridge Provinces

If you fly over western Maryland, you will see an 80-mile-wide region of alternating valleys and ridges, mostly parallel, with over a dozen in the sequence, oriented northeast to southwest. If you drive from Frederick to Cumberland on I-70 and I-68, you will find yourself continually climbing and descending, and only seldom driving on level ground. This region includes two provinces, the Valley and Ridge Province of sedimentary rock and the Blue Ridge Province of mostly metamorphic rock. For roadside geology purposes, the two provinces will be treated together because they resemble each other in topography.

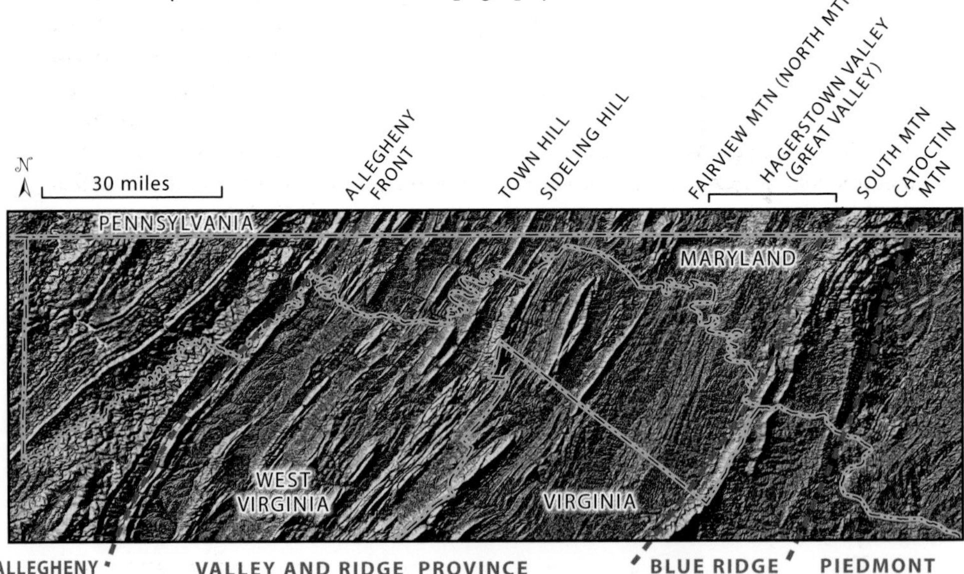

The shaded relief map of western Maryland clearly shows the northeast-trending ridges and valleys of the Valley and Ridge and Blue Ridge Provinces. —Compiled at Mountain Press from U.S. Geological Survey, National Elevation Database

The rock layers of these provinces have been tightly folded into wavelike structures called anticlines and synclines by the great tectonic pressure exerted when the African Plate shoved into the North American Plate during the Alleghanian mountain building event between 320 and 250 million years ago. The rock layers on the surface today lay deeply buried during those times, under so much pressure they were plastic or bendable—thus the folding. Over 200 million years of erosion have removed thousands of feet of covering rock to expose what we see today—the lateral limbs, or flanks, of the folds, with the tops removed.

Why were ridges and valleys rather than a level plain produced by weathering? Because of the folding, adjacent layers of different kinds of rock were exposed to the elements, and different rocks weather at different rates. Of the three most common sedimentary rocks, limestone and shale weather much faster than sandstone. As a result, the sedimentary region is today composed of relatively resistant sandstone ridges alternating with shale or limestone valleys—the more rapidly weathering rocks losing elevation more quickly. In the metamorphic Blue Ridge Province, metabasalt and gneiss eroded into a valley, leaving quartzite ridges standing to either side. In each case, the landscape produced is strikingly varied and beautiful.

The Blue Ridge Province lies east of the Valley and Ridge Province, from the eastern base of Catoctin Mountain, which flanks Frederick Valley, to the western base of South Mountain, which flanks the Great Valley. It is composed of rock that was once the Cambrian-Ordovician offshore wedge: Grenville basement gneiss, Catoctin lava flows, and an overlying sedimentary package that included Weverton sands, Harpers shales, and many limestone formations. Being on the colliding edge during the tectonic collision with the African Plate, many of these layers were metamorphosed and folded into a huge, upward-arching, overturned fold called the South Mountain anticlinorium, a regional-sized anticline. This large structure, measuring about a dozen miles across, defines the Blue Ridge Province.

The term *Blue Ridge* has several applications, which can be confusing. The Blue Ridge Province, stretching from Georgia to Pennsylvania and bordered on the east by the Triassic Border Fault, is the easternmost expression of the Appalachian Mountains. It is sometimes called the Blue Ridge Mountains, particularly in Virginia, or just Blue Ridge. In Maryland and Pennsylvania, however, South Mountain and Catoctin Mountain are the names of these continuous elongate ridges. In Maryland, a third ridge in the province is called the Blue Ridge. This thrust-faulted section ends in Washington County where it slopes into the Great Valley due east of Antietam Battlefield.

The Valley and Ridge Province stretches over 60 miles from the western base of South Mountain all the way to the foot of the Allegheny Front. From airplane or satellite, its most visible feature is the Great Valley, called Hagerstown Valley in Maryland—a 20-mile-wide, mostly limestone valley flanked by South Mountain to the east and North (Fairview) Mountain to the west. Its carbonate rock is discussed below.

West of the Great Valley, weathering of tightly folded sedimentary rock has produced a series of closely spaced ridges. From the top of one ridge, you can usually

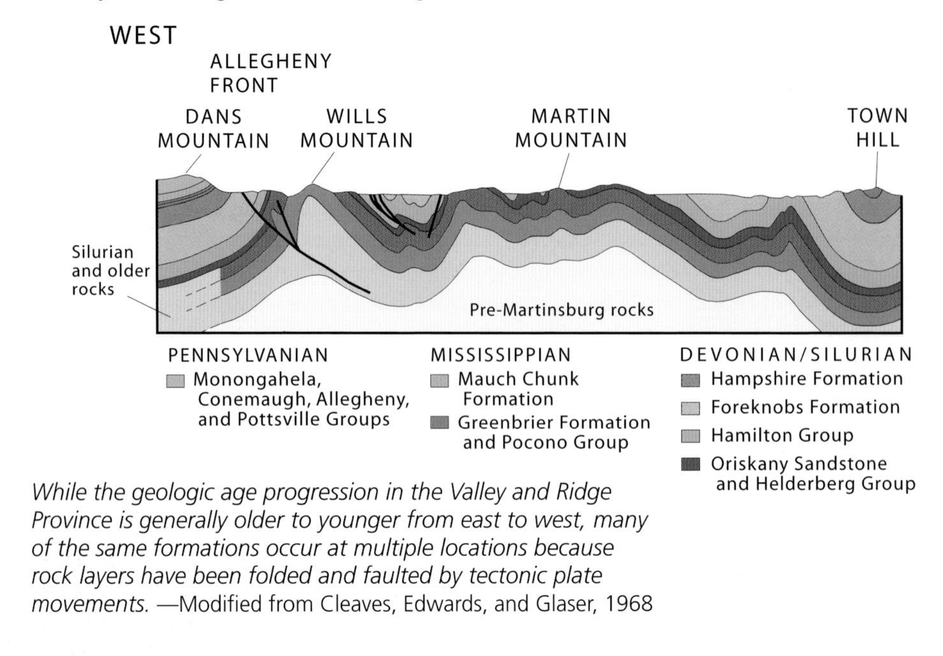

While the geologic age progression in the Valley and Ridge Province is generally older to younger from east to west, many of the same formations occur at multiple locations because rock layers have been folded and faulted by tectonic plate movements. —Modified from Cleaves, Edwards, and Glaser, 1968

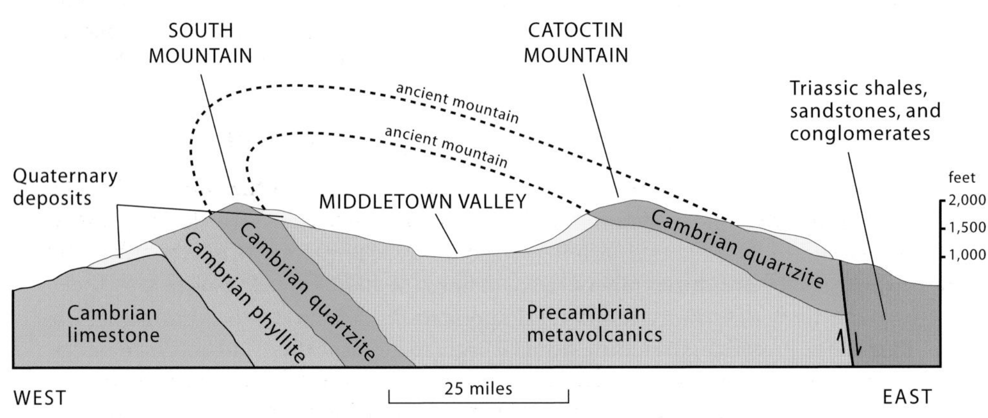

Rock of an ancient North American craton lies at the surface in Middletown Valley between the long-weathered limbs of a giant anticline that once reached alpine proportions. —Modified from Cloos, 1947

see one or more ridges in the distance, with valleys in between. Most of the ridges are 1,500 feet in elevation, more or less, while the intervening valleys are 500 feet or more below the ridge crests. Here the collision with Africa folded the thick sedimentary packages of the Queenston Wedge, eroded from the Taconic Mountains, and the Catskill Wedge, eroded from the Acadian Mountains. These thousands of feet of rock were composed primarily of clastic sediments—sandstones, siltstones, and shales. The tighter folding of the Valley and Ridge as opposed to gentler folds of the Allegheny Plateau is thought to be due in part to the greater resistance to compression put up by the thicker sections of these sedimentary wedges.

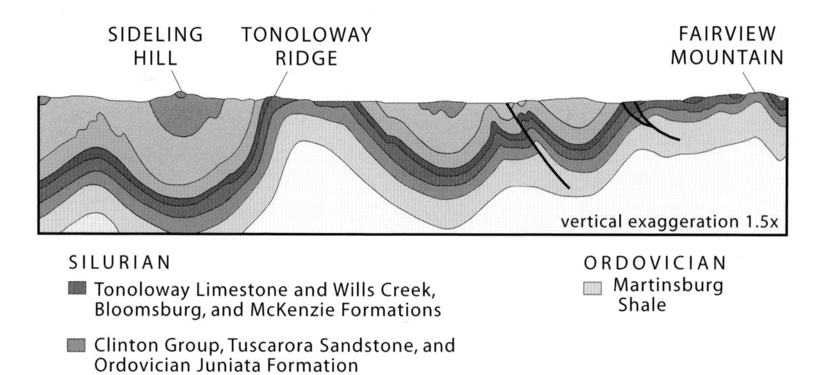

The progression in geologic age from east to west across these two provinces is generally older to younger, but this may not be true for a given sequence. Because of the multiple folds, and because of thrust faulting that also occurred, many of the formations repeat themselves as limbs of folds, successively dipping in opposite directions, across many miles. For example, you can see the red Hampshire Formation five times in a 20-mile stretch between Licking Creek and Fifteen Mile Creek.

GREAT VALLEY OF CARBONATES

The 20-mile-wide Hagerstown Valley is just a small piece of the Great Valley, the remains of a once-offshore carbonate bank that today stretches from New England to Alabama. Carbonate rocks, the most common being limestone, form very differently from the way other sedimentary rocks do. Clastic rocks like conglomerate, sandstone, and shale form by the cementing of fragments that have eroded from preexisting rocks—fragments that have often been deposited on seafloors. Carbonate rocks, however, form directly from the sea and may or may not contain clastic materials. They can form from precipitation of minerals in solution or from the accumulation of wave-broken fragments of the remains of marine organisms such as algae, corals, and shelled creatures.

Stream water and seawater are never pure. They contain ions of minerals, chemical compounds that have dissolved from rock due to weathering and erosion of rock. One of those many ions is calcium, Ca^{+2}. Water, particularly precipitation, also contains gases that have dissolved from the atmosphere. One of these is carbon dioxide, CO_2, the same gas that comes out of solution to form bubbles in cola and beer. Chemists know that removing CO_2 from water allows an increase in dissolved carbonate, the ion $(CO_3)^{-2}$. When Ca^{+2} and $(CO_3)^{-2}$ combine, calcium carbonate, $CaCO_3$, is formed. Limestone is composed of this chemical compound.

Therefore, any process that removes CO_2 from seawater will promote the precipitation of limestone. Three conditions will do this: raising the temperature, reducing the pressure, and agitating the water. These same three will remove CO_2

TIME	VALLEY AND RIDGE FORMATIONS		BLUE RIDGE FORMATIONS
Miss.	Pocono Group		
	Purslane Sandstone		
Dev./Miss.	Rockwell Formation		
Devonian	Hampshire Formation		
	Foreknobs Formation	Formerly Chemung, Woodmont, and Jennings Formations	
	Scherr Formation		
	Brallier Formation		
	Harrell Shale		
	Hamilton Group		
	Mahantango Formation		
	Marcellus Shale		
	Needmore Shale		
	Oriskany Sandstone		
	Shriver Chert		
Sil./Dev.	Helderberg Group		
	Keyser Limestone		
Silurian	Tonoloway Limestone		
	Wills Creek Formation		
	Bloomsburg Formation		
	McKenzie Formation		
	Clinton Group		
	Rochester Shale		
	Keefer Sandstone		
	Rose Hill Formation		
	Tuscarora Sandstone		
Ordovician	Juniata Formation		
	Martinsburg Shale		
	Chambersburg Limestone	carbonates of Hagerstown Valley	
	St. Paul Group		
	New Market Limestone		
	Row Park Limestone		
	Beekmantown Group		
	Pinesburg Station Dolomite		
	Rockdale Run Formation		
	Stonehenge Formation		
	Conococheague Formation		
Cambrian	Elbrook Formation		Tomstown Formation
	Waynesboro Formation		Chilhowee Group
			Antietam Formation
			Harpers Formation
			Weverton Formation
			Loudoun Formation
Proterozoic			Catoctin Formation
			Swift Run Formation
			Middletown Gneiss

Rocks of the Valley and Ridge and Blue Ridge Provinces.
—Compiled from Cleaves, Edwards, and Glaser, 1968; Edwards, 1978; Darling and Slaughter, 1962; Fauth, 1977; Fauth and Brezinski, 1994; Brezinski, 1993; Glaser, 1994b; Brezinski, 1992; Southworth and Brezinski, 1996; Sando, 1957

from your cola: heat it in a pan and watch the bubbles come out; pop the top and hear the release of gas; shake it and see it fizz.

For limestone formation, we need a tropical location for warmth, shallow depths so that underwater pressures are not too great, and wave and tidal movement for water agitation. These conditions were met from early Cambrian through middle Ordovician time, a period of about 100 million years, when the then-smaller North American continent straddled the equator. The sea had risen and submerged the continent edge. Offshore, the Chilhowee Group of clastics had been deposited on top of the Catoctin Formation lava flows. A continental-length offshore carbonate bank was then deposited on top of the Chilhowee Group in warm, shallow, tidal flats.

Other factors affect carbonate deposition. Marine organisms use seawater to produce their calcium carbonate shells and skeletons, which accumulate on the seafloor when the organisms die. Plants remove CO_2 from water, and because light promotes plant growth, equatorial positioning and shallow water optimize exposure. Muddy, sediment-filled water interferes with both plant and animal health, but because the Grenville Mountains had pretty much been eroded, the offshore waters were receiving very little sediment.

Classification schemes for different varieties of carbonate rock are highly complex and reflect different environments of deposition as well as different marine organisms living during a certain geologic period. After limestone, the most common rock of this carbonate bank is dolostone, often called after its principal mineral, dolomite. Dolomite is a rock in which part of the calcium has been replaced by magnesium. $CaCO_3$ becomes $CaMg(CO_3)_2$. The process of dolomitization is still not fully understood, but possible factors include periods of global warming, extreme evaporation and hypersaline waters, groundwater reflux through limestone, and excretions of ancient bacteria. What is known is that today very little dolomite is forming, and the proportion of dolomite in carbonate rock increases with geologic age. In the carbonates of Washington County, the limestone to dolostone ratio is 3 to 1. In addition, the Cambrian and Ordovician formations contain dolomite, while the later (younger) Silurian and Devonian formations of the Queenston Wedge do not. Interbedded gray limestone and tannish brown dolomite, often in thin laminated layers, reflect cyclical changes in depositional environments from shallow subtidal for limestone to supratidal for dolomite. The supratidal zone is above high tide, immersed only during unusually high tides or storm surges. Subsequent evaporation from these flooded areas created conditions under which dolomitization occurred. Today you can see interbedded layers in many Great Valley formations.

The 100-million-year-long deposition of the 2-mile-thick carbonate bank ended in middle Ordovician time with the tectonic convergence from the east of an offshore island arc and the subsequent collision with it—the Taconic mountain building event. Tectonic downwarping of the carbonate platform and deepening waters became the depositional environment for the influx of clays and sands from the emerging Taconic Mountains. These fine sediments covered the last limestone, the Chambersburg, and solidified into the thick Martinsburg Shale.

About 200 million years later, the African Plate converged and collided with the eastern shore of North America, causing the Alleghanian mountain building event. The compression shortened the east-west width of the carbonate bank by about 50 miles, or 30 percent, through folding and fracturing. Tiers of buried layers were driven horizontally over underlying layers in sheetlike fashion along slip zones. When these tiers bent toward the vertical and thrust faulted, they brought up to the surface carbonate layers from many miles to the east and from many feet belowground. In addition, the overall syncline in the carbonate bank placed the younger Martinsburg Shale near the middle of the older carbonates.

Calcium carbonate reacts chemically with acidic rainwater and groundwater, causing limestone to weather rapidly compared to, say, sandstone. Thus, carbonates have weathered to valleys while sandstone remains as higher-elevation ridges. Within the valleys, though, there is much local relief—an up-and-down, rolling landscape reflecting the different resistances to weathering of individual carbonate formations, often related to the presence of interbedded sandstone layers or to sand within the carbonate layer.

Also, limestone is very different from clastic rock hydrologically. In limestone, groundwater dissolves underground solution cavities and channels, often along fractures and bedding-plane partings. As a result, surface drainage is of relatively low volume because much of the water moves down and through the permeable subsurface rock as groundwater. Thus, limestone terrains often have fewer surface streams and creeks than are found in other rock formations but more springs—places where groundwater emerges onto the surface from an underground channel. More than 190 springs have been identified in the carbonates of Hagerstown Valley.

Underground cavities and channels, if large enough, are known as caves or caverns. Ones that are not currently filled with groundwater were dissolved in the limestone when the whole landscape and water tables were higher in elevation, at a less advanced stage of erosion. There are more than fifty known caves in Hagerstown Valley, as well as about two hundred cavernous zones that have been intersected by wells. Sinkholes—depressions where ground has collapsed into underground cavernous openings—are very common. Areas with sinkholes and cavern systems are called karst topography. Because of the interconnectedness of passageways, groundwater in karst terrain is more susceptible to contamination than it is in most other rock types.

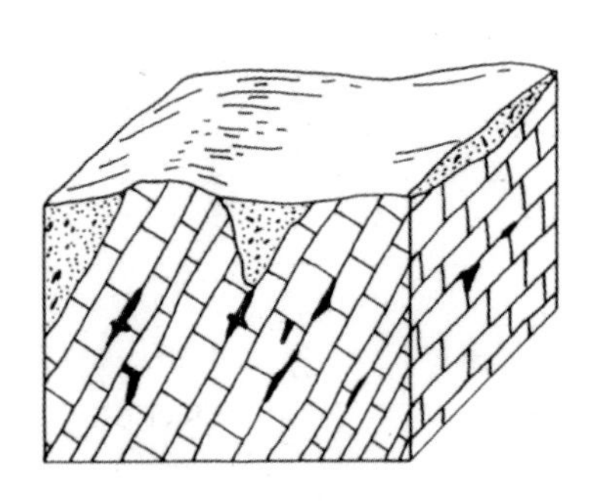

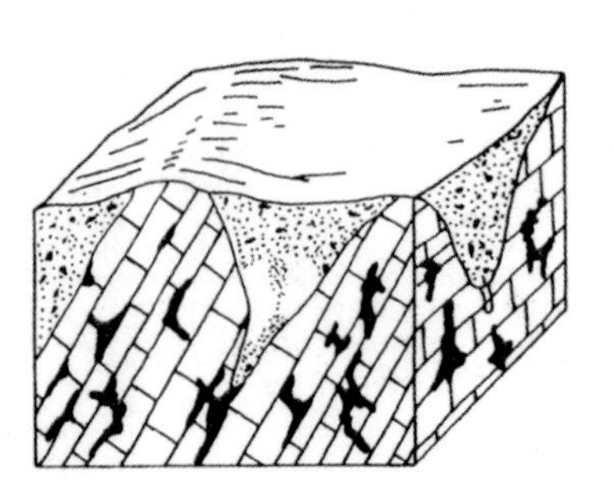

Limestone is basic, and groundwater is acidic. The resultant acid-base reactions create caverns and channels in limestone country. The dark zones are caves. —From Geyer and Wilshusen, 1982

Road Guides to the Valley and Ridge and Blue Ridge Provinces

Interstate 70 and US 40
Braddock—Hancock
50 MILES

The I-70/I-68 east-west connection between Braddock (west of Frederick) and La Vale (west of Cumberland) provides an excellent synopsis of the Blue Ridge and Valley and Ridge Provinces. Although many interstates are not good for roadside geology, this route is an exception, with numerous excellent roadcuts. Although stopping on the shoulder of this very busy route is extremely dangerous, you can walk onto the shoulder and look closely at exposed rock in some places if you want to take the risk. The National Road, old US 40, parallels I-70 and I-68 under different road numbers: US 40, US 40 Alternate, Scenic 40, and MD 144. It is accessible from most interstate exits and can be used for a slower paced trip. While National Road roadside stops are probably safer, the roadcuts are usually not as large and are more weathered due to their earlier construction.

I-70 exits 48 and 49 at Braddock are less than 1 mile west of the Triassic Border Fault. In fact, if you drive about 1.5 miles toward Frederick on US 40 through a congested mall area, you can see exposures of Potomac "marble" on a slope in front of a retail store. This is not a true marble but a conglomerate that consists of marble pebbles eroded from the high fault block—Catoctin Mountain—down into the basin that became the Frederick Valley.

West of the shopping area is the eastern slope of Catoctin Mountain—the eastern limb of the South Mountain anticlinorium. At the tops of the long, northeast-to-southwest-trending South and Catoctin Mountains is the highly resistant quartzite of the Weverton Formation of Cambrian age. This quartzite, a hard, ridge-forming metamorphosed sandstone, runs north into Pennsylvania and south into Virginia. On a clear day you can see from one mountain to the other across the intervening Middletown Valley. At maximum elevations of about 1,700 to 1,800 feet, these two ridges stand well above the carbonate elevations of about 400 to 500 feet of Hagerstown and Frederick Valleys, and noticeably above the 500- to 700-foot elevations of Middletown Valley. From Catoctin Mountain to South Mountain, a distance of about 10 miles on I-70, you travel from the Weverton Formation down onto the Catoctin Formation of metabasalt (greenstone) and metarhyolite and then back up and onto the ridge-forming Weverton again.

How can the Weverton Formation repeat on the landscape? These two ridges are the limbs of a giant anticline called the South Mountain anticlinorium. Compression during the Alleghanian mountain building event buckled the rock layers into a giant arch and metamorphosed them into crystalline rock. The limb farthest away from the colliding African continent was tilted over by the great compressive force.

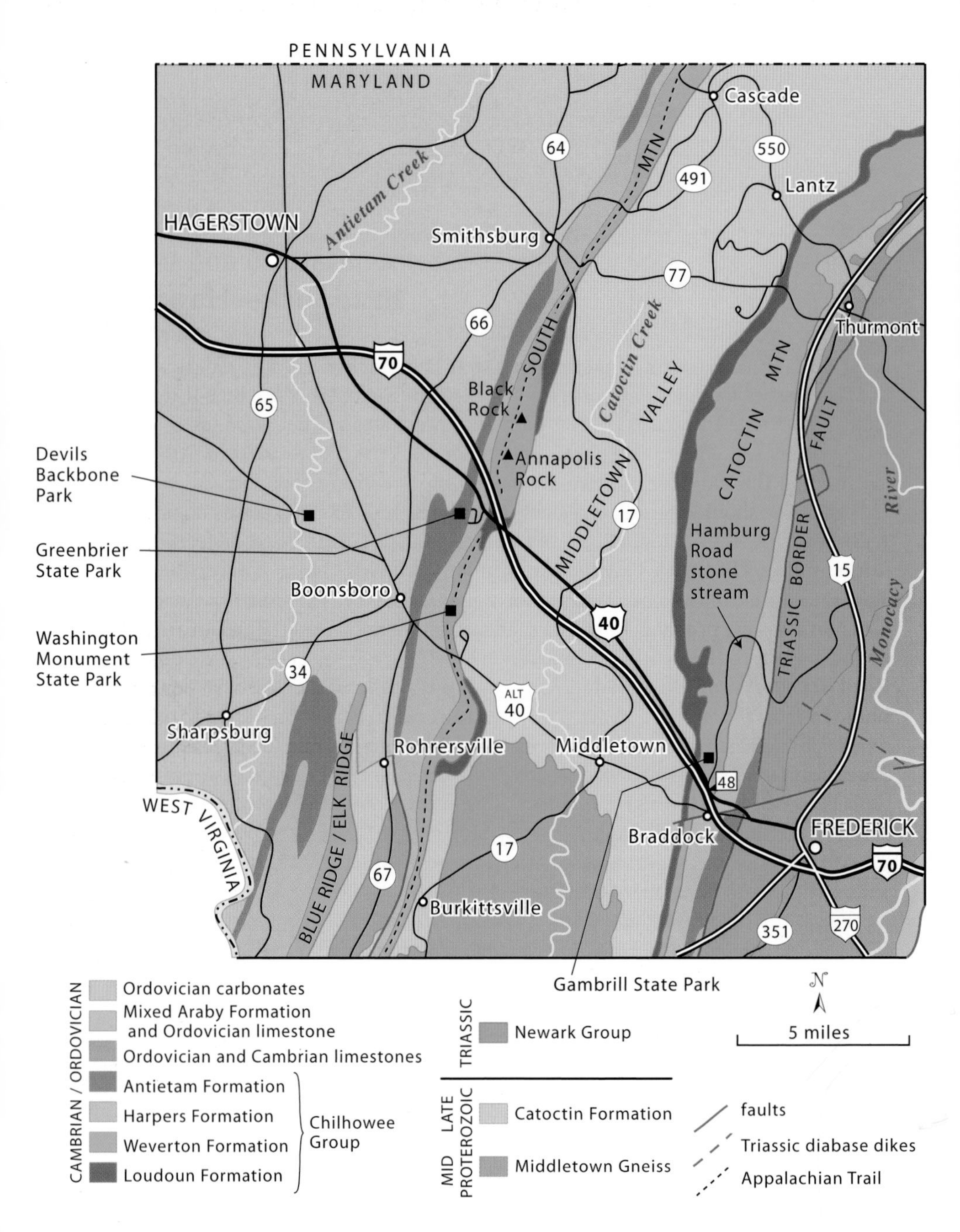

Geology along I-70 and US 40 between Frederick and Hagerstown.
—Modified from Cleaves, Edwards, and Glaser, 1968; Edwards, 1978; Fauth, 1977; Reinhardt, 1974; Fauth and Brezinski, 1994; Brezinski, 1992; Southworth and Brezinski, 1996

More than 200 million years of weathering and erosion have removed the top of the giant arch, leaving a valley between ridges—the rock in the middle weathering faster than the hard quartzite on the limbs.

I-70 over Catoctin Mountain

West of exit 48 at the top of Catoctin Mountain you see not the Weverton Formation, which is thinly exposed here, but a deep roadcut in the Catoctin Formation, originally a basaltic lava that poured out episodically about 600 million years ago through cracks in the 1.1-billion-year-old Grenville basement rock. Here it dips to the east as part of the eastern limb of the anticline. The Catoctin Formation runs across the valley to the eastern base of South Mountain, and you'll see the metamorphosed basalt, aptly called by its field name, *greenstone*, in roadcuts.

Gambrill State Park from US 40

On top of Catoctin Mountain is a sign for Gambrill State Park. Drive up about 1 mile and see a great view of Middletown Valley and the two ridges that form the limbs of the anticline. The bedrock you see in Gambrill State Park is the ridge-forming Weverton Formation.

Not far away is a rare opportunity to drive through a large boulder field. Go about 3.5 miles north on Gambrill Park Road, turn east (right) onto Hamburg Road, and descend Catoctin Mountain. Soon you will be among a slope of quartzite boulders, which probably slid down the mountain in a water-saturated soil layer over a buried permafrost layer during the Pleistocene Ice Age. See **Maryland 77: Thurmont—Smithsburg** for a detailed explanation of this process. As in Catoctin

You can see this impressive stone stream, a relict from the Pleistocene Ice Ages, from Hamburg Road near Gambrill State Park.

West of Ridge Road on US 40, you can find dark green Catoctin metabasalt with white quartz veins, formed when water that was superheated by magma dissolved quartz and flowed through fractures, later cooling and solidifying.

Mountain Park to the north, the source of these large stones is an upslope scarp or cliff in the middle member of the Weverton Formation, the ledge-maker quartzite. You usually have to do some strenuous hiking to view boulder-covered slopes, but here you can see them from your car, and you may be able to find a shoulder by the road for parking.

West of the Gambrill Park Road exit near Ridge Road and Shookstown Road are good exposures of east-dipping metabasalt of the Catoctin Formation. You cannot miss its characteristic dark green color and occasional white quartz veins.

I-70 over South Mountain

Rock exposures at the top of South Mountain are poor, but if you are westbound, you can see the broad carbonate valley to the west. If you are eastbound from the Hagerstown Valley floor, you can see west-facing cliffs and talus of Weverton quartzite at the very top of South Mountain to the northeast.

US 40 over South Mountain

An excellent exposure along US 40 is on the eastern slope of South Mountain west from Easterday Road. Near the top of this slope, near the state highway facility, you can see the rock change from the metabasalt greenstone to the light gray of a finely laminated metarhyolite. This rhyolite was also part of the Catoctin volcanic outpouring, but rhyolite differs chemically from basalt. Rhyolite is felsic lava, having

Metarhyolite along US 40 near the state highway facility on the eastern slope of South Mountain.

Freeze-and-thaw erosion of exposed, overturned quartzite beds of the Weverton Formation created this huge talus field below Black Rock.

a high percentage of silicon, while basalt is mafic lava, containing magnesium and iron, and lower percentages of silicon. You can spot the difference easily because the rhyolite is lighter in color and has many layers. Farther west, upslope and west of Pleasant Walk Road, you can see more of the metarhyolite.

At the top of South Mountain in the roadcut next to the parking area for the Appalachian Trail is yet another exposure of the Catoctin metarhyolite. From this convenient US 40 access point to the Appalachian Trail, you can hike north 2 miles one way to Annapolis Rock or 3 miles one way to Black Rock, each an escarpment of high cliffs of the hard, ridge-forming quartzite of the Buzzard Knob Member of the Weverton Formation. If you look closely at the quartzite rock of the clifftops, you will find much crossbedding. Below each cliff you can see talus fields of boulders, an especially extensive one at Black Rock.

If you work your way around the cliff and follow the boulder field down the mountain, you will find that it goes well into the woods. Freeze-and-thaw weathering, in which water seeps into cracks, freezes, and expands, is probably responsible for this mass of angular quartzite boulders. From either vantage point you can see all the way across the carbonate Hagerstown Valley to the sandstone ridges to the west.

Washington Monument State Park

You can reach the ridgetop vista at Washington Monument State Park from US 40 by taking Boonsboro Mountain Road—the first turn west of the US 40 overpass of I-70—to Boonsboro and then following US 40 Alternate east up South Mountain,

You can see crossbedding in the Weverton quartzite stones of the tower on South Mountain at Washington Monument State Park.

or by taking Monument Road from US 40 near the bottom of the eastern slope of South Mountain. The monument, which shares its name with the one in D.C., is a tower built of Weverton quartzite boulders on top of South Mountain in 1827 by the residents of Boonsboro as the first monument to George Washington. It is probably the most easily accessed point on top of the mountain for seeing a view of the Great Valley (here called the Hagerstown Valley) and for seeing one of the large talus fields of angular boulders that are common on the mountain. You can climb the tower, and in the stones of the tower you can see crossbedding.

On US 40 Alternate between the mountaintop and Boonsboro, you pass east-dipping exposures of the Harpers siltstone.

Greenbrier State Park and Vicinity

At Greenbrier State Park, off US 40 just west of the Appalachian Trail crossing, is one of the few exposures of sandstone of the Antietam Formation of Cambrian age, which forms ridges lower than and to the west of South Mountain but higher than those in the limestone valley. If you walk the trail to the dam, you can see an extensively weathered outcrop along the shore and a boulder sticking out of the lake. A gap cut in the Antietam sandstone ridge here by two converging streams afforded the location for the earthen dam. If you look northeast from the dam, you can see Annapolis Rock composed of Weverton quartzite on top of South Mountain.

On US 40 just downslope and west of the entrance to Greenbrier State Park is a roadcut in the siltstone, shale, and phyllite of the Harpers Formation, here overturned on the west limb of the South Mountain anticlinorium so that younger

This US 40 roadcut in the Harpers Formation is on South Mountain just below the entrance to Greenbrier State Park.

layers are above older layers. Remember that the Harpers Formation is found over a dozen miles east from here on the other limb of the giant fold.

Hagerstown Valley

At the bottom of the western slope of South Mountain, you are at the eastern edge of a 20-mile-wide, limestone and dolomite, thrust-faulted synclinorium. At the regional scale it is called the Great Valley, but in Maryland it is called the Hagerstown Valley. These carbonate formations weather at slightly different rates because of differing component mineral composition, so you will encounter a rolling terrain of many low hills and valleys.

For many decades, the limestone valley was largely rural. Weathering of carbonates produces good soil for growing crops. You will still see many crop and pasture fields as you cross the rolling landscape, but rapid development in the last couple of decades has drastically cut down on this acreage. Hagerstown is built on limestone and exhibits many springs, pinnacle outcrops, and sinkholes. Through the Hagerstown area on the eastern limb of the Great Valley synclinorium, you cannot see much geology (but you can go shopping).

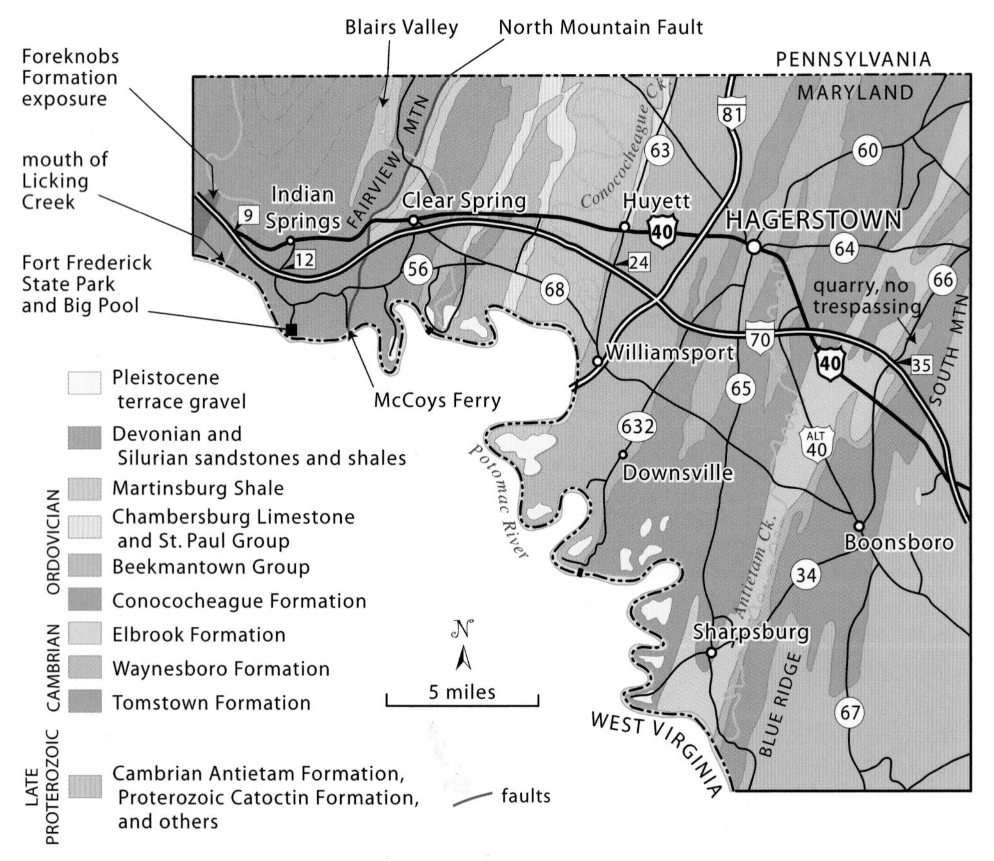

Geology along I-70 and US 40 across the Hagerstown Valley and west to Indian Springs. —Modified from Cleaves, Edwards, and Glaser, 1968; Edwards, 1978; Glaser, 1994a; Brezinski, 1992

Sinkholes can open into underground caverns and are dangerous places where children or adults can become trapped.

Hagerstown Valley on Interstate 70

Two familiar features of limestone country are located just off exit 35 at Beaver Creek: a spring with large discharge (at the fish hatchery) and a large quarrying operation. Because limestone reacts with naturally acidic rain and groundwater, solution cavities form underground and serve as groundwater discharge conduits—springs. As a mineral resource, limestone gravel is excellent for road building and many other construction jobs.

Most of the Hagerstown Valley is floored by Cambrian and Ordovician carbonates of the pre-Taconic offshore carbonate bank, but a couple of miles west of the junction with I-81, I-70 crosses a 2-mile-wide band of Ordovician-age Martinsburg Shale, which lies at the center of the synclinorium. In the Queenston Wedge, this formation represents some of the first clastics—mostly clays—from the newly rising Taconic Mountains, eroded into the inland Queenston basin when it was deepest. At exit 24, Huyetts Crossroads, on the westbound entrance ramp, there is an excellent exposure of this shale and a relatively safe stopping place. You can see that it is very crumbly and weathers into the clay of which it was made.

West of Huyetts, I-70 crosses Conococheague Creek, which meanders in the Martinsburg Shale from Pennsylvania to the Potomac River. From I-70 you might be able to see the difference between the gentle rolling contours of limestone terrain and the gullied, dissected shale topography.

You can see a superb exposure of the St. Paul Group limestone in the roadcut west of the overpass for Cedar Ridge Road, but this is usually not a safe place to stop. Like many of the carbonates in the valley, this exposure contains a mix of dolomite and limestone. While the carbonates are vulnerable to chemical weathering by rainwater, they are very difficult to excavate, and this roadcut undoubtedly required substantial blasting. This exposure is the western limb of the syncline, and between here and Fairview Mountain west of Clear Spring, you cross the same carbonate formations, in reverse order, that you crossed east of the Martinsburg Shale. Proceeding west, you can see the precipitously steep eastern slope of Fairview Mountain (the one with towers), topped by the highly resistant Tuscarora Sandstone of Silurian age.

The thick Martinsburg Shale is exposed on the westbound entrance ramp at exit 24.

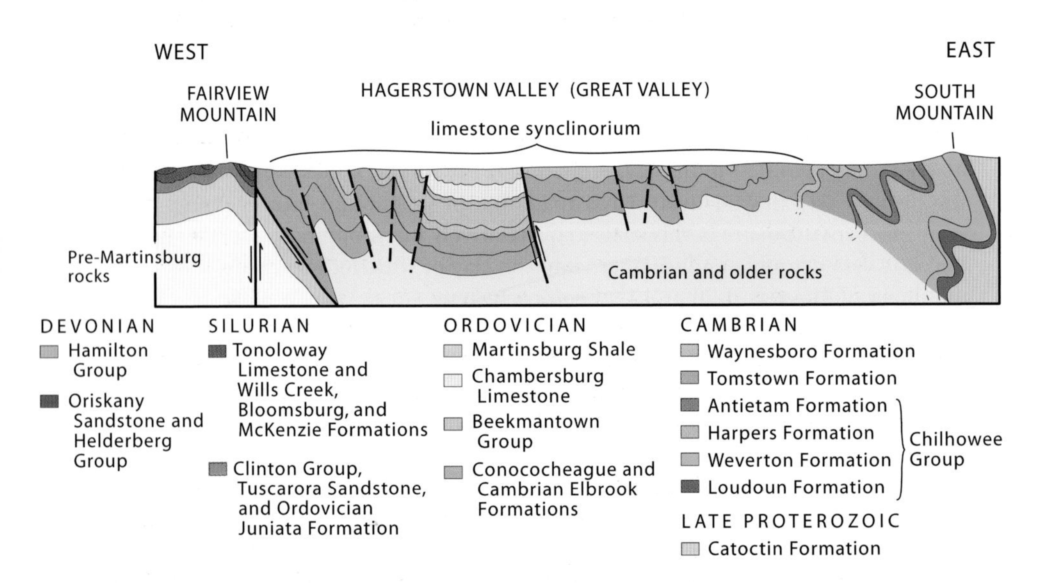

The synclinorium that floors Hagerstown Valley is a continuation of the large fold of the South Mountain anticlinorium. —Modified from Cloos, 1958

A broad limestone exposure of the St. Paul Group on I-70 westbound, west of the Cedar Ridge Road overpass, was formed in a warm ocean just before sediments from the Taconic Mountains began washing in. This is not a safe place to stop.

Hagerstown Valley on US 40

If you are traveling through Hagerstown Valley on US 40, just east of the MD 66 junction (and just west of the drag strip) there is a small ridge in the Red Run Member of the Waynesboro Formation of Cambrian age. This ridge is one of the many gentle hills that exist in the carbonate rock of the Hagerstown Valley. Here you can see why—the formation contains fine-grained sandstone and sandy dolomitic limestone, and the sand makes it more resistant to weathering than pure limestone.

West of MD 66 is another slope and exposures of the Elbrook Formation. Here you can see laminated dolomite (thin brownish layers) and thinly bedded limestone that reflect cyclical changes in tidal coverage. Cutting across the bedding on some sections of limestone in cleaved fractures are strands of white marble—metamorphosed limestone.

At Cool Hollow Road, US 40 crosses Beaver Creek, fed in part by one of the many springs that emerge from limestone solution channels. Topping the next hill west of Beaver Creek is an outcrop of massive, thick-bedded limestone of the Elbrook Formation. Of particular interest here are the cavities or holes typical of the

The white you see here is sand in the carbonates of the Waynesboro Formation along US 40 just east of the MD 66 junction.

underground solution channels and passageways that are dissolved in limestone by acidic groundwater. The Elbrook ranks fairly high among valley limestone formations in the categories "caves per square mile" and "wells penetrating cavernous zones per square mile"—which means that its calcium carbonate composition is probably higher than many of the other, less soluble formations.

Just east of where US 40 changes from two lanes to four is a roadcut in the Conococheague Formation where you can see massive layers as well as ribbon rock—alternating thin layers of brown-weathering dolomite and light-weathering limestone, formed in quiet waters marked by intermittent evaporation and exposure to air.

From I-70 exit 32 southeast of Hagerstown to I-70 exit 24 west of Hagerstown, it is probably better to follow I-70 instead of going through the congested downtown. From exit 24 of I-70 you can follow MD 63 north about 1 mile and rejoin US 40. From Huyett, US 40 climbs into the thick, 2-mile-wide band of Martinsburg Shale. Here you will see gullied terrain and two human features for which shale and clay are uniquely qualified: landfills and race tracks. At Earth Care Road you can drive down and see the grass-covered mountains of human garbage. This landfill was sited in shale because its impermeability helps prevent groundwater contamination from any possible landfill leakage. Groundwater in nearby limestone formations with underground channels could be quickly contaminated.

Conococheague Creek, locally called the Jig, flows almost entirely within this shale because it is bordered on either side by more resistant limestone formations. At the creek you can turn off at Hagerstown Speedway and drive down to the 1819 Wilson Bridge, which carried traffic until damaged by the 1972 Hurricane Agnes flood. You can walk up on it and look down at the wide, shallow creek that lacks rocks. Also nearby is the speedway, one of the best stock car racing dirt tracks in the country. It owes much of its reputation to the clay that results from the weathering of the Martinsburg Shale.

West of Beaver Creek, this exposure of Elbrook Formation along US 40 demonstrates in miniature what underground caverns and solution cavities look like in limestone.

Calcite veins in fractured Elbrook Formation on US 40 Alternate at Dead Man's Curve.

This ribbon carbonate rock of the Conococheague Formation has interbeds of darker dolomite and lighter limestone.

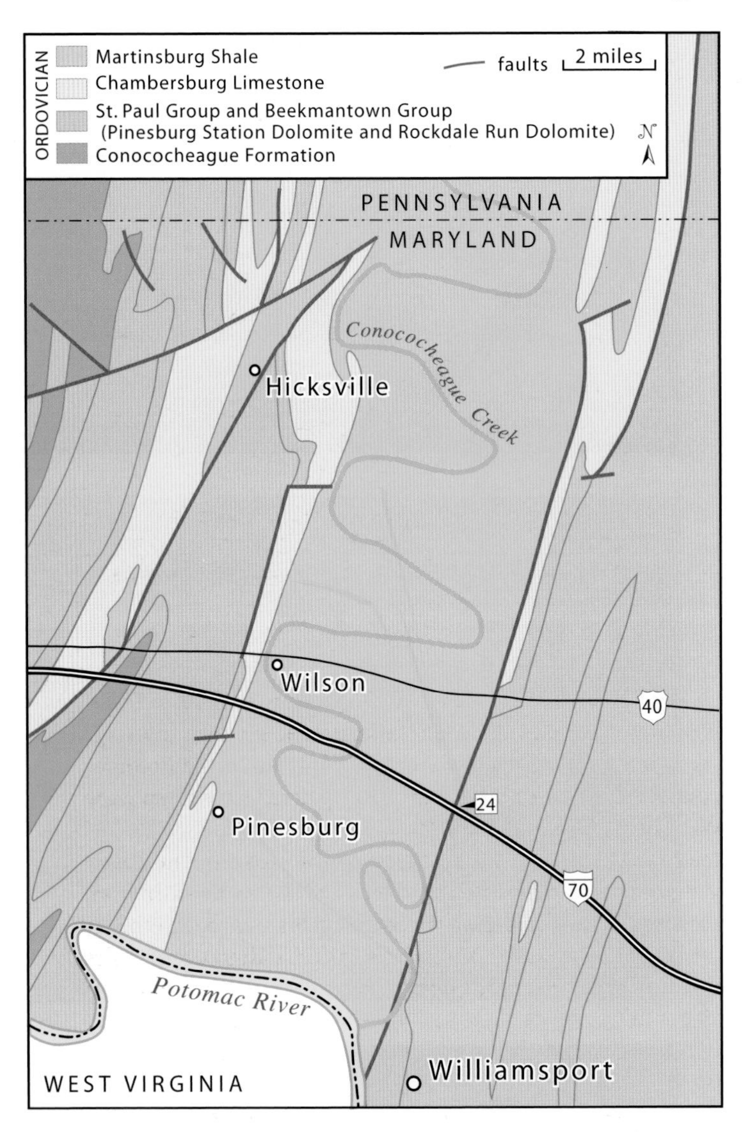

Conococheague Creek has cut meanders into the Martinsburg Shale between resistant limestone. —Modified from Cloos and others, 1951

West of the Jig, the road cuts through the limestones that contain the meanders of the creek. East of Clear Spring, you can look northwest and see the water gap in the ridge cut by Little Conococheague Creek. The creek drains the shale-bottomed Blairs Valley to the west. On the east slope of Fairview Mountain you can see a barren of the Martinsburg Shale. A thrust fault, the North Mountain Fault, transported this slice of shale to the west side of the Great Valley.

I-70 from Clear Spring to Hancock

West of Clear Spring, near the Boyd Road overpass, I-70 crosses the North Mountain Fault, where Cambrian and Ordovician limestones are thrust onto Silurian and Devonian formations. You can see Devonian shale along the road for several miles.

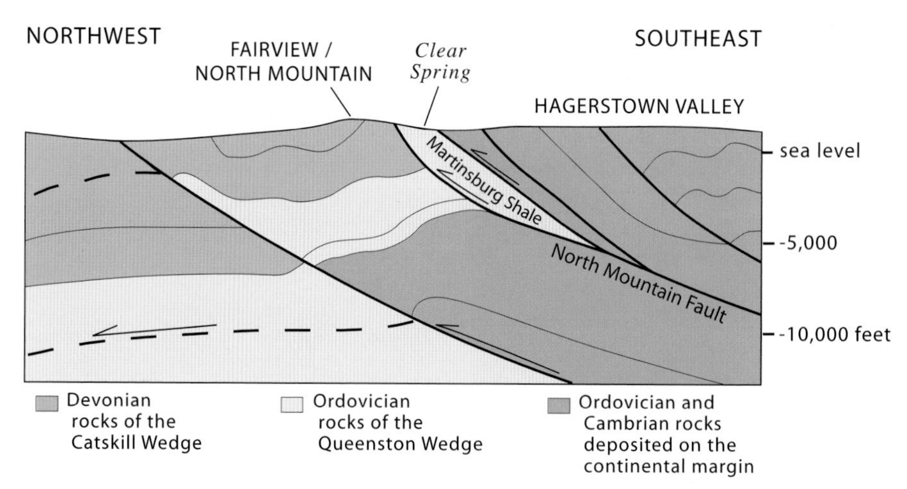

This cross section of the North Mountain Fault Zone (drawn just south in West Virginia) shows how Cambrian and Ordovician continental-margin rocks were thrust onto the inland-sea layers of the Queenston and Catskill Wedges. —Modified from Cardwell, Erwin, and Woodward, 1968

If you get off at exit 12 (Big Pool), drive east on MD 56, and turn at the sign for McCoys Ferry, you can drive 1 mile through shale exposures of the Devonian-age Mahantango Formation (formerly the Romney Formation) down to the Potomac River and towpath. From the boat ramp, you can see a quartzite rock exposure upstream that on the West Virginia bank marks the fault.

Near the Boyd Road overpass on I-70 you also begin crossing the many alternating anticlines and synclines, in mostly sandstones and shales, that run from here to Cumberland. Because the tops of the wavy series of folds have been eroded, you cross the same formations, alternately dipping in opposite directions, miles apart. These formations were deposited in the changing environments of an inland sea with clastic sediments eroded from the Taconic Mountains (Queenston Wedge) and Acadian Mountains (Catskill Wedge) to the east. They were then folded during the Alleghanian mountain building event.

At Fort Frederick State Park on MD 56, near I-70 exit 12, is Big Pool of the C&O Canal. This 2-mile-long lake was constructed out of floodplain swamps. In the park you can walk a trail through a remaining wetland area near the campground. Large sandstone boulders from ancient river floods were used to build the original fort in 1756, during the French and Indian War. The Western Maryland Rail Trail begins, as of 2010, at the I-70 exit 12 for Big Pool. See **I-68: Hancock—La Vale** for discussion of the trail.

On I-70, about a half mile west of the westbound entry ramp for exit 9, Pectonville Road, you can see a several-hundred-foot-long, interbedded sequence of gray, green, red, and brown sandstones and shales of the Foreknobs (formerly Chemung) Formation. Red rock here and in thicker sequences to the west indicates terrestrial, alluvial deposits—a time when water levels in the inland basin dropped

For protection from Native Americans, Fort Frederick was constructed in 1756 from sandstone boulders that had been deposited on the nearby Potomac floodplain. Did George Washington sleep here? As a matter of fact, he did.

Different colors and compositions of rock layers in the Foreknobs Formation, west of Licking Creek (at exit 9) along I-70, reflect changes in the depositional environment.

or deposits had filled the basin. The change back to darker colors indicates the return of submarine deposition.

Another couple of miles west, between exits 5 and 3, you pass the almost 4,000-foot-thick, red Hampshire Formation of Devonian age. It was deposited directly next to the sourceland by numerous streams draining from the Acadian Mountains to the east. It decreases in thickness westward—with increasing distance from the former highlands. It is easily recognizable and well exposed, and you will see its folded, dipping limbs again if you cross the synclinal ridges of Sideling Hill and Town Hill to the west on I-68.

Just west of exit 3 for Hancock, I-70 climbs a ridge with brown and red Brallier siltstone and shale on the east slope and dark gray Mahantango siltstone and shale on the west slope. These are some of the early inland-basin sediments derived from the rising Acadian Mountains. Due to folding, these formations recur several times between Clear Spring and Hancock.

Interstate 68 (Including MD 144 and US 40)
Hancock—La Vale
40 MILES

Western Maryland Rail Trail

The Western Maryland Rail Trail, as of 2010, begins at I-70 exit 12 for Big Pool and parallels the C&O Canal Towpath and the Potomac River for 22 miles up to Pearre (pronounced PAH-ree) in the Sideling Hill water gap. Suitable for cycling, rollerblading, and walking, this nearly flat, asphalt passage offers better access to geologic sites than the towpath because the original Western Maryland Railroad bed had to be cut into the mountains above the towpath in many locations. It also provides access to some classic structures that are not easily reached by automobile. Hancock is a convenient access point, with plenty of parking and services.

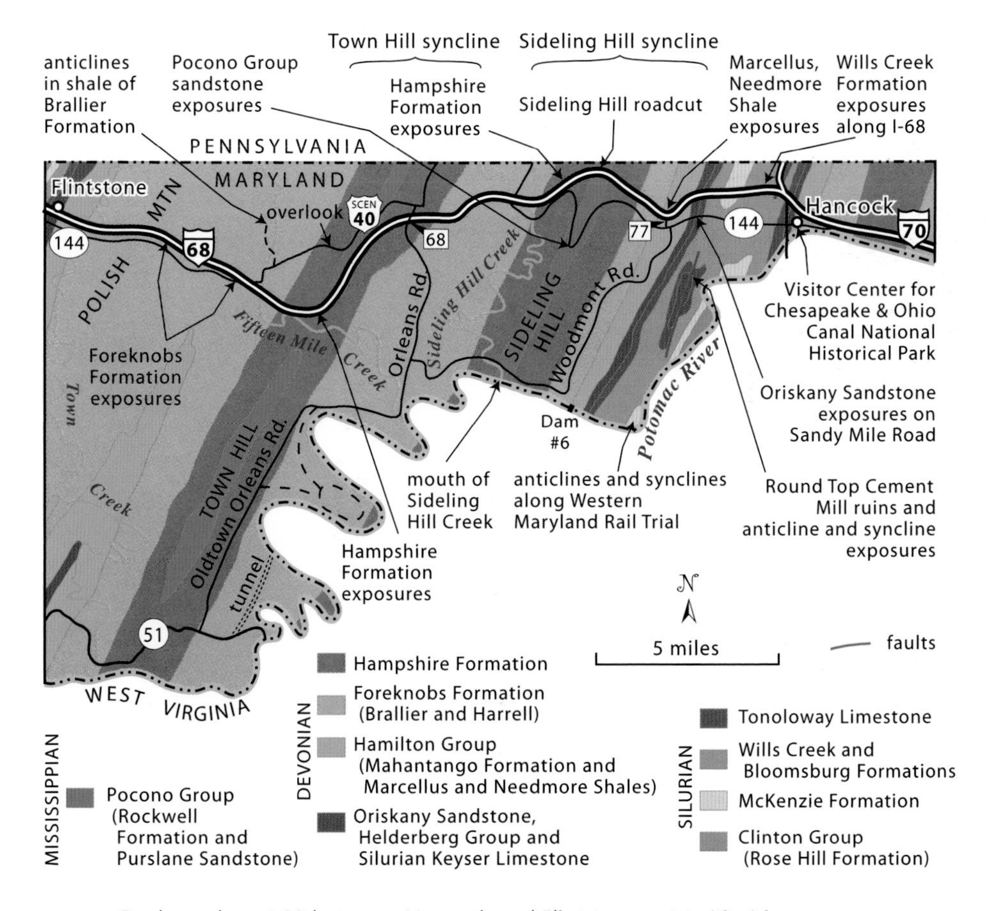

Geology along I-68 between Hancock and Flintstone. —Modified from Cleaves, Edwards, and Glaser, 1968; Edwards, 1978; Darling and Slaughter, 1962; Glaser, 1994a

At milepost 13 west of Hancock, you can see the remarkable anticlines at Round Top. West of Hancock on MD 144, you can turn onto Locher Road and drive down to the Western Maryland Rail Trail. Walk or bike the asphalt trail about 1 mile west to Roundtop Hill to see what geologist John Glaser called "the best section of folded rocks in Maryland." In the easternmost cut along the old rail bed is an anticline in shale and sandstone of the Bloomsburg Formation, and west of there a syncline and anticline in the Wills Creek Formation. A little farther west, in a second railroad cut, is another anticline in the Wills Creek Formation. Also, you can see numerous openings in Wills Creek limestone beds where rock was mined for the old cement mill, the furnace ruins still in place below the railroad bed.

A third anticline, known as Devil's Eyebrow, is down the hill, about 25 feet above the towpath. An informal, unmarked, steep trail runs down from the east side of the easternmost railroad cut. Calcareous shale and limestone in the Bloomsburg Formation have weathered out in the middle of the arch to form a cave. From the towpath you can also view the kilns of the former, nineteenth-century Round Top Cement Mill.

Between mileposts 18 and 19 are numerous intricately folded Silurian shale beds, small-scale anticlines and synclines within the larger Cacapon Mountain anticline. Near the 20-mile point you can see a classic example of creep in vertically dipping shale of the Devonian Brallier Formation. Creep is an imperceptibly slow form of

At an old railroad cut, now on a bike trail, about 4 miles upstream from Hancock, you can see this symmetrical anticline in the Wills Creek Formation.

Trees along the Western Maryland Rail Trail near milepost 20 lean or bend as the slope of Brallier Formation that they grow on creeps downhill.

The Devil's Eyebrow anticline, with eroded layers of Bloomsburg Formation, is uphill from the C&O Canal Towpath at Round Top.

downslope mass wasting due to gravity and frost heaving. Layers of shale intersecting the steep slope have been bent downslope. If you stop and walk up the dirt road that passes right next to the rail trail here, on the steep slope you can see trees that have been bent as the underlying shale and clay have crept downward.

Wills Creek Formation

Just west of the I-68/I-70 split is an excellent exposure of the Wills Creek Formation along the westbound lane. This roadcut is a relatively safe place to stop and see the many different colors of these Silurian-age sandstones and shales and even limestone. These deposits were eroded from the Taconic Mountains, which were built in middle Ordovician to early Silurian time and nearly completely eroded by late Silurian time. The presence of limestone indicates the slowing of clastic deposition.

The variety of beds of the Wills Creek Formation just west of the I-70/I-68 split is a beautiful coda to the ancient Taconic Mountains.

Sideling Hill Syncline

At a sweeping turn and the overpass of Sandy Mile Road is roughly the eastern extent of the 3- to 4-mile-wide Sideling Hill syncline. Look to the southeast side of the overpass to see an exposure of the Oriskany Sandstone of Devonian age. At exit 77, you can go a half mile east on MD 144, turn onto Sandy Mile Road and drive to the Oriskany Sandstone exposure. Here you may be able to find samples of fossilized Devonian brachiopods—bivalves that resemble scallops.

Exit 77, Woodmont Road, affords a safe place to stop and inspect an excellent exposure of the Marcellus and Needmore Shales. The roadcut is right beside the westbound exit and entry ramps.

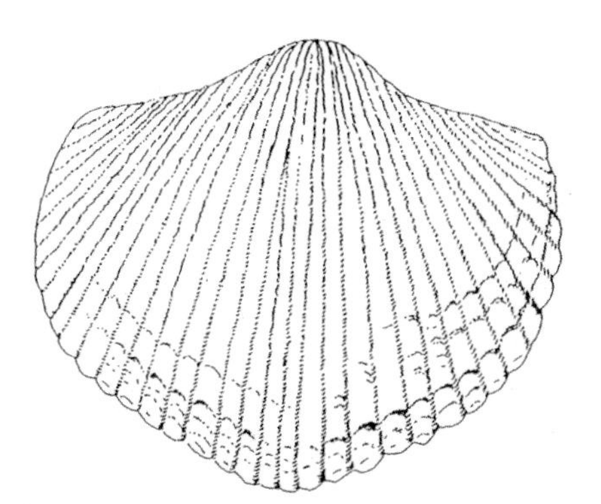

Some of the fossils you can find on Sandy Mile Road include the gastropod (snail) Platyceras (left) *and the brachiopod* Costispirifer (right). —Modified from Glaser, 1979

Along I-68 between the Sandy Mile Road overpass and the top of Sideling Hill, you will cross nearly all of the formations assigned to the Catskill Wedge, which eroded from the Acadian Mountains in Devonian and Mississippian time. On the east side of Sideling Hill, all of them dip to the west and form the eastern limb of the syncline. You will see various shales, but most are poorly exposed. Note the increase in slope near the top of the hill, a result of the resistant Purslane Sandstone there.

At the top is Sideling Hill Visitor Center, where excavation for I-68 extended from the ridgetop at 1,620 feet down to 1,280 feet at the base of the roadcut, exposing the center of the syncline—with 10 million tons of rock removed. The visitor center was closed in 2009, but you can still use the walkways to view the roadcut. The geology exhibits that had been here have been moved to Hancock. You can see them there.

The I-68 roadcut through Sideling Hill in 1988 shortly after it was made.

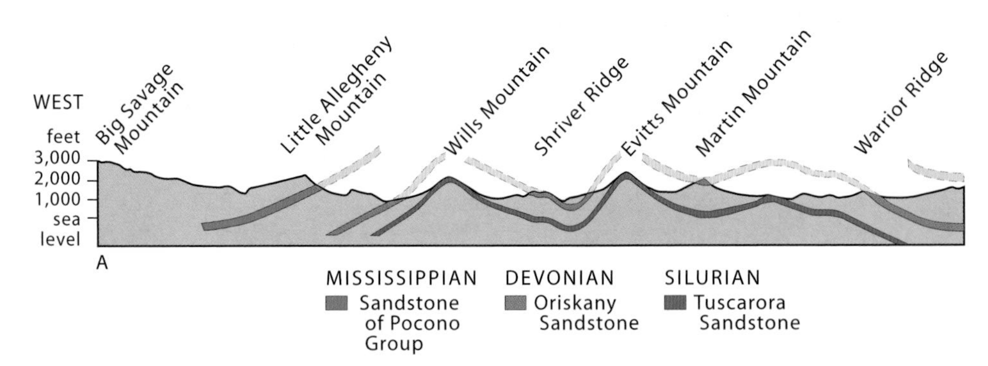

I-68 crosses numerous sandstone ridges between Hancock and La Vale. —Modified from Vokes, 1961

Sideling Hill and Town Hill to the west are parallel ridges topped by the resistant sandstones and conglomerates of the Mississippian Purslane Sandstone. Each ridgetop is at the center of a syncline formed during the Alleghanian mountain building event between 320 and 250 million years ago. Immediately after folding, the sandstones of today's ridges were structurally low axes in the troughs of the wavelike series of folds. After more than 200 million years of erosion, they are topographically high ridges because the exposed center axis is hard sandstone, contrasted to the less resistant shales on the limbs.

The original deposition of the coarse quartz sandstones and pebble conglomerates of the Purslane Sandstone represents the almost complete erosion of the Acadian Mountains because the most resistant mineral, quartz, lagged behind as less resistant minerals were washed away. Most of the rock layers in the upper, central part of the Sideling Hill syncline were alluvial-plain channel deposits dropped by

Look carefully for the ancient stream channel in the north-facing highway cut.

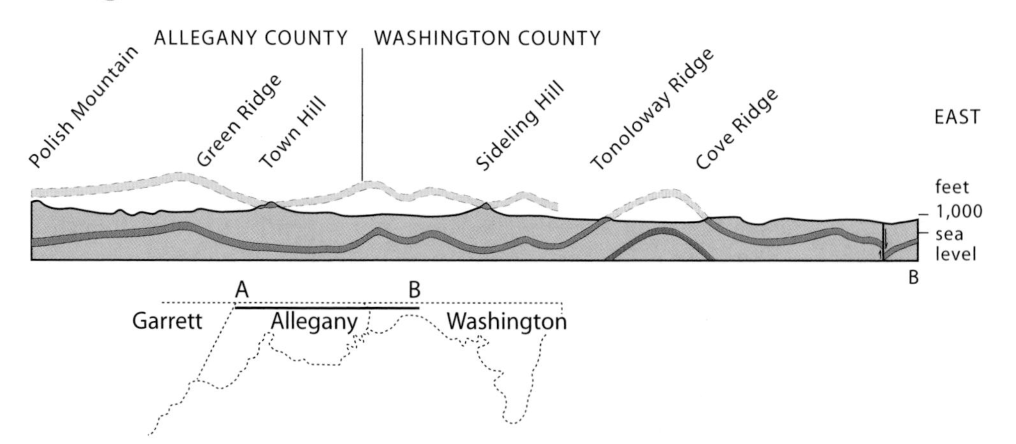

rivers washing off the low sourceland, with interbeds of shales. On the north-facing cut, at the eastern edge you can see an ancient stream channel. The layers of semi-anthracite coal that you see near the top are rare for this early date and may have formed in vegetated swamps along the rivers. Underlying the Purslane is the older Rockwell Formation (both Mississippian) of interbedded sandstones, siltstones, and shales—probably deposited in an alluvial plain or delta. These two formations were deposited not long before the beginning of the Alleghanian mountain building.

Since the cut was completed in 1984, differential weathering has gone to work on it. Today you can see where less resistant shale interbeds have eroded more

Differential weathering of shale and sandstone has created a ridge and valley contour across the cut at Sideling Hill.

rapidly than the sandstone layers, producing grooves or small gullies, some 2 to 3 feet deep. This same process generated the ridge and valley topography on a larger scale. When in shadow, the grooves define the contour of the fold from a great distance. In addition, you can often see groundwater discharging or ice forming at the axis of the syncline from above the clay layers. Water percolates readily through the porous sandstone layers, but when it reaches the nearly impermeable clay layers, it runs out to the edge of the open cut and drips down.

To the east from the visitor center, you can see across the Great Valley all the way to South Mountain, and you can see several water gaps in the parallel ridges. To the west, after you drive through the roadcut (in which no stopping is permitted), you can pull off and look back for a different angle, and you can also look west at the many parallel ridges and water gaps.

At the bottom of the western side of Sideling Hill, just below the westbound runaway truck ramp, is another spectacular roadcut—the less resistant red Hampshire Formation, deposited by streams on alluvial fans. This roadcut is a fairly safe place to stop and explore the exposure. You can see crossbedding, many thicknesses of interbeds, and slickensides—polished, scratched surfaces resulting from friction between beds that occurred during folding or faulting. This roadcut marks the western edge of the east-dipping limb of the Sideling syncline. The Hampshire Formation runs down deep under the mountain and emerges on the eastern slope. Just west of this roadcut is a small anticline that leads into the Town Hill syncline.

Scenic 40 also climbs up and over Sideling Hill. About a half mile from the top on either side, you can see good exposures of the ridge-forming Purslane Sandstone

Were it not for the roadcut at the top of Sideling Hill, this cut through the ubiquitous red Hampshire Formation in the east-dipping, west limb of the syncline would be considered spectacular.

and Rockwell Formation, and on the very top a coal bed. Scenic 40 West loops back up Mountain Road to the I-68 entry ramp, where you can get a great view to the west of the Ridge and Valley and a unique western viewpoint for the Sideling Hill roadcut.

Town Hill Syncline

On the eastern slope of Town Hill, a roadcut on the eastbound side shows red Hampshire Formation rock clearly dipping to the west—the limb of another syncline. Westbound travelers on I-68 should look at the eastbound lane when they see the sign that says exit 68 for Orleans Road is 1 mile ahead. West of here, there are few exposures for several miles until westbound travelers reach the 1-mile sign for exit 64, MV Smith Road. Here are two roadcuts in the red Hampshire Formation, this time dipping east as part of the western synclinal limb.

On the western slope of Town Hill at exit 62, Fifteen Mile Creek, is a superb exposure of the Devonian Brallier Formation on MD 144, just off the exit and parallel to I-68. This roadcut is a safe place to examine the small folds, slickensides, and papery thin layers of fissile shale. Clay molecules have a sheetlike structure and when buried beneath hundreds of feet of sediment, the compaction flattens clay particles into parallel layers. Over 2,000 feet of shales of the Brallier Formation are thought to have fallen out of suspension in a deep, quiet basin west of the Acadian Mountains.

Paper-thin, fissile shale of the Brallier Formation was created by compaction and is very fragile. Please do not trample—leave it for others to see.

Town Hill by Way of the National Road

Scenic 40 also crosses the Town Hill syncline. Just above Orleans Road are exposures of the red Hampshire Formation and, higher up, the resistant Purslane Sandstone—a repetition of the layers in the Sideling Hill syncline. On top of Town Hill you can stop at the overlook and see the Sideling Hill roadcut and the Potomac water gap to the east.

On the Scenic 40 slope down to Fifteen Mile Creek, you can see shale of the Brallier Formation along the road. If you don't mind a short trip on a dirt road, turn into the Green Ridge State Forest camping area, just east of where Scenic 40 crosses the creek, and drive about 1 mile to a swimming hole. If you explore the upper part of the cut in the Brallier Formation, you will see some very interesting anticlines. Do not climb on the fragile fissile shales—save them for others to see.

Shale that is now brittle was once plastic enough to be folded into these compact, well-exposed anticlines in the Brallier Formation along Fifteen Mile Creek.

From the old hotel overlook at the top of Town Hill, you can look east and see the gap cut by the Potomac River through Sideling Hill.

Polish Mountain

A 2,000 to 3,000-foot-thick submarine deposit, the Foreknobs Formation, continues west past the top of Polish Mountain. This ridgetop is hard quartz conglomerate, originally clay, sand, and gravel washing off the Acadian Mountains onto a submarine fan of steep, nearshore slopes. There are "No Stopping" signs at the huge I-68 roadcut, but you can see it much better from parallel MD 144, which provides a spectacular, almost aerial view. You can see extensive shale erosion at the base of the cut.

The vertical grooves here are due to human excavation, but in this almost aerial view of the I-68 roadcut through Polish Mountain from MD 144, you can see shale of the Foreknobs Formation eroding to the clay of which it is made.

Martin Mountain and Vicinity

About a half mile east of exit 56 for Flintstone is a roadcut in the Tonoloway Limestone, deposited in late Silurian time prior to Acadian mountain building, when little clastic sediment was available. At Flintstone the limestone is part of the eastern limb of an anticline, while the western limb lies on top of Martin Mountain, exposed in a large roadcut and quarry visible from many miles to the east.

On I-68 west of Flintstone at the foot of the eastern slope of Martin Mountain you will see barren, crumbly Silurian shales in roadcuts. Nearby are shale barrens where few plants grow. Shale barrens are fairly common in the rain shadow areas lying east of the Allegheny Front on drier, south- or east-facing slopes. Crumbly shale fragments create unstable soil conditions, and the clay does not absorb or hold water very well.

Mountaintop removal? The roadcut and quarry in the Tonoloway Limestone atop Martin Mountain are visible from many miles to the east.

Near the top of the eastern slope of Martin Mountain, you see a large quarry and the huge roadcut in the Tonoloway Limestone. Seemingly violating the rules of differential weathering, this limestone does not occupy a valley. If you look on the very top, you will see why—a cap of resistant Oriskany Sandstone, which is particularly visible between the eastbound and westbound ramps of exit 52. Along MD 144 west of the entrance to the quarry, you will encounter a roadcut in the same limestone that is being mined. It contains calcite veins, which resulted when fractures were filled with minerals precipitated from a water solution of calcium carbonate dissolved from the limestone itself. You can stop at the MD 144 overpass above I-68 on top of Martin Mountain and look back down at the roadcut.

From the top of Martin Mountain on MD 144 or I-68, on a clear day you can see the Allegheny Front to the west and the wind gap and the Narrows above Cumberland. As you descend the west slope of Martin Mountain, look for the unnamed fault—a fractured offset of rock layers—in the roadcuts across I-68 from the westbound ramp for exit 50, Rocky Gap State Park.

Calcite veins in the Tonoloway Limestone.

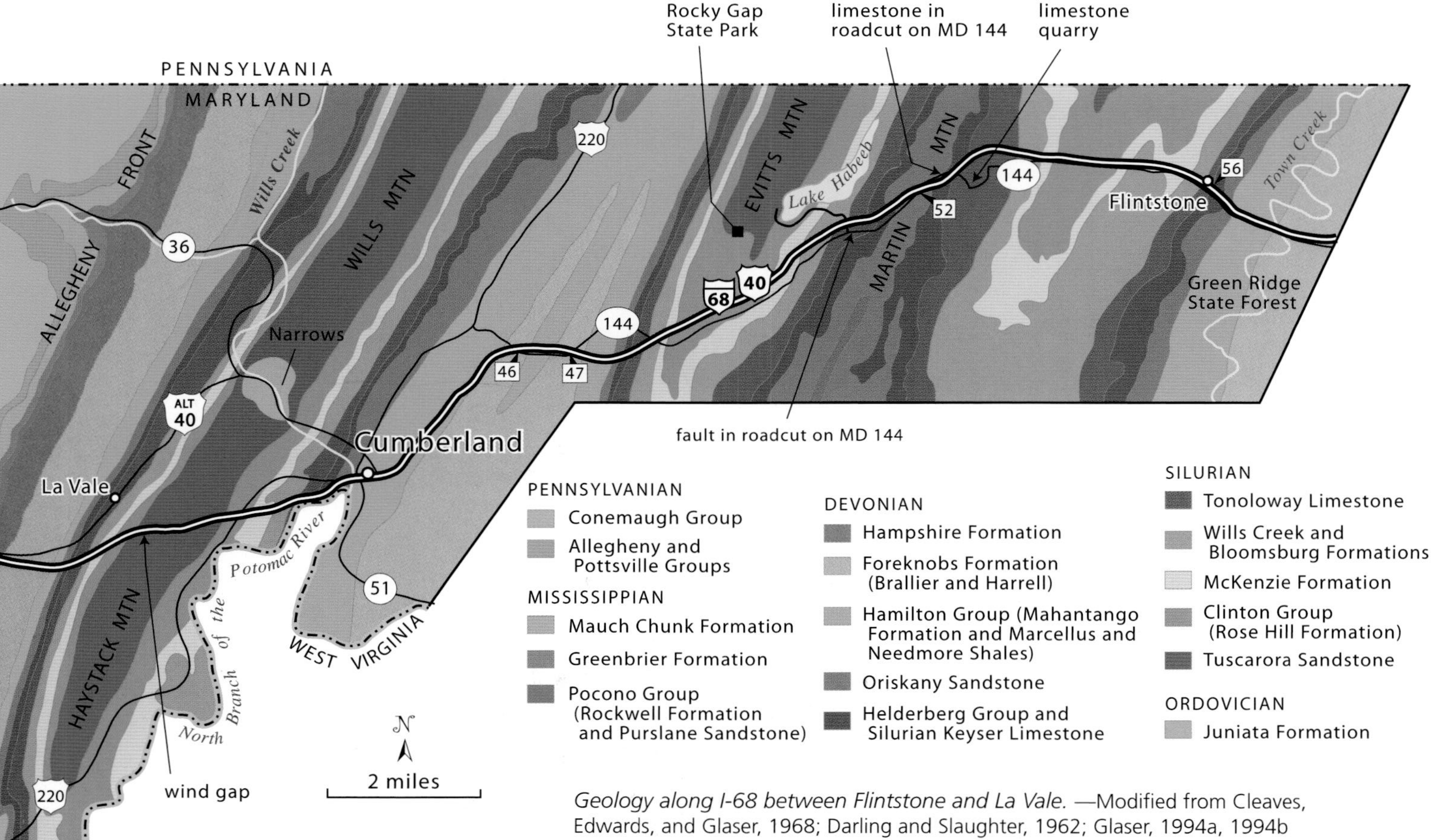

Geology along I-68 between Flintstone and La Vale. —Modified from Cleaves, Edwards, and Glaser, 1968; Darling and Slaughter, 1962; Glaser, 1994a, 1994b

West of the westbound entry ramp, you can see the recurring Silurian shales flaking off in sheets, and at the bottom of the descent from Rocky Gap, just west of milepost 49, you will see a steep slope of another shale barren. Between exits 46 and 47 is a roadcut in gray Devonian shale, and another is visible at exit 45. At exit 46 you can see the Allegheny Front looming to the west.

Rocky Gap State Park

You can take exit 50 and visit Rocky Gap State Park. Behind Lake Habeeb are the steep slopes of Evitts Mountain, topped by the resistant Tuscarora Sandstone of Silurian age. This formation also crops out about 35 miles east of here in the Bear Pond Mountains. The best hike here for seeing the Tuscarora Sandstone is probably the first mile or so of the Evitts Mountain Homesite Trail. Follow it down into the stream valley and back up the other side, and where it intersects the trail from the campground, you can turn right (back toward the lake), descend to Lakeside Loop Trail, and return. On the way, or if you are just walking to the dam, be sure to look at the rock exposure in the spillway. The short Canyon Overlook Nature Trail is also worth a look.

Evitts Mountain Homesite Trail at Rocky Gap State Park has accessible exposures of the Tuscarora Sandstone, with shale interbeds eroding more rapidly.

Wills Mountain

Wills Mountain, the last western ridge of the Ridge and Valley Province, lies above and just west of Cumberland. I-68 follows a wind gap through the Haystack-Wills Mountain anticline. A wind gap is a notch in a ridge formerly cut and occupied by a stream when the entire landscape was at a higher elevation and less eroded. It is thought that this gap was cut by an ancient Braddock Run, but at some point its flow was pirated or diverted to the northeast, where today it joins Wills Creek to cut through the Narrows. This type of direction change occurs when natural erosion cuts through a drainage divide and one stream captures another. A wind gap no longer has water in it. The gap here is about 300 feet below the ridge, which is topped

From the top of Martin Mountain, to the west you can see the Allegheny Front (Dans Mountain), the Narrows, and the wind gap above Cumberland.

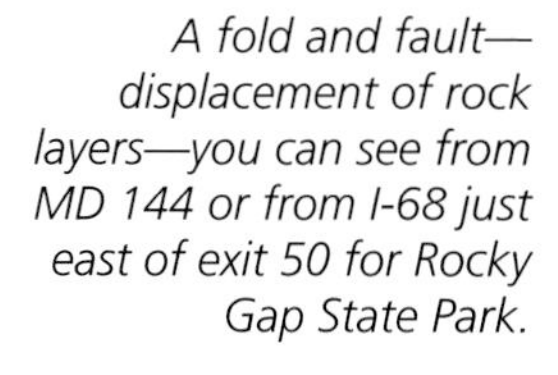

A fold and fault—displacement of rock layers—you can see from MD 144 or from I-68 just east of exit 50 for Rocky Gap State Park.

In the lower-precipitation rain shadow east of the high Allegheny Plateau, the clay soils derived from shale support little vegetation.

The wind gap through Wills Mountain, cut by the stream that once flowed through it, retains much of the shape of a valley. Photo taken from US 40 Alternate in La Vale.

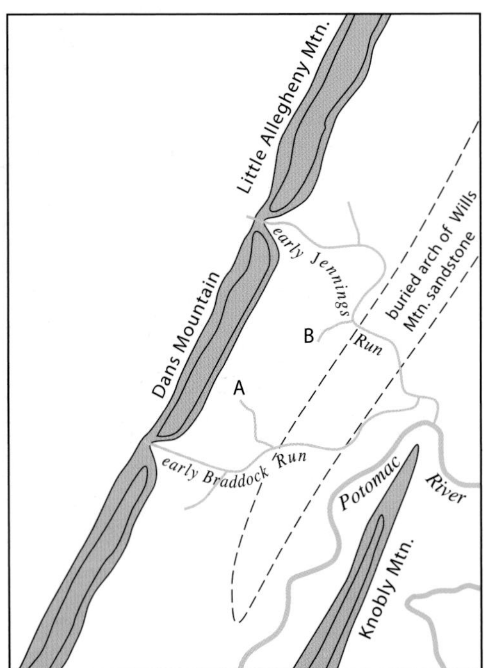

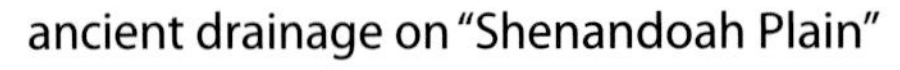

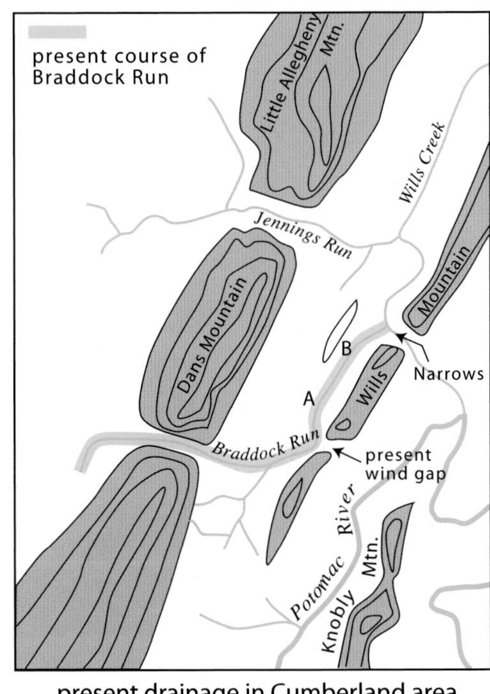

present drainage in Cumberland area

The wind gap in Wills Mountain, today occupied by I-68, was cut by a former course of Braddock Run when landscape and streams were at less eroded, relatively higher elevations. In the valley between Dans Mountain and Wills Mountain, tributary B of Jennings Run and tributary A of Braddock Run eroded headward toward the drainage divide between them. Eventually B reached A and pirated its flow from south to north, causing Braddock Run to follow and occupy a new course in the valley. Today Braddock Run enters Wills Creek west of the Narrows, while the former course of Braddock Run is left high and dry as a wind gap in the erosion-resistant Tuscarora Sandstone of Wills Mountain.
—Modified from Abbe, 1900

with Tuscarora Sandstone and was a convenient route for locating a highway. As you drive over the mountain, near the top you can see the white sandstone. Immediately to the west, La Vale lies at the foot of the Allegheny Front.

The Narrows

US 40 Alternate (the National Road), between I-68 exits 40 and 44, passes through the Narrows, a water gap cut through hard rocks by a stream that flowed thousands or even millions of years ago. The present stream bed is about 1,000 feet below the ridge cliffs to either side. Streams plunging down the Allegheny Front have converged into Wills Creek and cut through the hard Tuscarora Sandstone—so that it forms just the rim of the Narrows. At the bottom of the gorge, the stream has cut down into the Ordovician Juniata Formation, the oldest rock exposed at the Narrows, which is the only place you can see it west of the Bear Pond Mountains. It is also the oldest formation exposed in Allegany and Garrett Counties. If you drive from the Narrows to the top of Dans Mountain on the Allegheny Front, you will traverse about 150 million years of rock—the top half of the Queenston Wedge and all of the Catskill Wedge, a slice through the sediments eroded from one and a half ancient mountain ranges.

MD 36 begins at the western end of the Narrows, a water gap through Wills Mountain. A cliff of Tuscarora Sandstone looms high above the road.

Maryland 17
Middletown—Brunswick
13 MILES

The section of MD 17 between Middletown and Brunswick, an area called Middletown Valley, affords a unique opportunity to drive over a landscape where the rock is over 1 billion years old. The road passes through the erosional remnant of the center of the 10-mile-wide South Mountain anticlinorium. This giant, upward-arching geologic structure formed during the assembly of Pangea about 320 to 250 million years ago and once reached elevations perhaps equaling those of the Rockies or Alps. Hundreds of millions of years of weathering and erosion reduced the land to the relatively low ridges of about 1,000 feet that you see today flanking the valley. South Mountain to the west and Catoctin Mountain to the east are the eroded but surviving limbs of this large, overturned anticline.

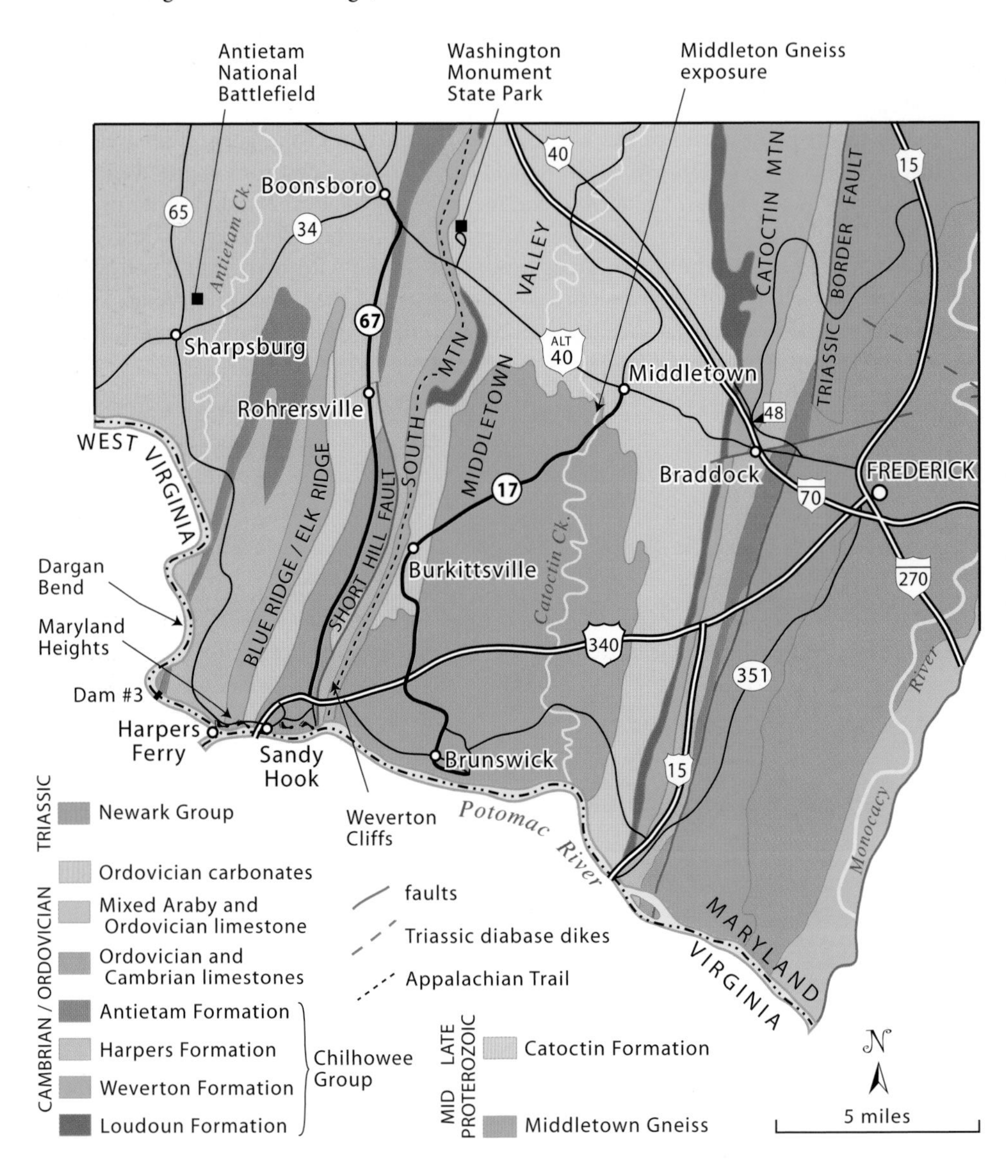

Geology along MD 17 between Middletown and Brunswick. —Modified from Cleaves, Edwards, and Glaser, 1968; Edwards, 1978; Fauth, 1977; Reinhardt, 1974; Fauth and Brezinski, 1994; Brezinski, 1992; Southworth and Brezinski, 1996; Southworth and others, 2007

At the MD 17 bridge over Catoctin Creek, you can examine excellent exposures of the 1.1-billion-year-old basement gneiss.

The middle of the anticline, where the oldest rock is always found, has worn down here to the basement rock of the ancestral continent. This 1.1-billion-year-old (that's 1,100 million!) granitic gneiss was once part of the Grenville Mountains, an ancient range that was formed and eroded in Precambrian time. You can take a close look at this rock, which is among Maryland's oldest, at the roadcut just north of where MD 17 crosses Catoctin Creek about 1 mile south of Middletown. It is also visible opposite the creek south of the bridge.

This rock, which was originally granite, made up part of the much smaller North American Plate over 1 billion years ago. At that time none of the other rock in present-day Maryland existed, and the shoreline of the ancient ocean, which was not even the Atlantic Ocean yet, was probably not too far east of Middletown Valley.

After this basement or crustal rock had weathered for hundreds of millions of years, plate divergence caused rifting and consequent outpouring of molten rock. Across Middletown Valley, for example, are many late Proterozoic metadiabase and metarhyolite dikes about 600 million years old, feeder dikes for the lavas of the Catoctin Formation. About 300 million years later, the Alleghanian mountain building event pushed up and metamorphosed the granite, lavas, and other formations to alpine elevations. Today the Catoctin metavolcanics, as they are known collectively, directly overlie the basement rock northeast of US 40 Alternate and occupy the slopes and ridgetops of Catoctin Mountain to the east and portions of South Mountain to the west. Because the anticline plunges to the northeast, the basement rock has been more readily eroded and exposed southwest of US 40 Alternate.

If you are on heavily traveled US 340, which passes under MD 17 north of Brunswick, you can see the Middletown Gneiss in a roadcut east of Catoctin Creek, between exits 4 and 8. Stopping is probably not safe. However, you can get off at exit 4 and go east on Jefferson Pike (MD 180) to the Catoctin Creek crossing and take a close look at the gneiss exposed there.

On MD 17 between Burkittsville and Brunswick, look for the beautiful stack stonewalls.

Maryland 34
Boonsboro—Sharpsburg
6 MILES

Maryland 34 passes through part of the Great Valley between Boonsboro and Sharpsburg. Just west of Boonsboro is the commercially operated Crystal Grottoes Caverns, an extensive system of underground caverns and solution channels in limestone and dolomite of the Tomstown Formation of Cambrian age. Naturally acidic groundwater has dissolved the carbonate rock, leaving behind caves you can walk through.

Just northeast of Antietam Creek is an excellent exposure of the Chewsville Member of the Waynesboro Formation of Cambrian age. This upper part of the Waynesboro Formation is composed of interbedded siltstone, sandstone, and shale. The exposure here can be climbed, but its steep face is covered with loose chips of

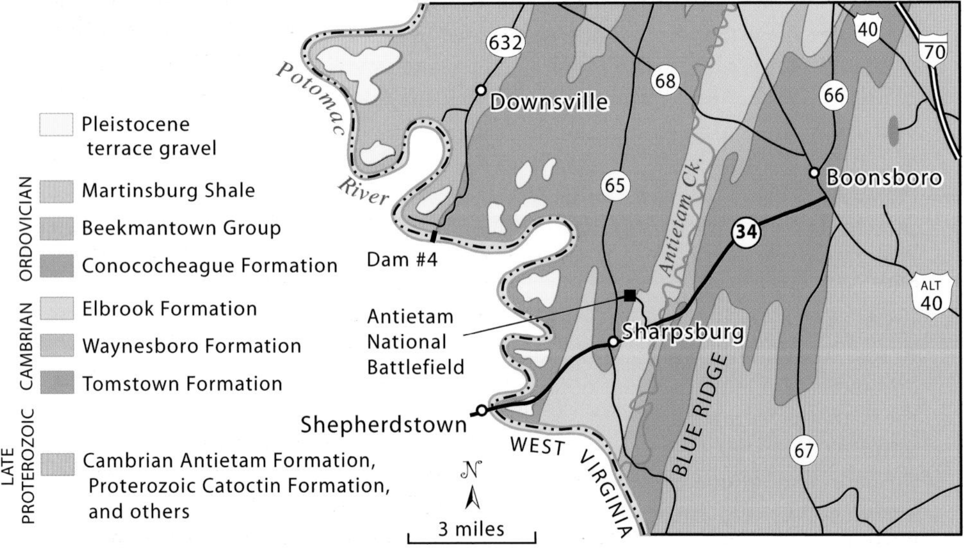

Geology along MD 34 between Boonsboro and Sharpsburg. —Modified from Cleaves, Edwards, and Glaser, 1968; Edwards, 1978; Brezinski, 1992

Just east of Antietam Creek on MD 34, you can see interbedded resistant sandstone and crumbly shale in the Waynesboro Formation.

shale. Still, you can walk along a concrete drainage channel and see ripple marks and mud cracks that indicate deposition in shallow water.

Just east of Sharpsburg, from the eastern edge of Antietam National Cemetery, you can get an excellent view of the northern end of Blue Ridge and its companion, South Mountain. You can also get a sense for the rate of chemical weathering on rock by observing the 100-plus-year-old headstones of soldiers killed at Antietam.

If you follow MD 34 to the headquarters of the Chesapeake & Ohio Canal National Historical Park above the Shepherdstown bridge over the Potomac River, you can get an idea of a former landscape. The flat area here is covered with gravel, cobbles, and boulders of sandstone and conglomerate that were deposited during the Pleistocene Ice Ages by the ancestral Potomac River as it was flowing at this level, which is over 100 feet above the present river. Over the thousands of years since, both the river and the landscape have been lowered through weathering and erosion. These terrace gravels are present on many high bluffs and necks of land above the Potomac in the Valley and Ridge and Piedmont Provinces, the sediments having been eroded from highlands to the west.

Just north of Sharpsburg, MD 34 passes Antietam National Battlefield. It may be disrespectful to speak of geology at a place where so many thousands were killed and wounded, but from the visitors center and from the tower at Bloody Lane, you can get exceptional views of the northern end of Blue Ridge and South Mountain to the east. As has often been the case throughout history, geology and the resultant topography were crucial in the planning and outcome of the battle. If you tour the battlefield, you will learn that high and low ground were critical. Lee chose Hagerstown Valley for his campaign into the North because the ridge of South Mountain afforded him protection from McClellan to the east. As McClellan fought through and over the gaps in South Mountain, Lee positioned his forces on ridges north of Sharpsburg and waited. A 2008 study compared the casualty rates for the Cornfield, an area of flat, open terrain in the nearly pure, uniformly weathered limestone of the Conococheague Formation with those of the area around Burnside Bridge, where the differential weathering of the interlayed limestone, shale, and dolomite of the Elbrook Formation had created an uneven, rolling terrain of hills and small ridges. Casualties were about three times higher in the Conococheague killing ground.

From Antietam Battlefield you can see the northern end of Blue Ridge (sloping down to the left), *topped with Weverton quartzite, with South Mountain in the distance.*

Maryland 51
Cumberland—Paw Paw, West Virginia
26 MILES

Maryland 51 between Cumberland and Paw Paw, West Virginia, passes through sandstones, siltstones, and shales deposited in Devonian time—400 to 360 million years ago. During this period the Acadian Mountains to the east were eroding sediments into a vast inland sea. The formations you see here represent different stages in the deposition of the Catskill Wedge, in environments ranging from inland sea basin and shelf through submarine fan and alluvial plain. After deposition, compaction, and cementation, these layers were compressed and folded into synclines and anticlines during the Alleghanian mountain building event of about 300 to 250 million years ago. You can see the results of this tectonic event in the tilted, dipping beds and wavy structures of these once deeply buried formations. You encounter these same rocks on I-68, but MD 51 is a quieter road and the roadcuts are perhaps safer (though not always) and more accessible.

Most formations here are primarily of shale, the most abundant sedimentary rock on the planet. Shale is formed when tiny clay- or silt-sized particles that are

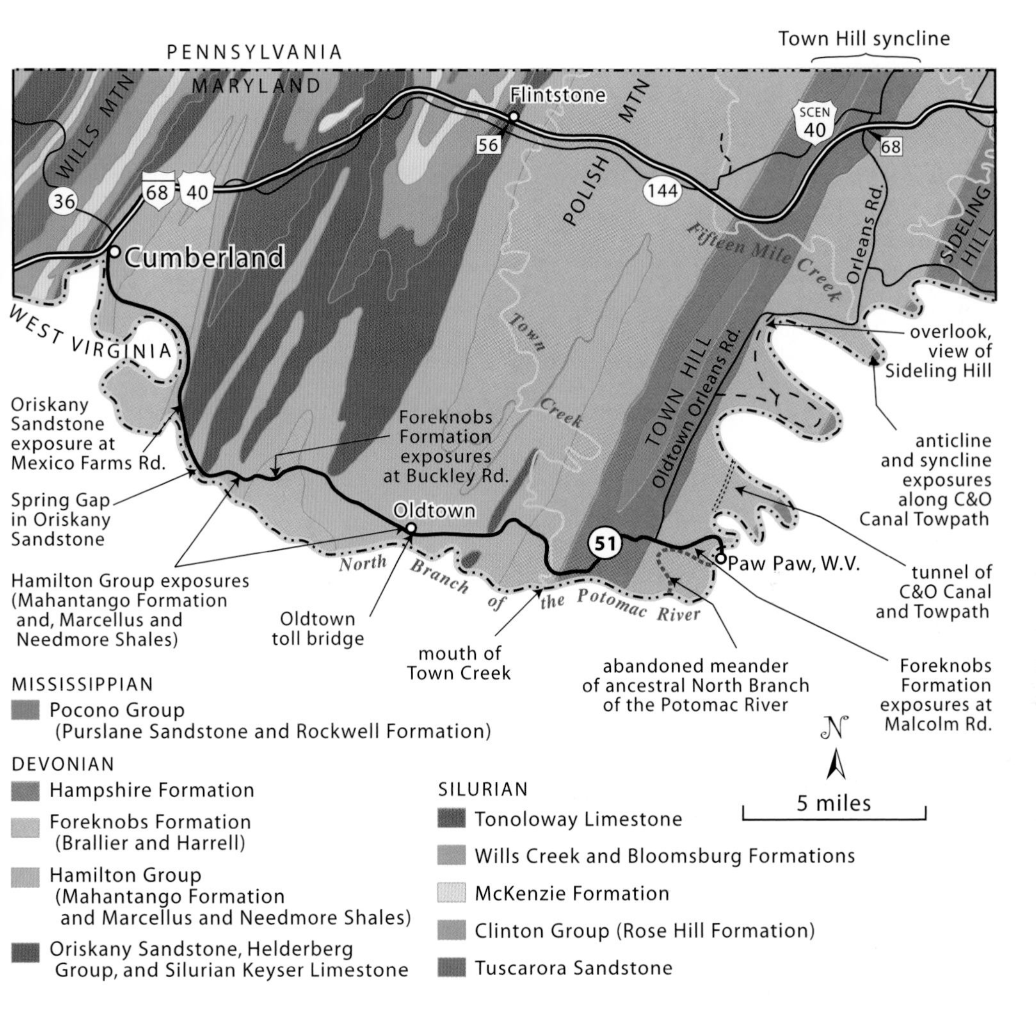

Geology along MD 51 between Cumberland and Paw Paw, West Virginia. —Modified from Cleaves, Edwards, and Glaser, 1968; Darling and Slaughter, 1962; Glaser, 1994a, 1994b; Edwards, 1978

held in suspension by moving currents settle to the bottom of quiet waters. In a deep basin or on a submarine slope or fan, currents can slow sufficiently to drop the clay. Once lithified to stone and uplifted, shale weathers into the clay of which it is made. You will always see clay at the bottom of a weathered shale exposure.

The names of the formations here can be confusing, depending on which map you consult. To keep matters simple, we will refer to four formations: from oldest to youngest, the Oriskany Sandstone, the Hamilton Group of grayish black laminated shales (the thickest of which is the Mahantango Formation), the Brallier Formation of interbedded shale and siltstone, and the familiar red shale, siltstone, and sandstone of the Hampshire Formation. Each reflects a different environment of deposition in the inland sea of the Catskill Wedge.

Interbeds of resistant siltstone and crumbly shale of the Brallier Formation erode to a steplike surface in a roadcut at the Mexico Farms Road exit off MD 51.

In Cumberland at I-68, exit 43B is the west end of MD 51. On MD 51, called Industrial Boulevard here, look to the west for a good view of the Narrows, a water gap through Wills Mountain. At the Mexico Farms Road exit, you can see a large roadcut of shale and siltstone interbeds of the Brallier Formation.

The flat Mexico Farms area and the floodplain neck in West Virginia where the Cumberland Airport is located are covered with alluvial gravel, cobbles, and boulders. Here and throughout the Valley and Ridge, the most extensive deposits of these terrace gravels are found upstream of ridges capped by resistant sandstone—possibly indicating that the former river was completely or partially dammed behind the ridge before a gap was fully cut through by water flow. Sure enough, downstream between Moores Hollow Road and Wheeler Road is a narrow water gap in the hard Oriskany Sandstone. In the bend just south of Collier Run you can see it again. You can go down to the C&O Canal at Spring Gap and look across at the West Virginia side of the water gap.

The massive, resistant Oriskany Sandstone, here at the bend south of Collier Run, may have dammed the ancestral Potomac River, resulting in the deposition of gravel upstream of the dam before the river cut a water gap through the rock.

Fissile shale of the Hamilton Group near Oldtown erodes as no other rock does.

Just across from the Spring Gap Post Office is an exposure of Hamilton Group shales. About 1 mile east of Spring Gap watch for a large shale exposure on the north side of the highway. Here the dark, 1- to 10-centimeter-thick, bedded shales have weathered rapidly to small chips. This shale may be part of the Mahantango Formation.

At Buckley Road is an exposure of the brown shale of the Brallier Formation, which contains siltstone interbeds. Between Buckley Road and Oldtown and east almost to Town Creek, you can observe the crumbly shales of the Hamilton Group in roadcuts. You can see the highly disintegrated, fissile, yellowish brown shale of the Hamilton Group in a roadcut across from Oldtown Volunteer Fire Company. Town Creek, however, flows in the red rock of the Hampshire Formation, and across

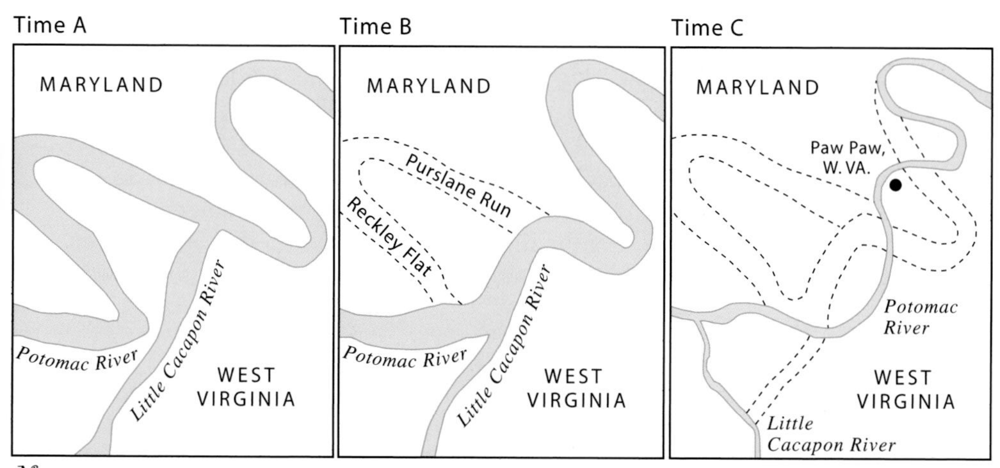

The former Potomac River meander in what is today Reckley Flat was abandoned in the past 1 million years when the Little Cacapon River captured, or pirated, the Potomac flow and redirected it into its already existing course across the meander neck. —Modified from Southworth and others, 2001

Gentle folds in this brittle siltstone of the Brallier Formation, in a roadcut on MD 51 west of Paw Paw, occurred when the rock was deeply buried and in a plastic, bendable state.

from the exit for the C&O Canal is a good exposure. The red rock is visible for a couple of miles east of Town Creek.

Across from David Thomas Road is a low field with a wooded point above it. This low area is the location of a former meander in the Potomac River. In the wintertime you might be able to perceive the old river channel, parts of which contain as much as 25 feet of gravel and sand deposited by the ancestral river. MD 51 parallels this old meander, and Purslane Run flows through part of it. The meander was abandoned when its neck was eroded through as the Potomac River cut into and occupied the former course of the Little Cacapon River in West Virginia.

Between David Thomas Road and the bridge over the Potomac River to Paw Paw, West Virginia, MD 51 passes through several interesting roadcuts in the Brallier Formation. Much of the rock is gently folded, and you can quickly identify the siltstone interbeds because of the angular blocks into which they weather. You can get out and look at some of these, but be careful because the road is narrow.

Paw Paw Tunnel

About 1 mile northwest of Paw Paw, you can exit MD 51 to a parking area for the C&O Canal and hike the towpath about a half mile down to the Paw Paw Tunnel. At over 3,000 feet in length, the tunnel saved the building of over 5 miles of towpath along bedrock cliffs of four incised meanders of the Potomac River. Progress on the tunnel was slow, however, because pick and shovel labor was extremely difficult.

The north portal arch of the Paw Paw Tunnel was located beneath the arch of an anticline in the Brallier Formation.

Funding problems also halted the work, which ran from 1836 to 1841 and 1847 to 1850. While shale of the Brallier Formation through which the tunnel was excavated might appear to weather readily, digging through it by hand was another matter. Above the north (downstream) portal, you can see an anticline, which figured in planning the location of the tunnel because engineers thought it would provide structural support.

You can walk through the spooky tunnel on the narrow, usually muddy towpath. Carrying a light is a good idea. The author once carried a canoe solo through the darkness without a light and swore he would never do that again. An alternative is Tunnel Hill Trail over the top, where you can see the barrenness of shale terrain and, near the north portal, spoil banks left from the digging.

Overlook of Paw Paw Bends

If you do not mind traveling on a well-maintained gravel road through the mountains and you want to visit a spectacular overlook and take a shortcut to I-68, you can turn off MD 51 onto either Malcolm Road or David Thomas Road and then turn right onto Oldtown Orleans Road. Follow this road through Green Ridge State Forest for about 5 miles, and stay on Oldtown Orleans as it turns right—do not follow Dug Hill Road. At Carroll Road, turn right and drive several hundred feet to the overlook. To drive through to I-68, stay on Oldtown Orleans Road until it becomes the paved Orleans Road, and follow it to I-68 exit 68.

From the overlook you can see the incised meanders of the Potomac River far below. The origin of these meanders has been much debated for over one hundred years. One key factor is the shale that lies between Sideling Hill to the east and Town

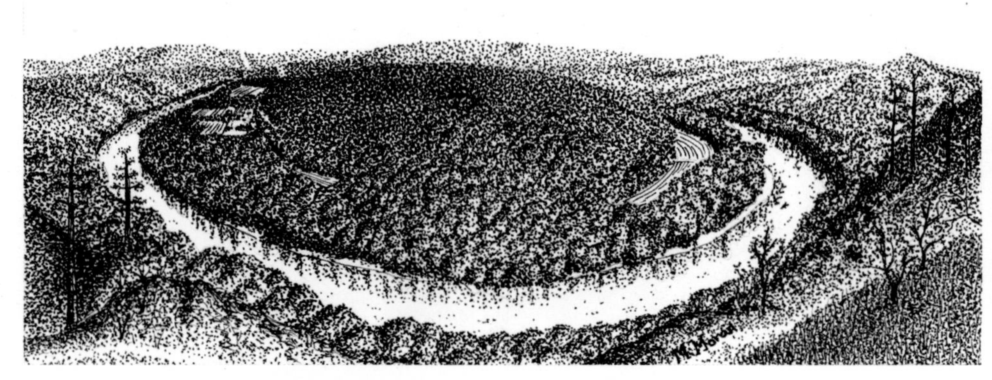

The view of a huge meander of the Potomac River from Green Ridge.
—Matt Moran sketch

Hill to the west. In the Valley and Ridge Province, many meandering streams are found in shale—the Conococheague, for example. The large scale of the meanders indicates that higher flow rates during Pleistocene time might have been responsible. Incision may be due to alluvial plain meanders maintaining their courses as the land rose because of isostatic rebound. Partial or complete impoundment behind Sideling Hill for a period may have also been a factor. From the overlook you can see the water gap cut by the Potomac in Sideling Hill. If you are a canoeist, a trip worth taking is from Paw Paw, West Virginia, to Little Orleans or on down to Hancock.

Maryland 67
Boonsboro—Sandy Hook
15 MILES
See map on page 88.

The north end of MD 67 lies at a T-junction with US 40 Alternate just south of a prominent hilltop of sandstone of the Antietam Formation of Cambrian age. To the east at the top of South Mountain are Turner's Gap and Fox's Gap, through which General McClellan's men fought and marched on their way to Antietam.

The first hill to the south of the MD 67 T-junction with US 40 Alternate is of Quaternary alluvium—a mix of sand, cobbles, and angular boulders eroded by freeze and thaw. Water expands when it freezes, and up on the quartzite ledges of South Mountain, rock was broken and wedged apart, and then eroded downslope. If you stop on the hill and look to the southwest, you will see the end of Blue Ridge, plunging from elevations of over 1,000 feet down into the 400- to 500-foot carbonate valley. This ridge, considered by some as the northern end of the Blue Ridge Mountains, is topped by the same ledge-maker Weverton quartzite that tops South Mountain to the east.

View to the southwest from MD 67 of Blue Ridge, plunging north into the limestone valley southwest of Boonsboro.

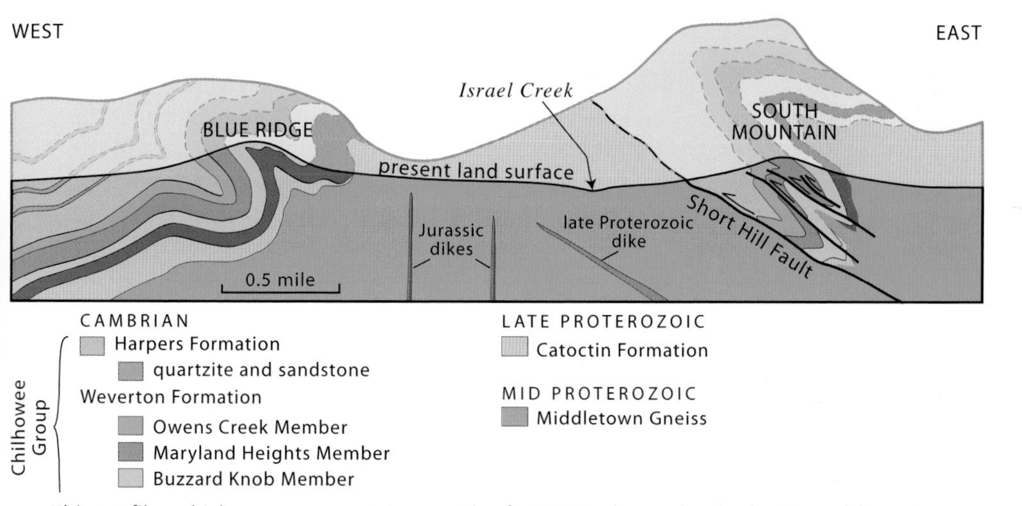

This profile, which runs east-west just north of US 340, shows the duplication of formations and ridges that resulted from tectonically induced slicing and thrusting. Note the present land surface and projected ancient mountains. —Modified from Southworth and Brezinski, 1996

How did this repetition of the Weverton quartzite occur? After the deposition of the Cambrian Chilhowee Group—to which the Weverton Formation belongs—and some Cambrian carbonates, continental rifting associated with tectonic divergence occurred about 500 million years ago. One of the downslippages that occurred with the crustal extension was along the Short Hill Fault. A block, or half graben, of Grenville basement rock topped with Catoctin volcanics, Chilhowee Group, and layers of carbonates slid downward, creating separated sections of these groups. Then, over 200 million years later, during the Alleghanian mountain building event, tectonic convergence reactivated the fault, this time compressing, folding, and thrusting the separated rock groups upward and westward into the South Mountain anticlinorium and the separately thrusted slice of Blue Ridge. Millions of years of erosion left a unique double ridge of the highly resistant Weverton quartzite. You will be able to see the ridges from many different perspectives farther south on the road.

At the MD 67 junction with Main Street in Rohrersville, you see something strangely out of place—the greenstone of the Catoctin metabasalt. This rock, located stratigraphically between the Grenville gneiss and the Weverton quartzite, lies buried in the core of South Mountain but tops Catoctin Mountain 7 miles to the east. Its exposure here is more evidence of the complex displacements associated with the Short Hill Fault.

South of Rohrersville, between Gapland and US 340, MD 67 passes between the two parallel quartzite ridges and over the gneiss of the Grenville basement rock, some of the oldest rock in eastern North America at over 1 billion years old. You can see an exposure of it in a roadcut at Brownsville along the northbound lane.

Just north of the south end of MD 67, you can turn east toward Weverton and at the end of the road join the Appalachian Trail for a 1-mile, 500-foot-elevation climb to the type location for the Weverton Formation—the Weverton Cliffs. No rock climbing is required and the view is spectacular. You look straight across the Potomac River water gap through South Mountain to Virginia. Far below, in the river rapids, you can see that the river is still cutting down through the hard rock. If you look closely at the rock under your feet, you can find crossbedding, testimony to the beach-environment origin of this sand-based rock. The Weverton quartzite is overturned here in the western limb of the South Mountain anticlinorium, so the older layers are on top.

Weverton Cliffs were created where the Potomac River cut a water gap through the highly resistant quartzite that tops South Mountain. The resistant rock creates rapids where the river flows over it. View is to the south from the top of Weverton Cliffs.

To the west is the gap in the companion Weverton-rimmed Blue Ridge. Harpers Ferry, West Virginia, is visible through the gap. The valley between South Mountain and Blue Ridge is floored by the Grenville basement gneiss. To the east you can see the Catoctin Mountain water gap and the intervening valley of Grenville rock. The youngest rock you can see from here is over 500 million years old.

US 340 Exit Ramp and Keep Tryst Road

About a quarter of a mile upslope from the Potomac River bridge, at the US 340 exit for Sandy Hook, you can look closely at some very old rock. Take the exit, find a place to park along the road, and walk back along the US 340 eastbound exit ramp. Watch out for traffic here—it is heavy. Toward the east end of the ramp, nearer Keep Tryst Road, you can safely find the 1.1-billion-year-old Grenville gneiss. For a riskier trip through the traffic, walk west on the ramp and cross US 340. Opposite the ramp is a roadcut adjacent to westbound US 340. Here you can see ancient folds in the gneiss, and if you look closely, small bodies of garnet and specks of graphite.

At even further risk, you can walk down onto the US 340 bridge for views of the river and the two water gaps. A safer bridge access is at Potomac Wayside just across the bridge in Virginia, but be sure to follow parking regulations.

Back at the exit ramp, continue east on Keep Tryst Road through cuts in the Grenville rock and pull off and park at the hairpin turn for a look at Weverton Cliffs

View looking upstream at Harpers Ferry through the water gap in Blue Ridge from the pedestrian walkway on the upstream side of the US 340 bridge.

to the east. You can walk down to the river from here and locate yourself in the very midst of the South Mountain water gap. Above the hairpin turn, the US 340 roadcut below Weverton Cliffs is an excellent exposure but much too dangerous to reach because of the traffic. Luckily, the same formations are repeated in Blue Ridge west of Sandy Hook, where you can see them safely.

As you drive back toward US 340, look for a dike near the second right-hand turn uphill from Israel Creek. This is one of the many feeder dikes—magma conduits—for the Catoctin lava flows, which oozed up through the extending Grenville crust about 600 million years ago.

Maryland Heights and Harpers Ferry

The road to Sandy Hook is the first right turn (toward the river) after the US 340 exit for Sandy Hook. West of the almost river-level village of Sandy Hook, the road travels about 2 miles along the base of the precipitous cliffs of Maryland Heights, the north end of the water gap through Blue Ridge. Parking places along this road are few, and you will have better luck getting one during a weekday or in early morning. From a parking place you can hike back along the road and look at the rock, or you can cross a footbridge and walk on the C&O Canal along the rapids of the Potomac down to its confluence with the Shenandoah.

You can cross the railroad bridge on a pedestrian walkway to Harpers Ferry, West Virginia. On the footbridge, if you look back at the cliffs to the left and up from the railroad tunnel, you will see a line at about 45 degrees where the vertical cliff meets a vegetated zone of more eroded material. This line marks the contact between the quartzite of the Weverton—the resistant cliff face—and the phyllitic metasiltstones of the Harpers Formation, which are weathering to soil that supports vegetation. The younger Harpers Formation lies under the older Weverton Formation because this is the overturned limb of an anticline.

Behind the young geologist, on the cliff of Maryland Heights, is the vegetated zone of the Harpers Formation below the sheer cliffs of the Weverton.

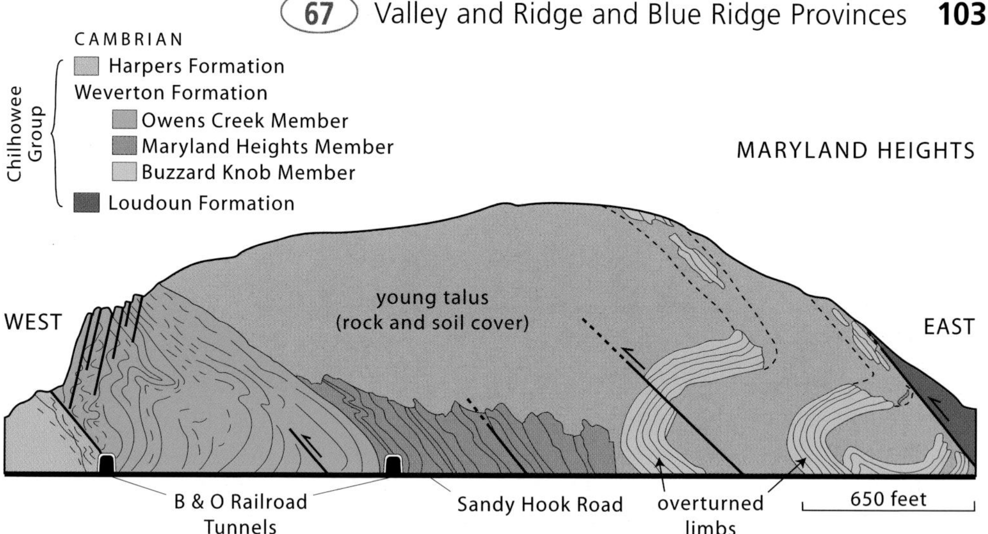

This diagrammatic sketch shows the overturned limbs below Maryland Heights.
—Modified from Southworth and Brezinski, 1996

On the Maryland side, you can take a moderately strenuous trail to the top of Maryland Heights. No rock climbing is required. The view from here may be better than the one from Weverton Cliffs, depending on your interest. This one reveals more of the river courses and of the Great Valley to the west. You can climb farther up to the Civil War redoubts and find good exposures of the Weverton quartzite.

Of hydrological interest in Harpers Ferry is the flood record on the side of the building across the street from the National Park Service bookstore. The flood crests are hard to believe. Of geological interest are the many stone walls and buildings constructed of phyllite of the Harpers Formation. If you walk up past Jefferson Rock to the upper part of the cemetery and look east, you will see a great view of the two Potomac River water gaps.

View looking downstream at the Potomac water gap and the confluence of the Potomac (left) *and the Shenandoah* (right) *from Harpers Ferry, West Virginia. US 340 bridge is in the distance.*

Maryland 68
Boonsboro—Clear Spring
17 MILES

MD 68 is a great route for observing the carbonate rock of the Great Valley. This rock formed in a carbonate bank in an ocean in Cambrian and Ordovician time. If you are just driving through and not making stops, one carbonate feature you can observe in fields is the uneven surface of grikes and pinnacles. Grikes are steeply dipping fractures or bedding planes that have been dissolved and filled with soil through many years of weathering. Protruding bedrock outcrops in between the relatively level soil areas are called pinnacles. Fields with many pinnacles are used more for pasturing than for planting.

At the intersection of MD 68 and US 40 Alternate and for the first mile and a quarter, you are on the Tomstown Formation of interbedded dolomite and limestone of Cambrian age. As you travel this route—if you arrive before the entire county has been developed—you will see that limestone soils are good for agriculture. In many places the road is lined with stack stone walls of limestone, hand built well over one hundred years ago to remove stone from fields and to enclose the land.

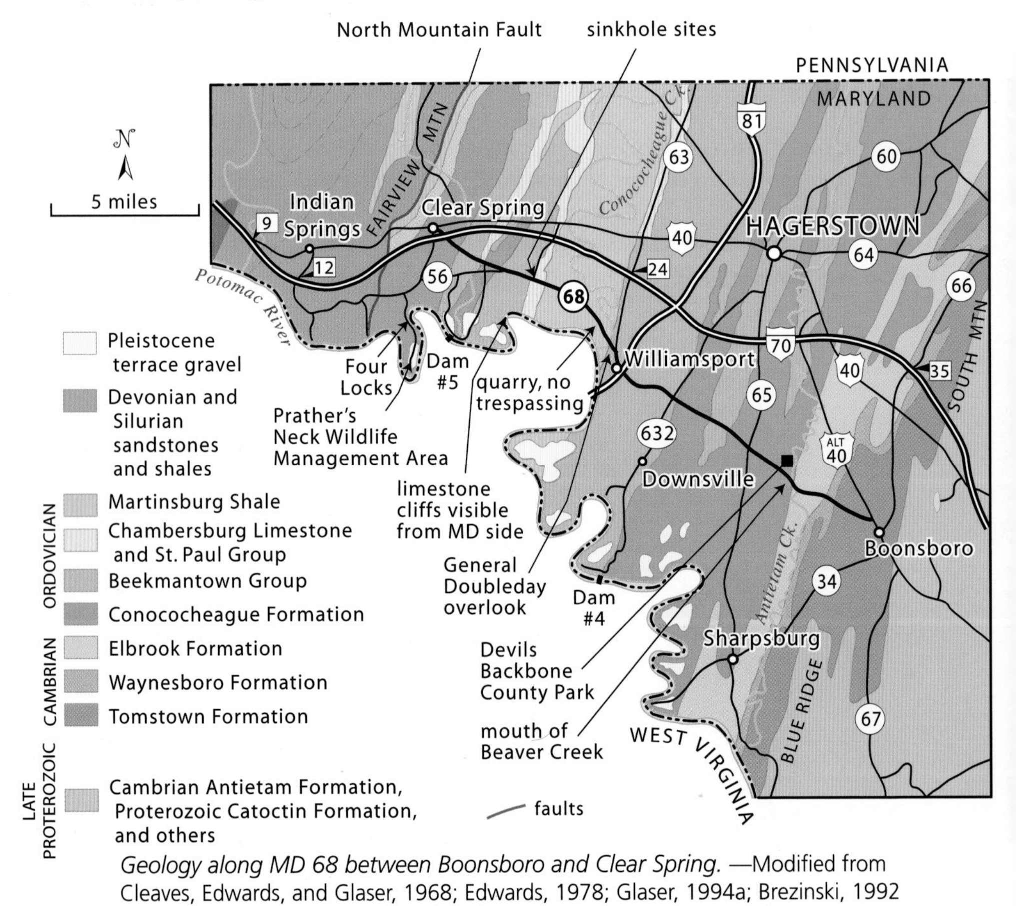

Geology along MD 68 between Boonsboro and Clear Spring. —Modified from Cleaves, Edwards, and Glaser, 1968; Edwards, 1978; Glaser, 1994a; Brezinski, 1992

For the first 1.5 miles west of US 40 Alternate, MD 68 passes over land of low relief—only gentle hills and valleys—and few outcrops. Just west of Millpoint Road is a ridge of interbedded limestone and sandstone. This sandy member of the Waynesboro Formation, not exposed here, is more resistant to chemical weathering than adjacent limestones but not nearly as resistant as the sandstones of South Mountain, visible to the east. You will cross many such small ridges in the Great Valley, and they usually contain either sandstone beds or sandy carbonate layers.

The slices of thrust faults and the many folds in the Great Valley carbonates bring repeated surface occurrences of the same formation. West of the ridge at Millpoint Road, you go onto the Elbrook Formation until you come to the bridge at the mouth of Beaver Creek, where it enters Antietam Creek. East of the bridge is an exposure of the laminated, well-bedded Elbrook Formation. West of the bridge are cliffs of the Conococheague Formation of Ordovician age, also with laminated dolomite. You may be able to walk down onto the floodplain of Antietam Creek here—check signs by the road. Notice cliffs on either side of the creek. Like the Potomac River and nearby Conococheague Creek, the Antietam has incised its meanders into a recently uplifting landscape. Acidity is usually not a problem in these creeks because it is neutralized by the dissolved carbonate ions, just as antacids neutralize excessive stomach acid.

Limestone cliffs in the Conococheague Formation at Devils Backbone County Park were created as Antietam Creek incised the land.

Farther west is the MD 68 bridge over Antietam Creek. Here you can stop at Devils Backbone County Park and, if the trail is open, cross the footbridge and take an easy hike of about a half mile up a small, incised cliff in the Conococheague Formation.

West of Devils Backbone County Park, MD 68 travels a fairly steep slope out of the creek valley. West of Lappans Crossroads (MD 65) is the rolling Conococheague Formation to Reichard Road, where in a little valley you can see pinnacle outcrops—the limestones of the Stonehenge Formation of the Beekmantown Group. Between the railroad crossing at St. James Run and Downsville Pike (MD 632) is, again, the rolling Conococheague Formation.

Side Trip to the Potomac River

If you want an interesting side trip, go south on MD 632 to the stop sign at Downsville and continue straight on Dam No. 4 Road to the dam on the Potomac River. Hearing the roar and seeing the great volume of water gushing over the dam will give you an appreciation of the great energy in this river that looking at its smooth surface in a quiet stretch does not. The dam, built about 150 years ago, provided water for the C&O Canal.

Between mileposts 88 and 89 on the C&O Canal Towpath, you can walk along the base of high cliffs made of interbedded dolomite (brown) and limestone (gray and white) of the Ordovician Rockdale Run Formation.

Near milepost 89, walkways from cabins down to the C&O Canal follow the dip of the sedimentary beds of the Rockdale Run Formation.

About a half mile south of Downsville, you can turn west on Dellinger Road, go another half mile, and turn south on Avis Mill Road. Go another half mile to the mill at the end of the road, where you can park in a gravel C&O Canal Towpath lot. On the towpath you can walk along the base of river-incised cliffs. The first 100 yards or so upstream is the Stonehenge Formation. Look carefully for paper-thin, folded laminations. About a quarter mile upstream is a cave and stream channel. When the entire landscape was at a higher elevation, this would have been an underground channel in the limestone. A little farther upstream is the Ordovician Rockdale Run Formation, which contains interbedded gray limestone and brown dolomite.

Williamsport Area

Between the Downsville Pike (MD 632) intersection and Williamsport, you will see a marked change in the lay of the land. Here the Rockdale Run Formation weathers almost uniformly to flat fields with many pinnacled outcrops. Because of the solubility of some limestones in precipitation, small fractures can become underground channels of subsurface drainage. Soil forms in the dissolved depressions and leaves a soil-bedrock interface that is pinnacled.

Aerial view of limestone pinnacles in a field near I-81.

From east of Williamsport you can look west and see the sandstone ridge of Fairview Mountain. Also on the horizon is the tall stack of a coal-fired electrical generating plant in Williamsport. The major north-south routes US 11 and I-81 run through this town, both built in the construction-friendly Great Valley.

In Williamsport you can take a three-block side trip to an interesting viewpoint at the General Doubleday site. Turn west onto West Salisbury Street at Town Hall, a solidly built limestone structure. Go two blocks and you will see a flag and cannon

From the General Doubleday high point in the cemetery above Williamsport, you can see areas that have been underwater during floods (left), *including the flood of 1936* (right). —1936 Photo courtesy of the Williamsport Town Museum

on a cemetery hilltop. Park and walk up for a commanding view of the Potomac River, the mouth of Conococheague Creek, the C&O Canal and one of its aqueducts (built of nearby limestone, probably from the St. Paul Group), and ridges to the west of the valley. The island that you see in the Potomac is probably the result of suspended sediment in high waters being dropped when currents were slowed on the inside of the river bend and by the downstream, almost head-on influx of Conococheague Creek. This area has experienced many catastrophic floods.

From the General Doubleday vantage point, if you look across the town, you can see a smokestack, which is on the other side of Conococheague Creek, and behind it you may be able to see the clay pit, or quarry, in Martinsburg Shale, long associated with brick works there. This more-than-2,000-foot-thick formation occupies a 2-mile-wide band in the middle of the general syncline of the Great Valley. Conococheague Creek has incised long, sweeping, fairly regular meanders in the shale. Thickly bedded and nearly vertical limestones to east and west restrict the flow of the creek to the Martinsburg Shale because the limestone is more resistant to this downcutting than the flaky, crumbly shale.

On the western edge of Williamsport, you cross Conococheague Creek and its floodplain and farther west encounter the deeply gullied and dissected terrain characteristic of shale. You drive over ups and downs that are steeper and more frequent than those of nearby limestone terrains. You will pass shale and clay banks right beside the road. Shale weathers to clay, and clay has low permeability—it

does not let water soak in readily. As a result, surface runoff is high. Gullies and steep-sided ravines are cut into the land—not the case with the highly permeable limestones.

In the shale are cedar trees, one of the plants hardy enough for the low permeability of the clay soil produced by the weathering of shale. Here are pastures rather than crop fields. In addition, you will pass the entrance to a landfill. All Washington County landfills are sited in the Martinsburg Shale. The low permeability helps contain any leakage of contaminants that might percolate into groundwater. In limestone, leaking pollutants travel quickly and readily through underground channels and contaminate groundwater. When you leave the shale, you should notice the change immediately. At Bottom Road, the relief to the west becomes much less pronounced in the limestone.

Sinkholes in Chambersburg Limestone

At Cedar Ridge Road, you can see in the pasture to the north not only numerous outcrops but also depressions and sunken areas. The Chambersburg Limestone is at the surface here, and it exhibits underground solution openings. Soil can collapse into funnel-like openings near the surface and form sinkholes. Water drains into them, down into the subsurface channels. If filled with dirt, the sinks might later collapse and reopen. You can drive up Cedar Ridge Road about a half mile and observe sinkholes immediately west of the road, some of them containing trees with their bases 5 to 10 feet below the rest of the field. Most of the sinkholes in Washington County are located in the Chambersburg and adjacent formations in a belt west of Conococheague Creek.

Western Edge of the Great Valley

West of Cedar Ridge Road you'll see the nearly flat and pinnacled limestone landscape of the Rockdale Run Formation. Folding produced this repetition of a formation—it runs in a more or less U-shaped syncline deeply under the Martinsburg Shale. To the west is another repeated layer of the syncline, the Conococheague Formation, which forms a small ridge at St. Paul Road, and then, just west of there, is the small valley of the Little Conococheague Creek. This creek flows through the gap in Fairview Mountain (North Mountain ridge) that you can see several miles to the northwest. The ancestor of this creek may be responsible for the small ridge at Ashton Road made of Quaternary alluvium that probably was eroded from the mountains to the west when the creek was larger and carrying more sediment. Clear Spring sits at the western edge of the Great Valley below a sandstone ridge. At the base of the ridge is the North Mountain Fault, where Cambrian formations have been thrust westward onto Devonian formations.

Fairview Mountain, regionally known as North Mountain, marks the western edge of the Great Valley in Maryland. Photo taken from MD 68, 2 miles east of Clear Spring. Note the gap cut by Little Conococheague Creek.

You can take St. Paul or Ashton Road south to Dam No. 5 and see the Potomac River roaring over the dam. For a few hundred feet upstream of the dam on the C&O Canal Towpath, you can see excellent exposures of ribbony laminations in the Conococheague Formation. These thin layers reflect the cyclically changing environments of tidal and shore/beach deposition.

Four Locks on the Potomac

South of Clear Spring, you can visit Four Locks on the C&O Canal. Here an incised meander of the Potomac River actually brings the flow toward the northwest. If you walk upstream on the towpath, you enter a shortcut across the meander that saved over 4 miles of canal construction. Elevated above the present river, this passage is an abandoned channel of a former Pleistocene tributary.

If you drive past the boat ramp and out Ankeney Lane onto the neck of the meander, you will come to a field that is part of the National Park Service C&O Canal property. The field is filled with rounded gravel and pebbles, indicating water transport. The river once ran along this terrace before it incised the meander into limestone of the Conococheague Formation. The stones here are from farther west in Maryland, and today they rest over 100 feet above the present river elevation. You can walk all the way out to the cliffs above the river on this park property, but take care because private hunt club land borders it. Many of the necks in the Potomac River meanders preserve gravel terraces deposited by the ancestral river.

At road's end you can also park at the newly opened Prather's Neck Wildlife Management Area and walk out to the cliffs above the river. There are no trails and this is a state-managed public hunting area, but you can see the same geologic features here.

Maryland 77
Thurmont—Smithsburg
10 MILES

Just west of US 15, MD 77 leaves the Triassic basin and heads up the eastern limb of the South Mountain anticlinorium, a steep climb up Catoctin Mountain along Big Hunting Creek. Near the western end of Catoctin Mountain Park, about 6 miles

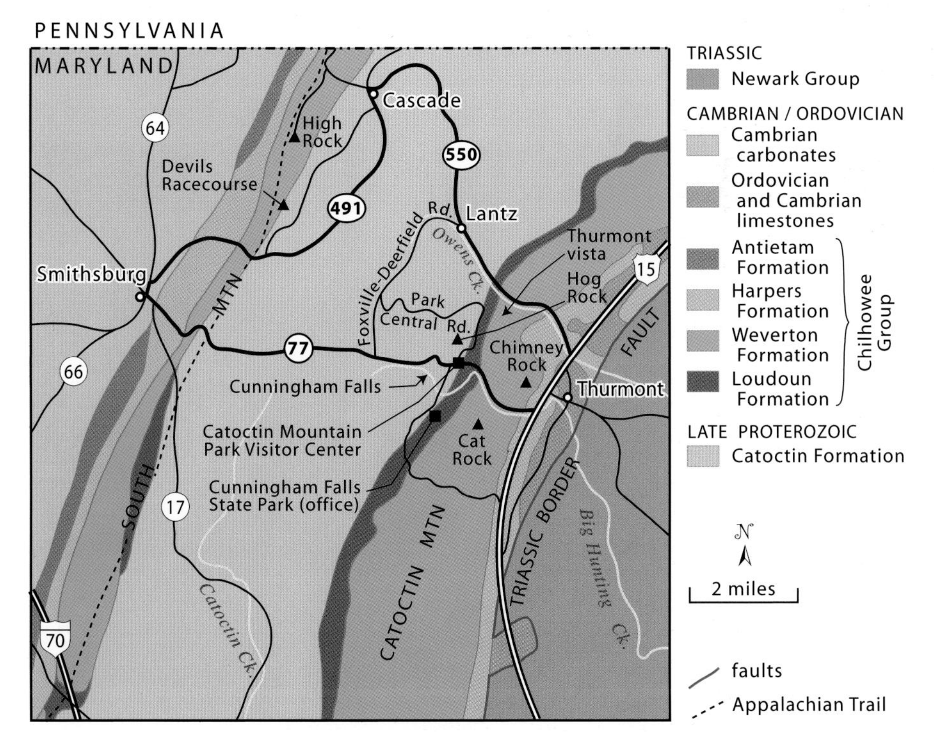

Geology along MD 77 between Thurmont and Smithsburg. —Modified from Cleaves, Edwards, and Glaser, 1968; Edwards, 1978; Fauth, 1977; Reinhardt, 1974; Brezinski, 1993; Godfrey, 1975; Brezinski, 1992

away, are the headwaters of the creek at elevation 1,450 feet. Near US 15 the elevation of the creek is about 550 feet. This change of 900 feet over 6 miles represents a stream gradient of about 160 feet per mile. The lower part of this steep gradient is due to the extreme downfaulting that occurred during continental rifting about 200 million years ago, but much of the upper part is due to thickly stacked layers of lava flows that occurred during rifting of the Grenville continent of 600 million years ago. Later metamorphosed into the metabasalt of the Catoctin Formation, this rock is highly resistant to weathering. You can see the steepest section of Big Hunting Creek at Cunningham Falls.

Stone Streams

About 1 mile west of US 15 you can begin to see remnants of erosion related to Pleistocene Ice Age climates of 30,000 to 10,000 years ago. Just below the first 90-degree right-hand turn, you can look up to your left and see a side-slope stone stream. You may be able to park right past the turn and walk back for a closer look. Side-slope stone streams are linear boulder deposits up to a half mile long and 50 to 500 feet across. Often bordered at their heads, or tops, by scarps, they can extend all the way down a mountain slope. Valley-bottom stone streams are linear boulder deposits in stream bottoms.

From the parking lot for Cat Rock and Chimney Rock Trails, you can walk or hike to other stone streams. One is on the mountainside just about 50 feet upstream from the parking lot, but the longest is across the road on the Chimney Rock Trail. How did these form? During the Ice Ages, continental glaciers did not reach quite this far south, but the colder climates generated freeze-and-thaw conditions for fifty to one hundred days each year. Much more snow and ice would have covered the

Big Hunting Creek flows over a valley-bottom stone stream, formed during Ice Age climates when ice-wedged boulders slid down from ridgetop outcrops over a water-saturated, underground permafrost layer.

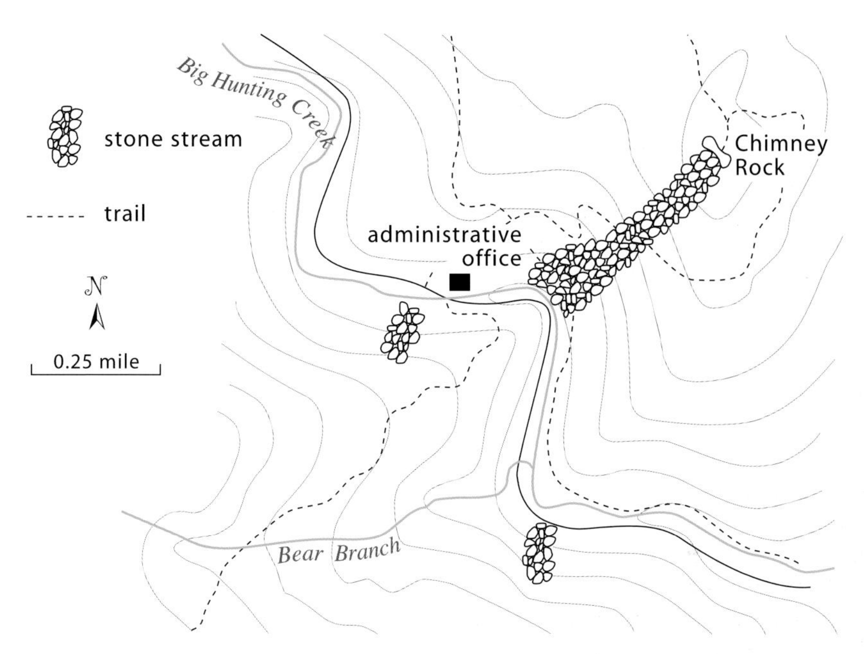

Several stone streams are located near the Chimney Rock Trail parking area. —Modified from Godfrey, 1975

ridgetop outcrops then. Sunlight during the day would have caused melting, and water would flow into cracks and crevices in rock. At night when temperatures fell below freezing, the water would solidify, expand its volume by over 10 percent, and wedge rock apart into the angular boulders you see. These boulders tumbled and collected below ridgetop cliffs into what are known as talus slopes. During warmer seasons, the ground was thawed to a depth of a foot or so, but deeper layers remained frozen. Water could not penetrate the impermeable frozen underground surface, so boulders in a slurry of saturated soil slid in a slow, glacial manner downslope in a process known as gelifluction. Then, after the last Ice Age, erosion cleared out soil, leaving behind the stone streams of angular boulders.

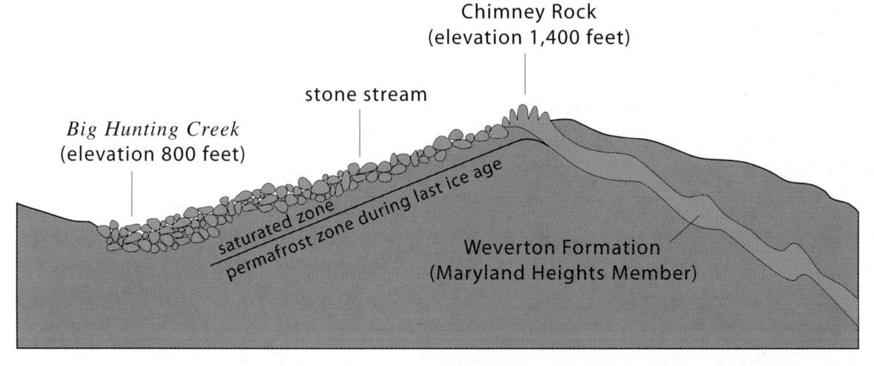

Freeze-and-thaw conditions during cold, Ice-Age climates wedged the hard Weverton quartzite into angular boulders that slid downward slowly into a stone stream from the outcrop of Chimney Rock. —Modified from Brezinski, 1992; Fauth, 1977; Means, 1995

Chimney Rock Trail is one of the shorter, relatively easier hikes to a Weverton quartzite overlook of the Frederick Valley. If you want to make it more challenging or even hazardous, you can descend from the Chimney Rock scarp, the rock source for the stone stream, down half a mile to Big Hunting Creek. The footing is not good, but you can see to either side the confines of the stone stream and experience firsthand (or first foot) the angularity of nonfluvial erosion. For a safer hike, just look for the stone stream where it crosses the trail about a third of a mile up from the trailhead or go down the less strenuous Crows Nest Trail.

On the opposite side of Big Hunting Creek from Chimney Rock, Cat Rock Trail leads up to another Weverton scarp. You can look north and see Chimney Rock, west to the metabasalt ridge where Camp David is located, and east toward Frederick Valley.

Catoctin Mountain Park Visitor Center

At the Catoctin Mountain Park Visitor Center on MD 77, you encounter a small bowl or valley in the conglomerate of the Loudoun Formation. Lying between the resistant Catoctin and Weverton Formations, this rock was less resistant to weathering and thus wore down faster—producing steep stream valleys to the north and south. The flat, open area opposite the visitor center is where the streams meet, level out, slow, and drop sediment, mostly during flood events, to form the floodplain along MD 77. You can see an exposure of the crumbly Loudoun Formation by walking to the end of the paved visitor center parking lot and then proceeding onto the old roadbed.

If you take some of the well-worn paths used by fishermen across MD 77 from the visitor center, you will see the characteristic erosion and deposition of a meandering stream. On the outside of a bend the running water—with its natural tendency to continue in a straight line—erodes the bank as it is deflected into a turn, while on the inside the slower moving current cannot hold its sediment in suspension and deposits it. You will find cutbanks on the erosional side, and sandbars or gravel bars on the depositional side. Most of this type of erosion and deposition occurs during high water.

From the Catoctin Mountain Park Visitor Center, you can get a park map and hike or drive in several directions. To the west you can hike to Cunningham Falls,

At Wolf Rock, fractures in the Weverton Formation are places where soil accumulates and trees grow.

and to the north and east you can hike up to the fortresslike Weverton outcrop of Wolf Rock.

Side Trip on Park Central Road

From the Catoctin Mountain Park Visitor Center, you can drive up Park Central Road (closed during winter) in the metabasalt to Thurmont Vista Trail, the shortest and easiest hike to an overlook of Frederick Valley. Higher up on the road, you can park and walk out to Hog Rock, an outcrop of the Catoctin metabasalt, with a view to the east of the Weverton quartzite ridge. The basalts are still over 1,000 feet thick today after millions of years of erosion, so the outpourings of about 600 million years ago must have been much thicker. Emerging from vertical fractures in the rifting Grenville continent, episodic lava flows oozed out in pulses separated by hundreds or thousands of years of erosion and deposition. Individual flows were almost 100 feet thick, and the entire event may have lasted up to 15 million years, completely blanketing the old landscape. Eventually, continued rifting took the Catoctin lavas below sea level, and then offshore clastics, most notably the Weverton Formation, were deposited on it in Cambrian time. The field name for metabasalt is greenstone, and you can see why in the many boulders and outcrops along the road.

Farther west on Park Central Road, you pass the entrance to Camp David—no admittance!—and then come to Chestnut Picnic Area. Here, on the periphery of the area, you can see plateaus or terraces below you. These are successive layers of extrusive lava outpourings, in this case a rock known as rhyolite. Unlike basalt and its intrusive equivalent diabase, which are high in iron and magnesium and low in silicon, rhyolite and its intrusive equivalent granite are high in silicon. This rock oozed out at about the same time as the basalt and was metamorphosed to metarhyolite. You can see more metarhyolite farther west in the bed of Owens Creek. If you tap two pieces together, you will hear a distinctive clink, and if you fracture a piece, you will produce a sharp, almost razor edge. The Indians used this rock for tools and arrowheads. The author has several pieces which he uses for cutting string and fishing line. You can return to MD 77 via Foxville-Deerfield Road.

When metarhyolite of the Catoctin Formation fractures, sharp edges are formed. Points and cutting tools of this rock have been found in Native village sites hundreds of miles from Catoctin Mountain.

Cunningham Falls State Park

From the Catoctin Mountain Park Visitor Center, you can drive a few hundred feet west and turn toward Cunningham Falls State Park. An excellent exposure of metabasalt of the Catoctin Formation is alongside the road up the hill from the bridge over Big Hunting Creek. This east-dipping rock is an inner layer of the South Mountain anticlinorium. Another exposure is in the spillway on the opposite side of the dam, accessible by a short walk. Looking downstream from the middle of the dam, you will see resistant Catoctin metabasalt on your left and resistant Weverton quartzite on your right, split by the valley of less resistant Loudoun conglomerate. On up the road from the dam you can enter Cunningham Falls State Park and take one of two trails to the falls. Depending on how it is measured, this is Maryland's highest waterfall, at about 80 feet of cascades. It is also the best exposure of the Catoctin metabasalt.

A waterfall begins at a nickpoint, or cliff, where there is a sudden dropping off. Then, over the course of many years it erodes its way upstream. The original Cunningham Falls may have begun at the contact between the resistant metabasalt and the softer Loudoun conglomerate or, more likely, all the way down at the Triassic Border Fault. Today Cunningham Falls flows over one of the thicker sections of the metabasalt, and as the landscape has been generally lowered over many years, this thicker section has created a steeper gradient relative to the rock downstream. Today running water is continuing to erode channels into the rock face at the falls, cutting its way upstream in small "canyons" that you can see either from the viewing platform or by climbing up onto the open rock area next to the channel.

On the trail back down to the park, observe the gorge and cliffs cut by Big Hunting Creek. Its path is similar to the water gaps cut by the Potomac River in resistant, perpendicular ridges, but here the downcutting has been accelerated by the sharp plunge down into the Triassic basin.

Cunningham Falls, originally known as McAfee Falls, flows over thick sequences of 600-million-year-old lava flows.

On MD 77 heading west, you will see a handicap access parking area for the falls, and then you begin a steep ascent up the same scarp down which the falls plunge a few hundred feet to the south. During winter you can see exposures of the green metabasalt as you climb.

Crossing South Mountain

West of Catoctin Mountain Park, MD 77 continues across metabasalt and metarhyolite that comprise the center of the anticline. Then as you descend South Mountain in a series of sharp turns, you may be able to catch a glimpse of the Great Valley to the west. Here you will cross the same formations you crossed driving up from Thurmont into Catoctin Mountain Park—in reverse order. Because the western limb of the anticline is overturned, the formations dip east, just as do the ones on the eastern limb above Thurmont. Few outcrops are visible from this section of road.

At the bottom of the mountain, where MD 77 intersects MD 64 at Smithsburg, you have entered the limestone of the Great Valley, here the Tomstown Formation. From here west lie about 20 miles of mostly limestone. Why is it a valley? Limestone is primarily calcium carbonate, and it reacts chemically with precipitation, which is acidic even in an unpolluted environment. This low resistance to weathering causes it to wear down much faster than most other rocks—creating low areas. The limestone valley runs from Pennsylvania to Georgia and is the site of the major north-south routes US 11 and I-81, and of major troop movements during the Civil War. The Tomstown Formation here may have once been continuous with the Frederick Limestone east of South Mountain, but they were compressed into the huge, arching anticline and eroded over millions of years.

Maryland 491
Smithsburg—Cascade
7 MILES

Just a half mile north on MD 64 from the western end of the MD 77 junction, you can turn onto MD 491, which with its wide shoulder and excellent outcrops, is probably the best locality for examining layers of the overturned limb of the huge South Mountain anticlinorium. Because the beds have been tectonically pushed up past vertical and flipped over, the younger layers now lie beneath the older layers.

Before you reach the outcrops, stop a few hundred yards after turning onto MD 491 and look west. On a clear day you can see all the way across the carbonate-floored Great Valley to the sandstone ridge of North Mountain. Even though this is not a ridgetop view, it is one of the best for appreciating the width of the carbonate bank and the results of long-term differential weathering.

Just as you enter the right-hand turn about 100 yards upslope from Fruit Tree Drive, you will see an exposure of sandstone and siltstone of the Harpers Formation on the north side of the road. In *The Physical Features of Washington County* (1951) geologist Ernst Cloos, who so well articulated the details of Maryland structural geology in the 1940s and 1950s, wrote, "All formations which participate in the South Mountain uplift and anticlinorium possess a typical, distinct, and regionally oriented cleavage." This cleavage transects bedding and is due to shearing and flow during the plastic deformation of metamorphism. It dips east in almost all outcrops

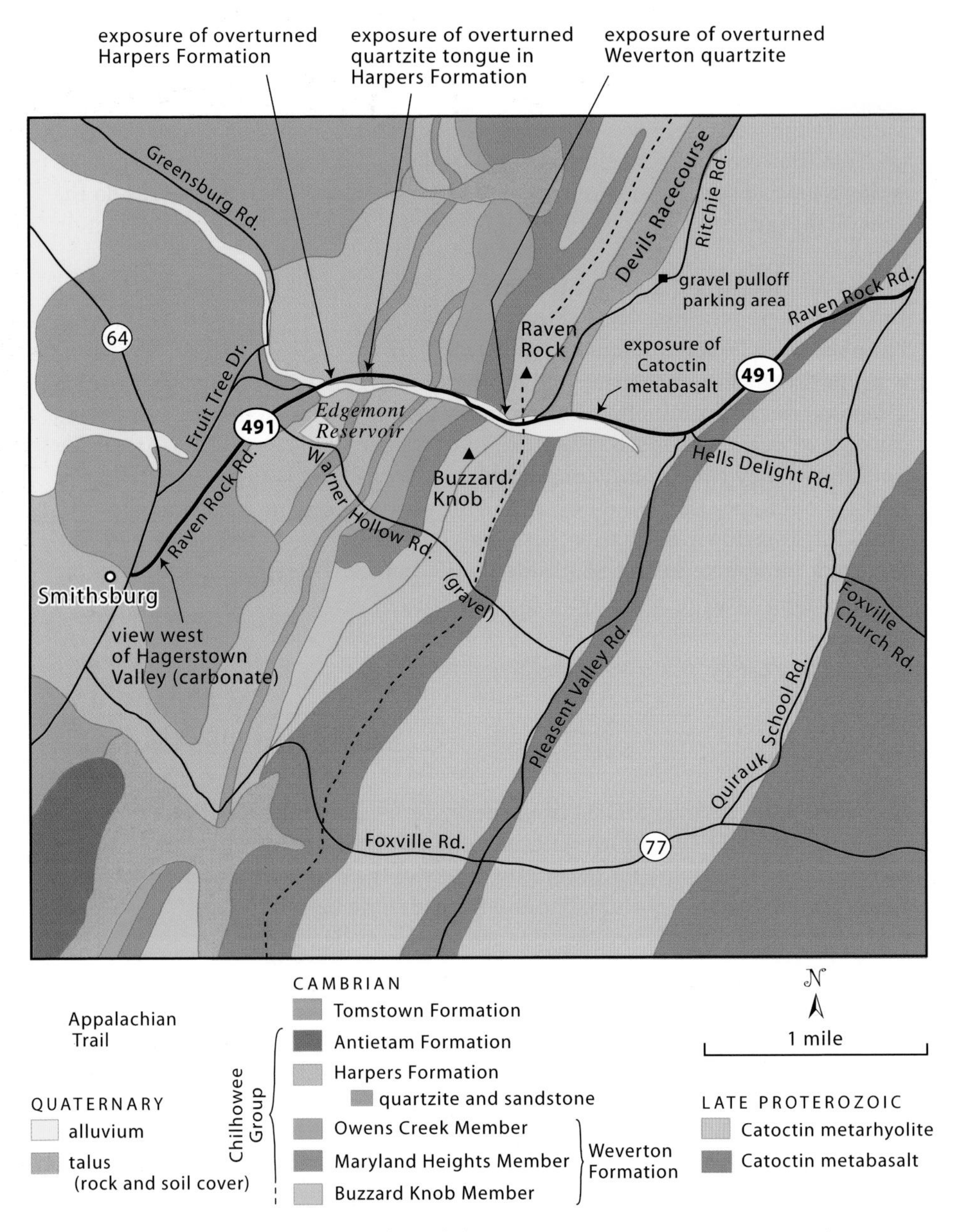

Geology along MD 491 east of Smithsburg. —Modified from Cleaves, Edwards, and Glaser, 1968; Edwards, 1978; Fauth, 1977; Brezinski, 1992; Brezinski, 1993; Godfrey, 1975; Cloos and others, 1951

and "can readily obscure bedding and has been mistaken for bedding." The east dip that you see in this first outcrop of the Harpers Formation is in the cleavage. You must look very closely to see the bedding, which is almost vertical. Moving upslope from the outcrop, across from a break in the guardrail and an earthen bank, you can find a crumbly section of the Harpers Formation. A bit farther east, across from the next earthen bank upslope, you will see a crystalline quartzite tongue of the Harpers Formation.

In this section of the Harpers Formation, you can clearly see the east-dipping cleavage transecting the more vertical bedding.

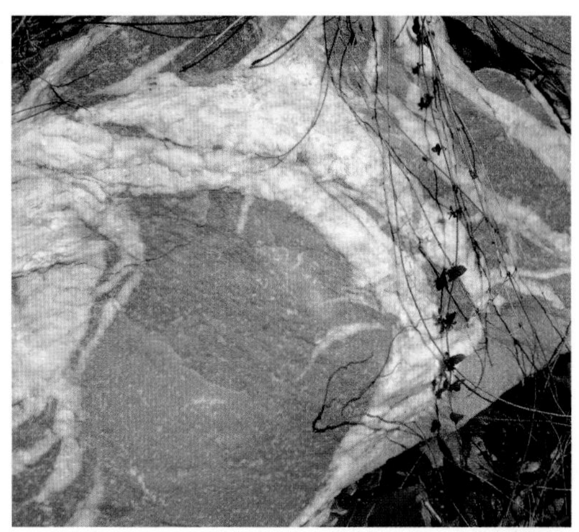

Quartzite of the Harpers Formation, looking very crystalline here, contains veins of quartz.

Raven Rock and Buzzard Knob

Just west of the Ritchie Road stop sign, Raven Rock rises above the north shoulder of MD 491 with cliff exposures of overturned Weverton quartzite. Look for the Appalachian Trail crossing of MD 491 downslope from Raven Rock. You can hike north about a half mile to Raven Rock and see the Great Valley to the west and Buzzard Knob to the south. These two high points are of the ledge-maker member of the Weverton Formation called the Buzzard Knob Member—hard, metamorphosed, high-silicon quartzite that resists erosion much more than the adjacent limestone to the west. The water gap between the two high points is quite marked here.

You can also hike south on the Appalachian Trail to Buzzard Knob. As you come to the top of the slope after less than a half mile, turn right onto a no-longer-maintained but still vaguely marked side trail through briars that heads toward the knob. You then climb several hundred yards over an unmarked rugged, rocky talus field. Just aim upward. At the top are well-defined, overturned beds of the Buzzard Knob Member of the Weverton Formation dipping east. The best time to go is during winter when there is no snow cover.

At the base of Raven Rock on MD 491, take note of the east-dipping rock. About 10 miles to the east on MD 550, you can see the same Weverton rock dipping in the

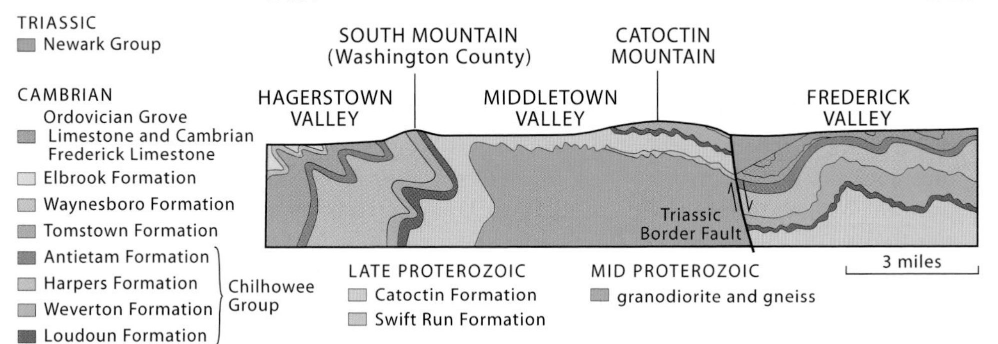

South Mountain and Catoctin Mountain are on the limbs of a large overturned anticline.
—Modified from Cleaves, Edwards, and Glaser, 1968

Separated by 6 miles, these Weverton Formation limbs of the South Mountain anticlinorium dip east at roughly the same angle. At Buzzard Knob (top) *off the Appalachian Trail is an outcrop of the western, overturned limb. At Isabella's Rock* (bottom) *in the Manor Area of Cunningham Falls State Park is an outcrop of the eastern limb. In the early growth of the alpine-scale Appalachians about 250 million years ago, these limbs were probably continuous across a towering arch.*

same direction but on the other limb of the fold—a characteristic of an overturned anticline. On route to Cascade, watch for outcrops of the Catoctin metabasalt and for the beautiful stack stone walls constructed with metabasalt.

Side Trip to Devils Racecourse

Turn north onto Ritchie Road from MD 491 (upslope a few hundred feet from Raven Rock), and go three-quarters of a mile to a rough gravel pulloff on the left where you can park and then walk about 200 feet to a huge block field known as Devils Racecourse. There are no signs, but this is public land. The block field is about 200 feet wide and 1 mile long.

Block fields, or boulder fields, are broad, mostly flat areas covered with angular boulders, their slopes usually less than 10 degrees. They resemble stone streams but are longer, wider, flatter, and usually unvegetated. Block fields like Devils Racecourse are common in the Arctic. Geologists disagree on the exact conditions necessary for their formation and movement, but some conditions include (1) chemically inert rock in ridgetop outcrops or cliffs as source material; (2) a cold climate with many temperature fluctuations between freezing and thawing so that water running into existing cracks in rock freezes, expands, and wedges the rock apart; or sustained temperatures below freezing during which water migrates chemically into rock and fractures it; and (3) subsequent slow, downslope sliding of a water-saturated layer of these fractured boulders and sand over an underground layer of frozen earth or permafrost—a process known as gelifluction.

How could this Arctic formation have met these conditions here? First, the Weverton quartzite on the ridge to the west of Devil's Racecourse and the Catoctin metabasalt on the ridge to the east are both inert to chemical weathering and form many high-elevation outcrops.

At Devils Racecourse, boulders from two different formations on adjacent mountaintops were transported into the intervening valley during Ice Age times.

Second, low temperatures would have been common here during Ice Age times. With temperatures cold enough a couple hundred miles to the north to sustain continental glaciers, near-glacial conditions and perhaps small valley glaciers existed here.

Third, the valley-bottom slope had a gradient sufficient for downward sliding. Because the limestone lay at a significantly lower elevation just a few miles west, any stream of interglacial times would have downcut this valley between the Weverton quartzite and the Catoctin metabasalt. Then, during glacial periods, with snowfields topping local ridges, ice-wedged boulders from the two formations would have slid slowly down the side slopes during seasonal warming and converged in the valley. A few feet beneath the surface of these side slopes and the valley bottom lay frozen, permafrost surfaces topped by a water-saturated, viscous, boulder-filled layer that slipped downslope over the water-lubricated basal layer. Movement was glacially slow, on the order of inches per year.

Be sure not to miss the obvious features of this mile-long deposit. The boulders are angular, not rounded—clear indication that they were not deposited by running water. Listen for the stream running today beneath the rocks. Take note that the boulders are both the dull gray, sandstone-textured quartzite of the Weverton Formation and the greenstone metabasalt of the Catoctin. If you walk up the side slopes, you will find them boulder strewn.

High Rock

You can drive all the way to High Rock for a panorama of the 20-mile-wide, limestone-floored Great Valley. This clifftop outcrop, on the overturned limb of the South Mountain Anticlinorium in the Buzzard Knob Member of the Weverton Formation, stands about 1,000 feet above the limestone valley. The quartzite is much more resistant to weathering than is the limestone. To reach High Rock, turn west just north of the lake in Cascade onto Pennsylvania Road toward Rouzerville. Drive about 1 mile and turn south (left) onto Pen Mar High Road, which heads up the mountain. Be careful at High Rock: it has no guardrail and the rock surface is slippery where it has been painted by "artists."

From the highly resistant quartzite of High Rock, you can see the more rapidly weathered limestone valley about 1,000 feet below.

Maryland 550
Cascade—Thurmont
9 MILES

See map on page 111.

From Cascade, MD 550 passes over a metabasalt-floored valley through Sabillasville to Lantz, where it joins the Owens Creek valley for the descent to the Triassic Border Fault at Thurmont. Owens Creek, like Big Hunting Creek along MD 77 to the south, has a steep gradient, about 115 feet per mile, and for the same reason—it cuts down to the faulted Triassic basin.

Between Lantz and Eylers Valley Flint Road, you can see several exposures of the metabasalt. Then, farther southeast, as you descend more steeply, you cross rock formations of Cambrian age. There are several places to park along the road by Owens Creek, and you can see the formations safely by walking behind the guardrail above the creek. At the first fairly hard left-hand turn if you are descending, you can see a large outcrop of the Maryland Heights Member of the Weverton Formation, a metasiltstone and metagraywacke (dirty sandstone). You can pull off and park right below the turn on the floodplain.

From a pullout located below the upper railroad bridge over MD 550 and above the lower railroad bridge, you can walk up to an exposure of the Owens Creek Member of the Weverton, a conglomeratic metagraywacke. If you want to climb up to the tracks—but not onto the bridge!—you can see a large, "slick rock," east-dipping exposure. All of these formations dip eastward, as they do throughout the overturned South Mountain anticlinorium.

From this same pullout, you can explore the depositional floodplain of Owens Creek and the erosional cutbank where it has incised the hard rock. Below the lower

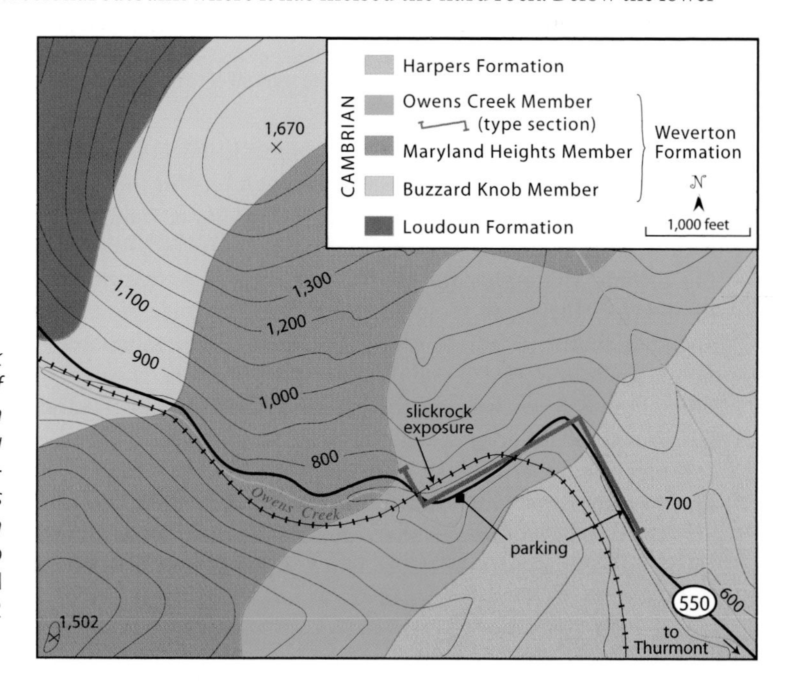

Along Owens Creek on the east limb of the South Mountain anticlinorium, you can see the east-dipping Harpers and Weverton Formations up close. —Modified from Brezinski, 1992

To put the railroad through, workers excavated hard rock of the Weverton Formation along its dip to make the job "easy."

railroad bridge and a right-hand turn, you will see the Harpers Formation, next to the "Falling Rock" sign. You can park right below the roadcut and walk back outside the guardrail.

Bear Pond Mountains

7 MILES

A side trip off I-70 or US 40 through the beautiful Bear Pond Mountains from Clear Spring to Pectonville will take you through terrain that is the product of three different mountain building episodes over the course of about 300 million years. The rock of these mountains is part of the Queenston Wedge of sediments eroded from the Taconic Mountains and part of the Catskill Wedge eroded from the Acadian Mountains. The many geologic structures—the folds of anticlines and synclines, and the giant thrust faults—resulted from the compressive forces of the Alleghanian mountain building event and the assembly of Pangea. The tight folding of the many different rock layers has caused alternating layers of mostly sandstone and shale to be exposed for millions of years to the elements of weathering. Sandstones, highly resistant to weathering, stay higher longer and form ridgetops, while shales, which are less resistant, weather down into low hills or valley bottoms.

You can drive through these mountains easily from either direction off I-70. If you are eastbound, take the Indian Springs exit (exit 9). Turn left (west) onto US

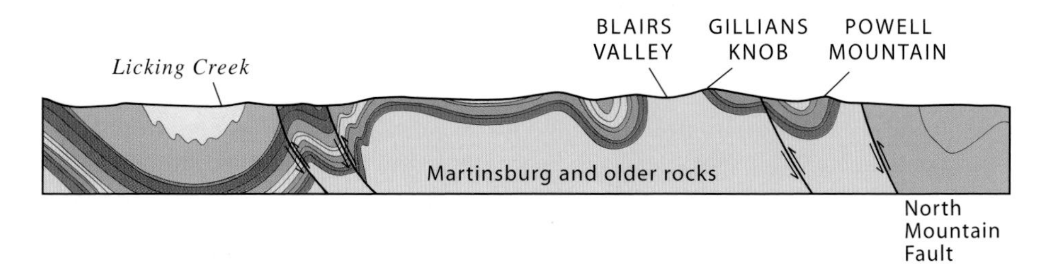

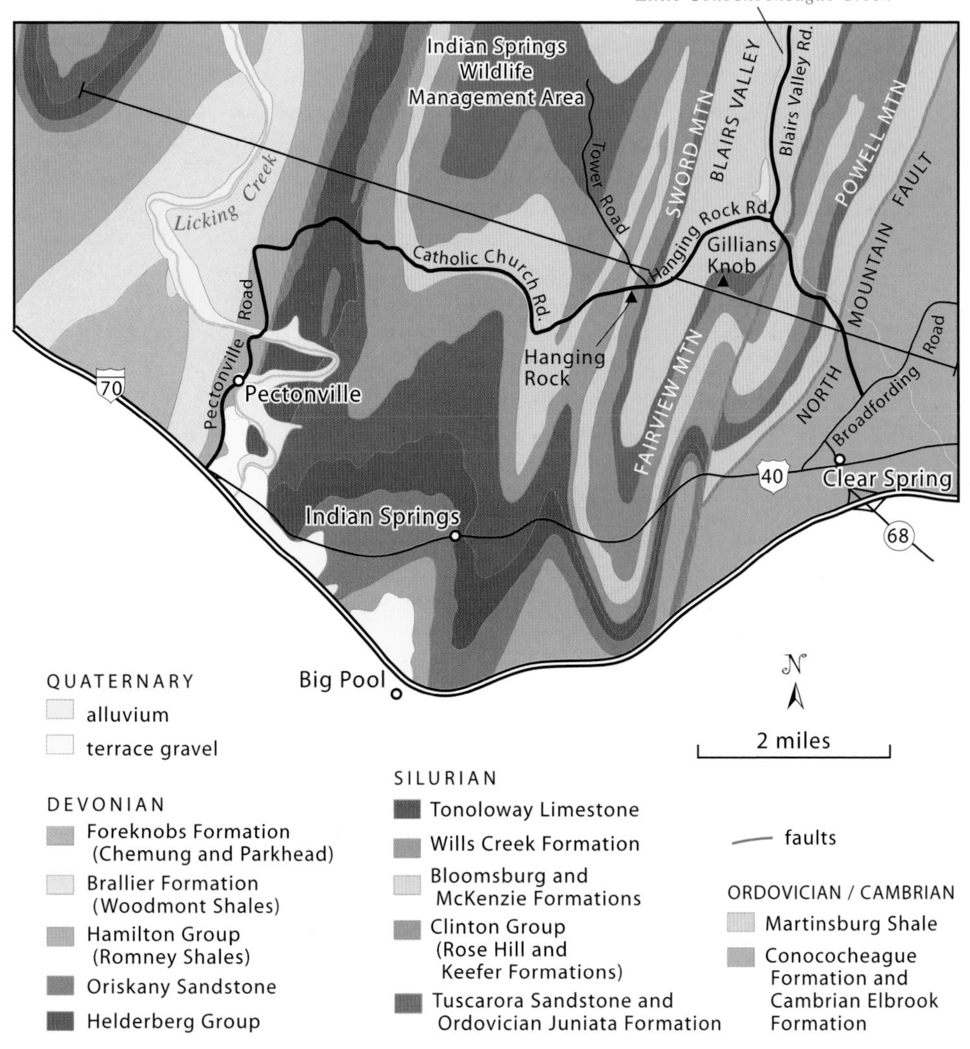

Geology of the Bear Pond Mountains. The transect through the mountains shows the anticlinal valley of Blairs Valley in Martinsburg Shale between ridges of resistant sandstone. Note the tight folding, the faulting, and the synclinal valley of Licking Creek. —Modified from Cleaves, Edwards, and Glaser, 1968; Edwards, 1978; Glaser, 1994a

40 and then take an immediate right onto Pectonville Road. If you are westbound, take the Clear Spring exit (exit 18), bear right to the traffic light, go straight across US 40, and drive out of Clear Spring to Broadfording Road. The text will describe the route from east to west, Clear Spring to Pectonville.

On Broadfording Road you drive on the westernmost occurrence of the carbonate rocks of the Great Valley in Maryland. To the immediate west is the steep-sided ridge of Fairview Mountain (regionally known as North Mountain), topped by the resistant Tuscarora Sandstone of Silurian age. This formation, which today runs from New York to Tennessee, was once mostly a beach deposit at the edge of an inland sea. The break that you see in this ridge is the water gap cut by Little Conococheague Creek. Take Blairs Valley Road right through the gap beside the creek. Before you get to the gap, you cross a Quaternary alluvial deposit eroded from the mountains and laid down on limestone.

West of the water gap is Blairs Valley. Pull into the parking lot at the lake and look west at the steep, almost vertical, east face of Sword Mountain, also topped by the Tuscarora Sandstone. Blairs Valley lies in the middle of a tightly folded, eroded anticline, and because the limbs of the anticline are of hard sandstone, ridges stand east and west of the less resistant shale bottom of the valley.

If you wish to attempt one of the toughest climbs in Maryland and experience major differential erosion firsthand, you can cross the dam and head up from the

Blairs Valley in the Martinsburg Shale lies between ridges of resistant Tuscarora Sandstone.

west side of the lake. The climb has no trail and is boulder-strewn, extremely steep, and a snarl of briars—best attempted during winter when there is no snow cover. From the top of Sword Mountain on a clear day you can see across the Great Valley to South Mountain.

Take Hanging Rock Road, which climbs the northern slope of Gillians Knob, an outcrop of Tuscarora Sandstone to the south of the lake. From the road you can see a great view of Blairs Valley and its enclosing ridges to the north. At Tower Road you can drive up 1 mile to a closed gate and see boulders of Tuscarora Sandstone by the road. Here you are on the western limb of a syncline of which the eastern limb is Sword Mountain. At the base of Tower Road you can park near the stop sign and walk carefully down Hanging Rock Road to Hanging Rock, which hangs almost over the road. Here is another water gap and another exposure of the Tuscarora. Partial clogging of this gap and siltation above it have created a pocket of wetlands, and you can see cattails on the other side of the road above Hanging Rock.

Continue west on Hanging Rock Road and then turn onto Catholic Church Road. Look along the road for exposures of red, interbedded sandstones and shales of the Silurian McKenzie and Bloomsburg Formations. These deposits, colored red by iron oxides, were laid down by streams as the inland sea basin filled with sediments from the ancient Taconic Mountains. Along these roads and farther west on Mooresville and Pectonville Roads, look for the hills and gullies of medium relief that are typical of erosion in shale formations. Also, look for gray and brown shales of the Devonian Hamilton Group and Woodmont Formation, associated with the inland sea basin of the later Acadian Mountains.

Just west of Indian Springs Pond on Mooresville Road and on Pectonville Road, you can see roadside pits or quarries where these thick shale formations have been mined. You can also see numerous small roadcut exposures along the way. Also in this area is a limestone of the Silurian/Devonian Helderberg Group, the solubility of which is responsible for a disappearing stream and for caves along Licking Creek just downstream from the Pectonville bridge on private property. Just north of the Pectonville bridge and at Camp Harding County Park at Pectonville, you can see clearly the floodplain laid down in modern times by the meandering Licking Creek—a characteristic of all streams and rivers.

Overall, this area is an excellent place to study the rock cycle. Ancient mountains generated by tectonic movement eroded sediments into low, inland basins filled with water. Here the sediments solidified into rock and then later were pushed and folded up into new mountains that have eroded again and are still eroding today, the streams carrying sediments to floodplains and on down to lower elevations. You can explore a large portion of this area on foot. The Indian Springs Wildlife Management Area comprises four tracts totaling about 6,800 acres. An office at Blairs Valley Lake has information about it.

The tough Tuscarora Sandstone hangs almost above Hanging Rock Road.

• — Piedmont Province — •

The Piedmont Province stretches from the Triassic Border Fault at the eastern foot of Catoctin Mountain to the Fall Line, where unconsolidated Coastal Plain sands and clays butt up against the hard rocks of the Piedmont. The Coastal Plain sediments are at least 400 million years younger than the Piedmont rocks on which they lie.

The Piedmont Province includes the metamorphosed remains of a volcanic island arc, sediments from the fore-arc basin, and ocean floor remnants, all of which collided with the eastern coast of North America in the Taconic mountain building event in middle to late Ordovician time. The intense metamorphism, folding, and faulting of that tectonic convergence are manifested in the many contorted rock structures you can see today. While relatively young mountains elsewhere are thousands of feet high, the long-eroded rocks here are only hundreds of feet above sea level. *Piedmont* means an area at the foot of the mountains, but the rocks here were once part of massive mountains. Today the Piedmont slopes gradually down to the east. The northeast-trending Parrs Ridge and Dug Hill Ridge form the drainage divide between Chesapeake Bay and the Potomac River, but Sugarloaf Mountain forms the high point at 1,282 feet.

Long after earlier mountains had eroded into hills and after Africa had collided with North America to form Pangea, the sutured continents were pulled apart by rifting. The Triassic Border Fault, a major normal fault, formed at this time, about 200 million years ago. Rocks on the east side of the fault dropped down thousands of feet and sediments filled the basin, known as the Gettysburg Basin in the Frederick Valley and north into Pennsylvania. Along the East Coast, a series of

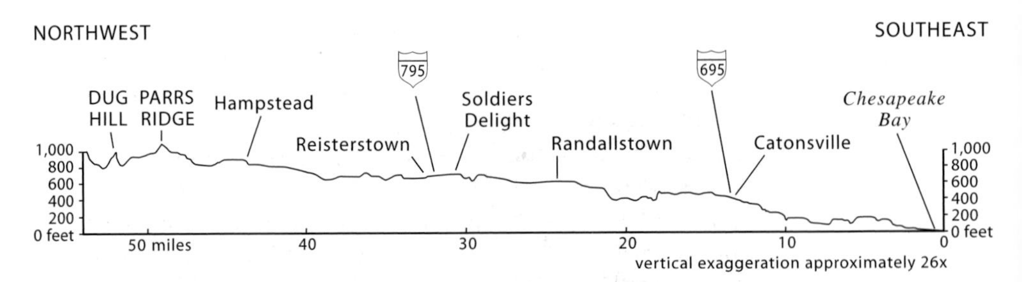

This profile, cut along the northwest-trending divide between the Patapsco and Gunpowder Rivers, reflects the general east-dipping slope of the Piedmont. —Modified from Maryland Geological Survey, 1929

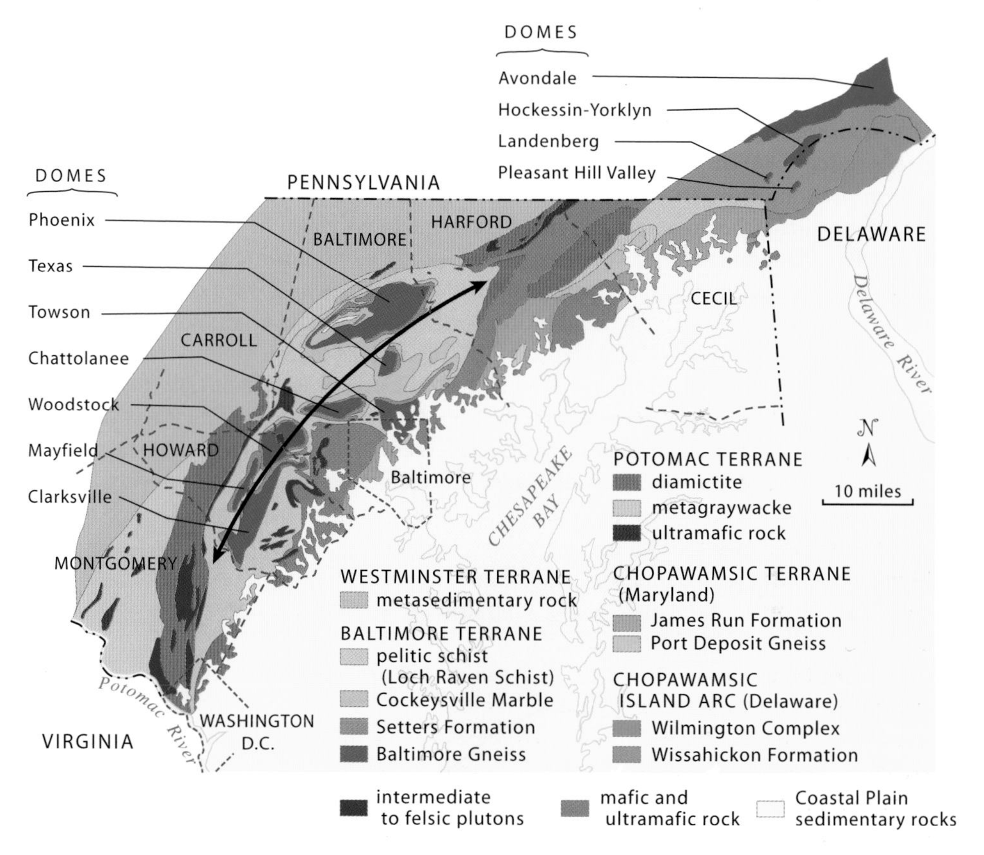

Piedmont rocks of different terranes—Chopawamsic, Potomac, Baltimore, and Westminster—run in a northeasterly direction. —Modified from Higgins and Conant, 1990; Plank and Schenck, 1998; Blackmer, 2004b, 2005

these basins formed from Massachusetts to North Carolina. The Culpeper Basin of Virginia extends north into Maryland near Poolesville. In Maryland, the sediments of Triassic age were deposited directly on an eroded surface of the metamorphic rocks of the Piedmont. Also, the divergence probably created an overall lowering or sinking of the crust from Frederick eastward as Africa drew away from North America. Today the Piedmont terrain is rolling, the ups and downs determined by various faults and by the different resistances to weathering of the bedrock. Parrs Ridge, for example, contains highly resistant quartzites.

Prior to Pangea times, the metamorphic rocks of the Piedmont originated in the island arc and adjacent ocean environments. Today they are grouped into four terranes in Maryland. Each terrane formed in a different location and under different conditions on the ancient seafloor and then was sutured or accreted to the North American continent during the Taconic mountain building event. Please note that these volcanic-arc and seafloor rocks are not called terranes in Delaware. See the Delaware Piedmont section for a detailed discussion of that area.

The Wilmington and Baltimore areas have gneissic rock that originally was part of an offshore volcanic arc, the Chopawamsic Terrane; in Wilmington this rock is called the Wilmington Complex and in Baltimore the James Run Formation. Ancient tectonic convergence and an eastward-dipping subduction zone on the ocean floor generated igneous activity that formed this chain of volcanoes about 488 to 470 million years ago, and these rocks are the metamorphic remnants.

Turbidite sediments that eroded from the volcanoes and were deposited in the fore-arc basin between the subduction zone and the island chain compose the Potomac Terrane of mostly Cambrian age. This mélange of sediments and fragments of ocean floor was later metamorphosed by tectonic pressures and burial. It is best exposed along the Potomac River upstream from D.C. and includes metagraywacke, schist, diamictite, and metavolcanic rock. The terrane is thousands of feet thick and includes the Mather Gorge, Sykesville, and Laurel Formations.

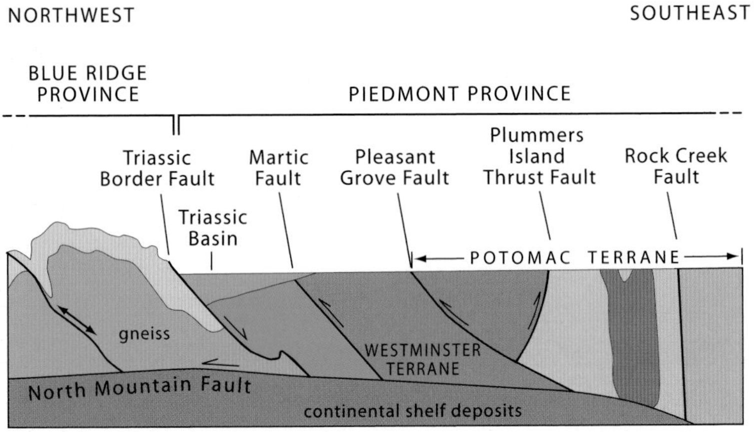

The rock from the ancient ocean trench was of considerable volume—today called the Potomac Terrane. This cross section runs from D.C. to west of Harpers Ferry and is mostly centered on the Potomac River. —Modified from Southworth and others, 2002

The quartz-rich schist and metagraywacke of the Mather Gorge Formation started out as turbidites—graded beds of sorted, coarse sands and clays derived from volcanic islands and deposited by underwater landslides in a deep ocean trench. Pieces of oceanic crust became incorporated into the Mather Gorge Formation during the converging and thrusting associated with the plate subduction that produced the deep ocean trench. The gneiss and schist of the younger Sykesville Formation contain clasts and fragments eroded from the Mather Gorge Formation. The Plummers Island Thrust Fault had brought the Mather Gorge Formation to the surface of the ocean trench, putting it in the position of being the source of material for the Sykesville Formation. The Sykesville is also intruded by igneous plutonic rocks originating from magmas produced by the subducting plate, magmas that

TIME	TRIASSIC BASIN AND WESTERN PIEDMONT	WESTMINSTER TERRANE	POTOMAC TERRANE	BALTIMORE TERRANE	CHOPAWAMSIC TERRANE	DELAWARE PIEDMONT
Triassic	dikes and sills Newark Group Gettysburg Shale New Oxford Formation					
Silurian				Ellicott City Granite Woodstock Granite		Iron Hill Gabbro Bringhurst Gabbro Arden Plutonic Suite
Ordovician	Grove Limestone	Peach Bottom Slate Cardiff Metaconglomerate	Kensington Tonalite Norbeck Intrusive Suite Georgetown Intrusive Suite Dalecarlia Intrusive Suite Bear Island Granodiorite			Wilmington Complex Brandywine Blue Gneiss Rockford Park Gneiss and others
Cambrian	Frederick Limestone Cash Smith Formation Araby Formation Sugarloaf Mtn. Quartzite	Peters Creek Formation Pleasant Grove Schist Urbana Formation	Sykesville Formation Conowingo Diamictite Laurel Formation	Baltimore Mafic Complex Mt. Washington Formation Hollofield Formation Aberdeen Metagabbro Perry Hall Gneiss	Gunpowder Granite Port Deposit Gneiss James Run Formation	
Prot./Camb.		Gillis Formation Wakefield Marble Silver Run Limestone Ijamsville Formation Marburg Formation Prettyboy Schist Libertytown Metarhyolite Sams Creek Metabasalt	Mather Gorge Formation Morgan Run Formation Soldiers Delight Ultramafic Complex other ultramafic rock	Glenarm Group Loch Raven Schist Oella Schist Cockeysville Marble Setters Formation		Wissahickon Formation Glenarm Group Cockeysville Marble Setters Formation
Proterozoic				Baltimore Gneiss		Baltimore Gneiss

Dates and correlations among different sources do not often agree about rocks in the Piedmont. This column for the Piedmont is a best fit of several different geological maps and reports from U.S. Geological Survey, Maryland Geological Survey, Delaware Geological Survey, and Pennsylvania Geological Survey. Igneous intrusions and extrusions are shown in red. —Compiled from Southworth and others, 2001, 2006; Southworth and others, 2007; Kunk and others, 2004; Southworth and Brezinski, 1996, 2003; Crowley, Reinhardt, and Cleaves, 1976; Reinhardt, 1974; Cloos and others, 1964; Edwards, 1993; Southwick and Owens, 1968; Higgins and Conant, 1986; Cleaves, Edwards, and Glaser, 1968; Gates, Muller, and Valentino, 1991; Schenck, Plank, and Srogi, 2000; Blackmer, 2004b, 2005

PHOENIX DOME

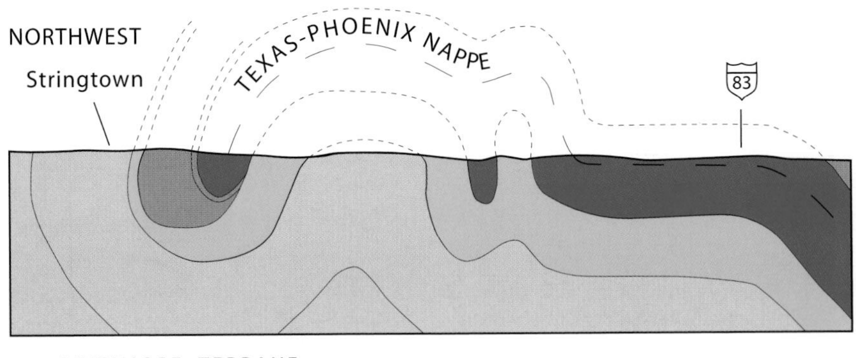

BALTIMORE TERRANE

CAMBRIAN / PROTEROZOIC
- Loch Raven and Piney Run Schists
- Cockeysville Marble
- Setters Formation
- Baltimore Gneiss

farther to the east were the source for the volcanic islands. Rocks that solidified from these magmas include the Dalecarlia and Georgetown Intrusive Suites.

The Sykesville Formation is a diamictite—a poorly sorted sediment in which pebbles and clasts "float" in a sandy or muddy matrix. The pressure of ocean-trench metamorphism altered the sedimentary rock into gneiss and schist with elongated or flattened clasts. The Sykesville diamictite has been described as a granitic-looking rock and was mapped as the Wissahickon diamictite on older geologic maps. The Wissahickon Formation in Delaware was also deposited in a fore-arc basin and may correlate with deposits of the Potomac Terrane of Maryland, but the formation is not assigned to a terrane in Delaware.

A third terrane, called the Baltimore Terrane, consists of thirteen nappes—large-scale folds that have been overturned by tectonic compression. This terrane runs intermittently from just north of D.C. to Philadelphia. In *The Geology of Cecil County* (1990), Michael Higgins calls it "the major structural feature of the eastern Maryland Piedmont." Regionally called domes, these folded structures contain 1.1-billion-year-old gneiss massifs of the ancient continent at their cores surrounded by metamorphic rock that was originally marine sedimentary rock deposited on the ocean floor. These massifs—giant crustal fragments—are thought to have been separated from the early North American continent during late Proterozoic rifting, inundated by the ocean, and then buried under ocean floor sediments. The cover rocks include the Setters Formation, Cockeysville Marble, and Loch Raven Schist. During the Taconic mountain building event, they were thrust back into and onto the continent.

Of course, the tops of the nappes have been eroded over hundreds of millions of years to reveal the 1.1-billion-year-old Baltimore Gneiss, the core of each fold. The different rates of erosion of the gneiss and the quartzite, marble, and schist that occupy the limbs or flanks of the domes are responsible for many of the rolling hills and marble-floored valleys of highly developed urban areas lying northwest of I-95.

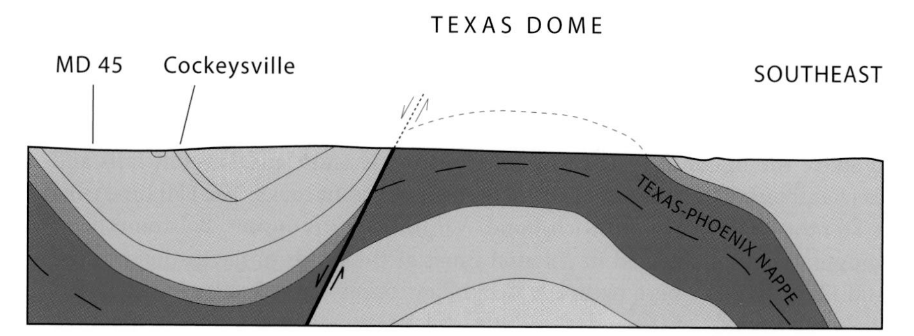

Even the 1.1-billion-year-old basement rock was contorted and deformed by tectonic collisions, regionally into folded domes, or anticlines, of which there are about a dozen from D.C. to Philadelphia. This section, from Stringtown through Cockeysville, features the Texas and Phoenix Domes north of Baltimore. Eroded portions of the folds are shown with dashed lines. —Modified from Crowley, Reinhardt, and Cleaves, 1976

Iron- and magnesium-rich mafic magmas of the Baltimore Mafic Complex intruded the Baltimore Terrane or possibly beneath the island arc. The complex includes the Hollofield Formation and the Mount Washington Formation of western Baltimore City, and it is difficult to determine their exact origin because subsequent mountain building events have thrust them about greatly.

The fourth terrane, the late-Proterozoic-to-Cambrian Westminster, originally deepwater ocean-floor sediments deposited on the continental margin landward of the Baltimore blocks, lies in Maryland northwest of suburban Baltimore and D.C. and northwest of suburban Wilmington in Pennsylvania. The western margin of the Westminster Terrane is the Martic Fault, a thrust fault along which the ocean bottom rocks were shoved upon the continental shelf rocks of the ancestral North American continent. The eastern margin of the terrane is the Pleasant Grove Fault, which separates it from the Potomac Terrane. The Pleasant Grove Fault formed in the Acadian mountain building event and was reactivated during the Alleghanian event. The Westminster Terrane includes gneiss derived from sandy sedimentary rocks, diamictite, pelitic schist, phyllite, slate, and marble. Over the years the rocks have been assigned various names (based on the geographic location in which they occur), including Marburg, Ijamsville, Urbana, Prettyboy, Peters Creek, and Peach Bottom.

Some terms used in Piedmont geology can cause confusion. The Baltimore Terrane is a collective term for the gneiss-cored nappes flanked by metamorphosed sedimentary rocks. The Baltimore Gneiss is the 1.1-billion-year-old basement core of the nappes, part of the ancestral North American continent more than 1 billion years ago. The Baltimore Mafic Complex, previously termed the Baltimore Gabbro, is mafic, intrusive igneous rock that is not considered a terrane and its exact origin is still being debated.

THE FALL LINE

Wilmington, Baltimore, and Washington lie on the Fall Line. It is not actually a line but rather a zone where the hard, crystalline metamorphic rocks of the Piedmont stand above the unconsolidated Coastal Plain sands and clays, and low falls and rapids characterize the passage of rivers over the resistant rocks. The Fall Line runs from Trenton, New Jersey, to Richmond, Virginia. Wilmington, Baltimore, and Washington were established in colonial times at the heads of navigation of tidal Coastal Plain rivers, where naturally deep, Pleistocene-carved shipping channels met the rocky rapids of the rivers. In each city, water-powered mills occupied the high-gradient rocky courses, putting them and their products close to the docks of the navigable waters. The rocky hills above the natural harbors and along the tidal rivers leading to the harbors made for excellent defensive positions.

In the D.C. area, Rock Creek, Cabin John Creek, and 17 miles of the Potomac River drop in rapids through rocky courses to the tidal Potomac River. In Baltimore, Jones Falls, Gwynns Falls, and the rocky stretch of the Patapsco River converge within a 1-mile radius to the tidal Patapsco River. In Wilmington, Brandywine Creek and Red Clay Creek tumble down rocky courses and form the tidal Christina River, which enters the wide, estuarine Delaware River. In Wilmington you can stand at the colonial head of navigation, the Market Street Bridge over Brandywine Creek, and see the change from the rocky upstream course to the tidal downstream section.

In each city the Coastal Plain sediments that have washed down from the western Piedmont and Appalachian highlands are at least 400 million years younger than the Piedmont rocks on which they lie. After the last glacial stage, rivers that had cut deeply into soft, unconsolidated Coastal Plain sediments were subsequently backflooded by a rising sea. The lower reaches became the wide, deep, tidal arms of estuarine waters. During colonial times, the low-lying Coastal Plain lands below each city's Fall Line were mosquito-infested marshes, and thus the wealthy settled on higher, drier Piedmont ground situated north and west of the tidal ports.

Baltimore was always the better harbor than Washington because its Fall Line lies just a little over 10 miles from Chesapeake Bay. The Fall Line of Washington lies over ten times that far from the bay. Washington is only about 30 miles from the bay by land, but the wide, deep Potomac River meanders through more than 100 miles of Coastal Plain deposits before reaching the bay.

Road Guides to the Maryland Piedmont

Interstate 70
Frederick—Baltimore
47 MILES

I-70 provides you with a good overall view and feel of the Piedmont terrain. Between Catoctin Mountain at Braddock and Mount Airy, a distance of only about 20 miles, I-70 traverses a large number of formations and structures from many different geologic periods, including the Martic Fault and rocks of the Triassic basin and the Westminster Terrane. While these features indicate this area has undergone tremendous changes in its long history, few of them are directly visible today because of soil cover and human structures. You cannot view the Triassic Border Fault or the Martic Fault on the surface anywhere, and the only visible evidence is in the topography: Catoctin Mountain above western Frederick Valley marks the Triassic Border Fault, and the phyllite ridge above the eastern part of the valley marks the Martic Fault.

Along the base of Catoctin Mountain just west of Frederick Valley is the northeast-trending, deeply buried, almost vertically dipping Triassic Border Fault. About 200 million years ago, the pulling apart of Pangea caused crustal fracture and downward slippage of thousands of feet along fractured sections of earth crust from present-day Connecticut to Georgia. The mountains to the west thus formed a ridge above the basin to the east. The basin structure is called a half graben, and its eastern margin is like the hinge of a trapdoor. Into this deep basin eroded sediments from the adjacent high mountains. Because this material moved down from the mountaintops to fill the basin over millions of years, the relief between ridgetop and valley floor was reduced from over 10,000 feet to about 500 feet. At the fault line, the Cambrian rock to the west is about 550 million years old, while the sedimentary red beds to the east, the basin fill, are about 200 million years old. Thousands of feet below the surface presumably lie the downfaulted Cambrian rock layers that were once continuous with those of the mountains before faulting.

Near the US 340 junction (at I-70 exit 52) but not visible on the surface is the eastern boundary of the filled basin, where Cambrian limestone meets Triassic red sandstones and shales. Millions of years before the downfaulting, this limestone would have occupied the eastern limb of the giant South Mountain anticlinorium of the Blue Ridge Province and would have made a continuous arching connection with the limestones to the west in the Hagerstown Valley. In Triassic time, the limestone was part of the half graben that sank downward, but being closer to the hinge, it did not sink as far as the rock to the west. Over the years, erosion exposed it, making the stone accessible on the surface for mining. Just south of Frederick, limestone quarry operations border I-70, and sometimes you can see clouds of dust. In November 1995, a huge sinkhole formed at the intersection of the exit 55 eastbound ramp and Reichs Ford Road. A car even plunged into the gaping cavity. Sinkholes form when the surface soil collapses into underground

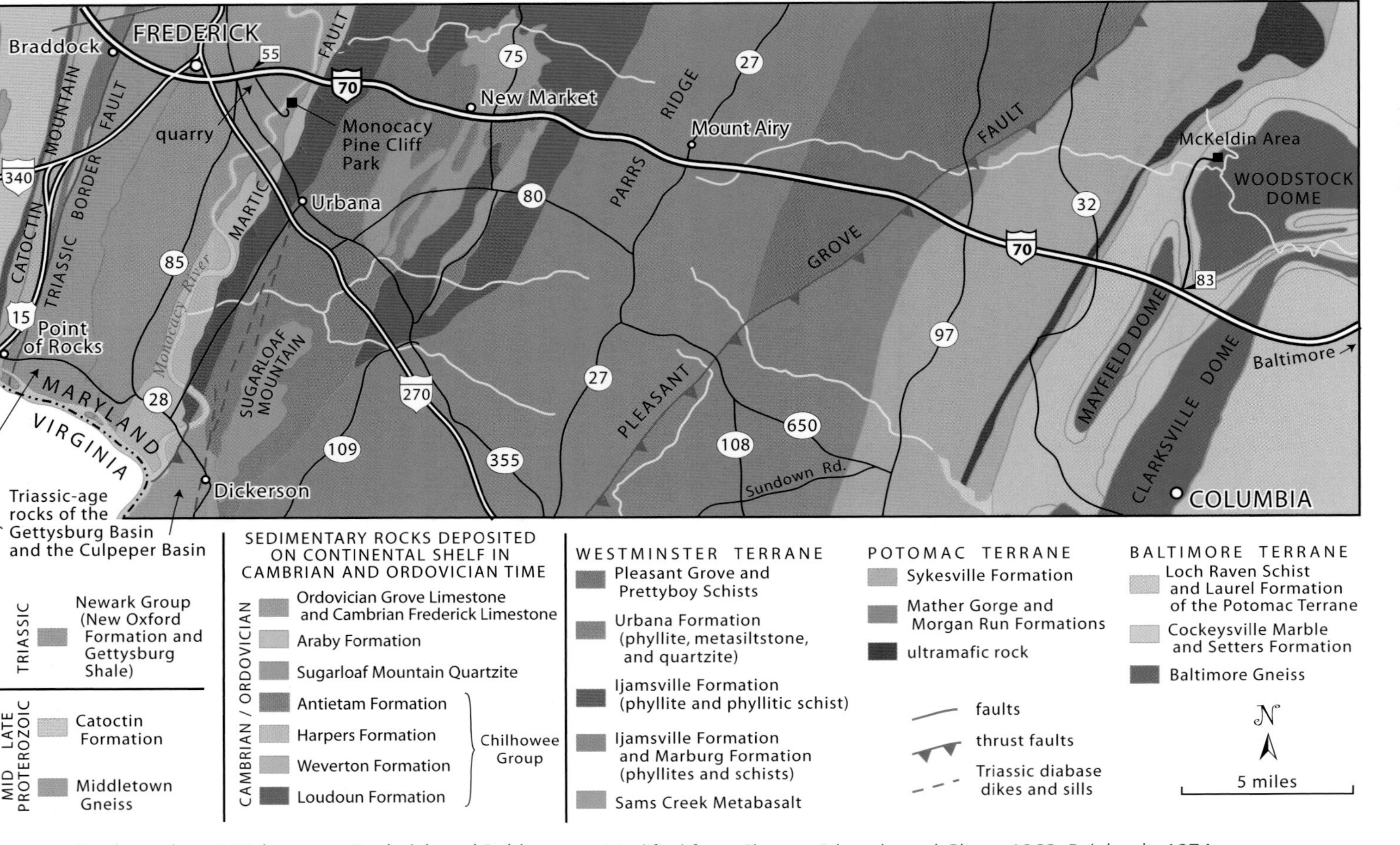

Geology along I-70 between Frederick and Baltimore. —Modified from Cleaves, Edwards, and Glaser, 1968; Reinhardt, 1974; Southworth and others, 2007; Meyer and Beall, 1958; Cloos and others, 1964; Edwards, 1993; Crowley, Reinhardt, and Cleaves, 1976

solution cavities or caverns in limestone. Fluctuations in groundwater levels often initiate the collapse.

On I-70 a couple of miles east of Frederick you cross the Monocacy River, which drains the large valley created by the downfaulted half graben in Frederick and Carroll Counties. To get a close look at the Monocacy River, take exit 55 and go south on Reichs Ford Road to Monocacy Pine Cliff Park. Across from the ball fields you can walk down to the river and look across at exposures of metasiltstone and phyllite of the Cambrian Araby Formation. Geologists think it correlates with the Antietam Formation of Washington County in the Valley and Ridge Province to the west.

About a half mile east of the Monocacy River is the Martic Fault, where geologists think Westminster Terrane rocks were thrust over continental shelf deposits during the Taconic mountain building event. Much later, during the Alleghanian mountain building event, rocks shifted and sheared laterally along the fault in a manner similar to the lateral slippage along the San Andreas Fault today. At the fault, metamorphic phyllite—deep ocean deposits of the Westminster Terrane—borders quartzite and limestone deposited on the continental shelf of North America in Cambrian time.

The I-70 bridge over the Monocacy River is at about 250 feet elevation. A steady 4-mile climb to the east, to the Mussetter Road overpass, brings you to an elevation of about 550 feet. On this slope are two diabase dikes, igneous intrusions of Triassic age (200 million years old), while just east of them near New Market is an igneous basaltic lava of late Proterozoic age (approximately 600 million years old). Each magma upwelling resulted from a different episode of plate divergence, when the crust stretched, thinned, and cracked under tension. Through these cracks oozed molten rock, which when cooled at or near the earth's surface, solidified into igneous rock. Because most igneous rock resists erosion well, topographic evidence of each episode remains. The diabase dikes of Triassic age form low ridges, and the Precambrian basalt forms hilltops and ridges next to other, less resistant rock into which it is interlayered.

Most of the rock east of the Martic Fault and down to the Fall Line is metamorphic. The high point along I-70 is Parrs Ridge, which you cross at Mount Airy at an elevation of about 850 feet. This northeast-trending high ground divides streams draining into the Potomac from those draining directly into Chesapeake Bay. Marburg schists and quartzites that resist erosion form Parrs Ridge.

Between Mount Airy and Baltimore, the Piedmont gradually slopes down to the east but resistant rocks form ridges here and there. At the Sand Hill Road overpass just east of the MD 32 exit, the elevation is about 600 feet.

Between MD 32 and the I-70/US 40 split, you drive gradually over Alpha Ridge, known to geologists as Mayfield Dome. It is a large fold of the Baltimore Terrane with hard gneiss at its core.

McKeldin Area of Patapsco Valley State Park

At the McKeldin Area of Patapsco Valley State Park, you can closely observe excellent exposures of the 1,100-million-year-old Baltimore Gneiss, the basement rock of the ancient Grenville continent. This park is located in the core of the Woodstock Dome, one of the dozen anticlinal nappes of the Baltimore Terrane. Millions of years of erosion have exposed the rock of the core. To reach the McKeldin Area take Marriottsville Road (exit 83 from I-70) north a few miles. Follow the park road to the last parking lot and either park there or drive on down the single-lane road to Rapids Trail. The trail passes by the falls, where the Patapsco River plunges 12 feet in about 60 feet of length over a broad, eroded surface of the ancient gneiss.

On the trail downstream from the rapids you can see floodplain deposits of considerable thickness. Such riverside sediments might be only decades old. During Hurricane Agnes of 1972, for example, many inches of mud were laid down. You can see these unconsolidated sediments, some of the geologically youngest, resting on the gneiss, Maryland's oldest rock. On this trail, after two right-angle bends below the falls, is a superb, water-scoured exposure of the gneiss with clearly visible folds

The clean, sculpted exposures of Grenville basement rock, the Baltimore Gneiss, along the Rapids Trail, may be the best in the state.

A lower part of the McKeldin Rapids in Patapsco Valley State Park is bordered by 1.1-billion-year-old Baltimore Gneiss kept free of vegetation by the erosive action of periodic high water.

and foliation. If you continue downstream from the rapids about a half mile, you will come to the confluence of the South and North Branches of the Patapsco.

I-70 crosses the Patapsco River in its deeply cut valley. Other areas of Patapsco Valley State Park farther downstream are discussed in the Baltimore section of this chapter.

Interstate 270
Frederick—Rockville
29 MILES

Many interesting and diverse geologic sites of the Triassic basin and the Piedmont lie between Frederick and the D.C. area. You can get to individual sites from exits off the heavily traveled and often congested I-270, but a combination of Maryland routes from Point of Rocks to Seneca Creek and Blockhouse Point gives you a much better tour. However, if you are driving I-270 without exiting, you can still get a general sense of the geology.

At the western end of I-270 in Frederick Valley, you can see Catoctin Mountain to the west and Sugarloaf Mountain to the east and south. Here called the Gettysburg Basin, this northeast-trending, low-lying region runs hundreds of miles along the East Coast between the Piedmont and the Blue Ridge Provinces. When Pangea began diverging into separate plates over 200 million years ago, the earth's crust was extended and fractured. At the eastern base of Catoctin Mountain, rock slid downward thousands of feet along the Triassic Border Fault, creating a trough that filled with sediments eroded from nearby highlands. The block tilted downward on its western side, and today, as you head east across the Monocacy River, you begin a climb up the slope of this downtilted structure.

It is a climb of over 3 miles from the river to the Urbana exit, and about halfway up you cross the Martic Fault, not visible on the surface but located near the ridge west of the Urbana exit. Here the Westminster Terrane—originally rocks from the offshore continental rise and slope—were thrust onto and over the nearshore, continental-margin rocks of Cambrian age during the Taconic mountain building event, in middle to late Ordovician time, over 200 million years before the Triassic downfaulting. If you are traveling west on I-270, about 2 miles west of the Urbana exit you can pull off at the scenic overlook, be very close to the Martic Fault, and see the downfaulted Frederick Valley.

Sugarloaf Mountain

Once you are east of the Martic Fault, you can catch glimpses to the south of Sugarloaf Mountain, made of an erosion-resistant quartzite of Cambrian age. The rock, here called the Sugarloaf Mountain Quartzite, is related in age and stratigraphic position to the continental-margin sands that formed the Weverton quartzite topping Catoctin and South Mountains to the west. The quartzite on Sugarloaf displays ripple marks and crossbedding like the Weverton quartzite—both characteristics of

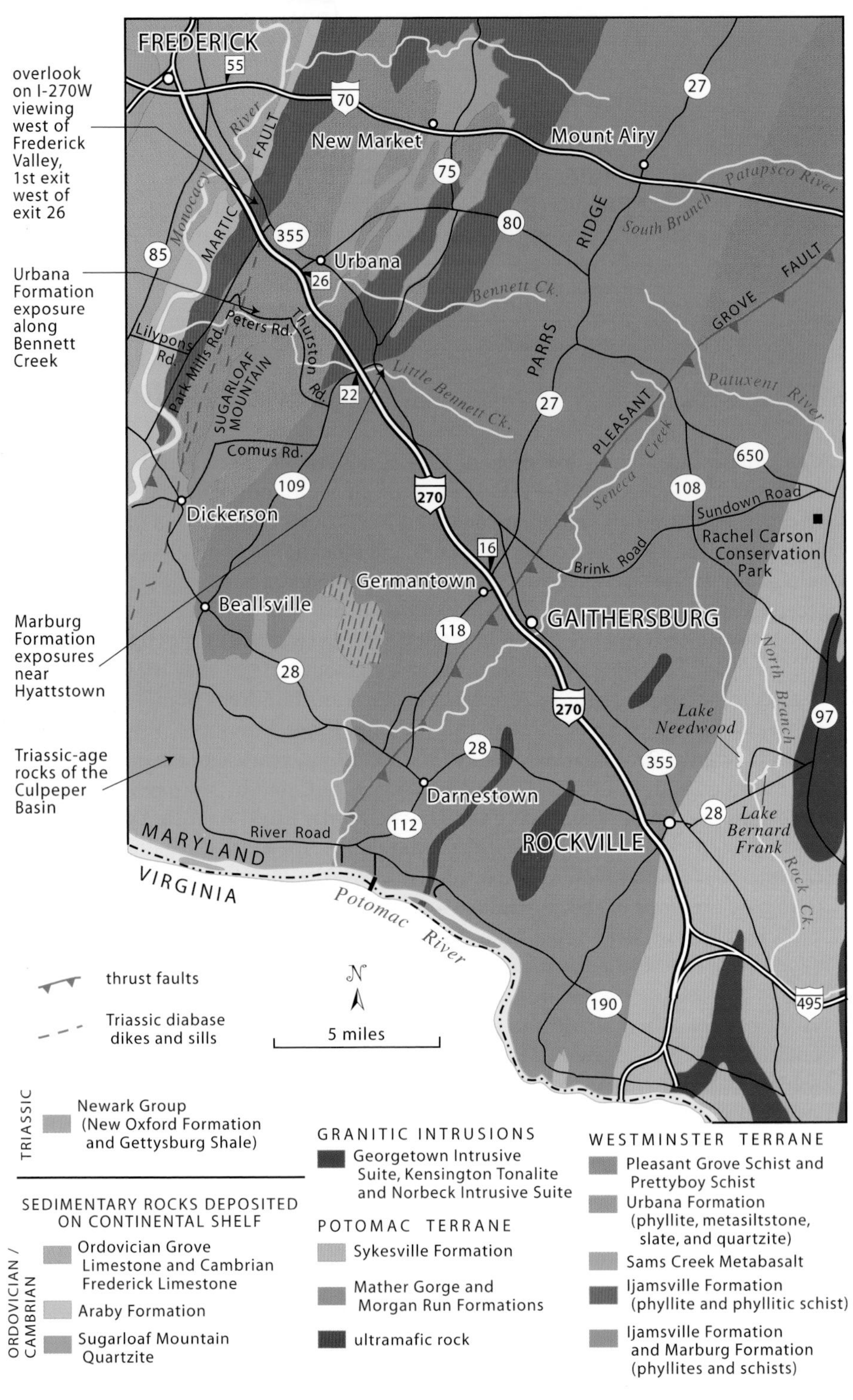

Geology along I-270 between Frederick and Rockville. —Modified from Cleaves, Edwards, and Glaser, 1968; Reinhardt, 1974; Southworth and others, 2007; Meyer and Beall, 1958; Cloos and others, 1964; Edwards, 1993; Southworth and others, 2001; Southworth, 1996; Kunk and others, 2004; Southworth and others, 2006

beach deposits. Also, the rock at Sugarloaf is part of an anticline overturned to the west, as is the larger South Mountain anticlinorium of correlative rock.

As part of the buried rock underlying the overthrusted Westminster Terrane, Sugarloaf did not, however, always stand as a monadnock, or isolated hill. The less resistant, overlying rock—the Urbana Formation—weathered down more rapidly over millions of years, leaving the more resistant quartzite to stand on its own. This type of erosion, which exposes rock hidden beneath an over-faulted block of different rock, is known as a tectonic window.

Though rock at Sugarloaf Mountain is now a hard, highly resistant quartzite, the original shore sands were deposited in waters that generated crossbedding and ripple marks. Note the ripple marks on the rock at right.

At 1,282 feet, Sugarloaf Mountain stands over 800 feet above the surrounding Piedmont landscape because it is made of hard quartzite.

At I-270 exit 22 at Hyattstown, you can take MD 109 south to Comus Road and follow it to Sugarloaf. Driving up Sugarloaf, you can see boulders of the quartzite all over the slopes and a concentration just above West View parking area at the A. M. Thomas Trail (green) trailhead. If you want to hike to the top, this one-quarter-mile green-blazed trail takes you alongside a talus slope of huge boulders, but the path is well-worn and does not require rock climbing. At the top look for the crossbedding and ripple marks, and also take in the magnificent views. To the west on a clear day, you can see parts of three water gaps on the Potomac River: the Catoctin Mountain water gap at Point of Rocks, and farther west, halves of the water gaps in South Mountain and Blue Ridge—all topped with similar quartzite. To the east you can see much of the Maryland Piedmont. If you want to explore further, there are over 15 miles of trails, all on the quartzite.

Slate Quarry and Other Westminster Terrane Exposures

Slate Quarry Road is less than 1 mile south on MD 109 from I-270 exit 22, and about three-quarters mile down this road, you can see slate of the Urbana Formation, Westminster Terrane, in roadcuts and an old quarry on private, posted land. Slate is a sturdy, metamorphic rock whose parent is the crumbly sedimentary rock shale, and it was often quarried in Maryland for roofing slate. The parent clay was deposited in deep offshore ocean bottoms during late Proterozoic to Cambrian time.

You can see excellent exposures of the Urbana Formation phyllite along Bennett Creek on Peters Creek Road not far from the old quarry. Peters Creek Road is a good dirt road that connects Thurston Road and Park Mills Road. The exposures are located right where the road runs along Bennett Creek.

These Urbana phyllite and quartzite exposures are on a gravel section of Peters Road along Bennett Creek.

Look for quartz veins in the exposures of Westminster Terrane Marburg Formation on gravel Hyattstown Mill Road.

At exit 22 for Hyattstown, you can go one block east on MD 355 onto Hyattstown Mill Road and see some good exposures of the Marburg Formation metasiltstone and phyllite of the Westminster Terrane at Little Bennett Regional Park. Go about 0.5 mile on the gravel road into the park to a sign for Dark Branch Trail. The exposures are on the small hill above the trailhead.

Parrs Ridge

At exit 16, Father Hurley Boulevard, you can get onto MD 27, Ridge Road, and follow Parrs Ridge north all the way to Westminster. This high ground, quartzite of the Marburg Formation, is the drainage divide for streams flowing west into the Monocacy River and on to the Potomac River, or east into primarily the Patuxent and Patapsco Rivers, which flow directly into Chesapeake Bay. Regionally, this ridge gives some indication of the eastern extent or erosional influence of the basin associated with the Triassic Border Fault. Southeast of Parrs Ridge, the land falls, more or less regularly, to the Coastal Plain.

Rachel Carson Conservation Park

From MD 27 at Brink (I-270 exit 15), you can take Brink Road east, which becomes Sundown Road. Turn onto Zion Road and turn in at mailbox number 22201 for Rachel Carson Conservation Park. At this 650-acre oasis of nature amid the sprawl of D.C. and Baltimore, you can walk 6 miles of trails over stream-dissected, wooded Piedmont terrain. On the Rachel Carson Greenway Trail and along Hawlings River, you can see excellent exposures of schist of the Morgan Run Formation of the

Potomac Terrane—the mélange sediments deposited in the ancient fore-arc basin that lay east of an offshore volcanic island arc in early Cambrian time. Eroded volcanic rock particles washed into a deep ocean trench, lithified into stone, and then were metamorphosed to the schist located in this park dedicated to the author of *The Sea Around Us*. The Morgan Run may be correlative with the Mather Gorge Formation (both contributing clasts to the Sykesville Formation), and the Plummers Island Thrust Fault and shear zone that separate these two from the Sykesville Formation may be near or in the park. The uncertainty here is due to the lack of agreement among several different geologic maps.

Lakes Needwood and Bernard Frank

At the Rockville exit for MD 28 East, you can follow MD 28 to Avery Road to Lakes Needwood and Bernard Frank, which are on two branches of Rock Creek in the Mather Gorge and Sykesville Formations, respectively, both of the Potomac Terrane. Each has hilly, wooded terrain with many hiking trails, and from the dam at Lake Needwood you can bike along Rock Creek all the way to Georgetown on the paved trail, a one-way distance of about 25 miles.

On the trail from the parking lot to the dam at Lake Bernard Frank, you can get a close look at gneiss of the Sykesville Formation.

At Lake Bernard Frank there are excellent rock outcrops of gneiss of the Sykesville Formation on the path from the parking lot off Avery Road to the dam. In the fields around Meadowside Nature Center off MD 115, you can get a good idea of the predevelopment, rural character of this area.

These two lakes were built in the 1960s to control flooding and sedimentation in downstream Rock Creek. However, sediments held in suspension by currents in a stream will fall to the bottom as the current slows in a quiet reservoir. Over several decades a reservoir can fill with sediment and lose its water-holding capacity. In spite of Montgomery County's very early implementation of measures to control erosion from construction sites (most erosion-control measures coming decades later), the Soil Conservation Service has estimated that by about 2020, the two dams will no longer be able to store enough water to prevent downstream flooding. Heavy rains in June 2006 pushed levels to historic highs. Water seeped through the earthen Needwood Dam, and park officials feared the dam might fail. About 2,500 residents were evacuated, but the dam held and was later pronounced safe.

MD 355 between Frederick and D.C.

Before I-270 was constructed, the main road to Frederick from D.C. was Wisconsin Avenue–Rockville Pike–Frederick Road, today MD 355. Although it is heavily developed, you can see some exposures here and get a good idea of the rolling Piedmont terrain. Between the Monocacy River and Clarksburg, you can catch glimpses of the late Proterozoic and Cambrian Ijamsville phyllites of the Westminster Terrane and the Urbana phyllites that surround the Sugarloaf tectonic window, but there are no safe places to stop. At the Monocacy River bridge you can pull off and cross the road to look at an exposure of the Frederick Limestone.

Jagged solution weathering of Frederick Limestone near the Monocacy River bridge on MD 355. Layers with higher percentages of calcium carbonate weather more quickly than those with higher clay content.

US 1/Maryland 273
Baltimore—Delaware Border (Cecil and Harford Counties)
50 MILES

I-95 is the main highway between Baltimore and Wilmington, but it is not the route from which to see much of anything natural. US 1 and MD 273 provide better views and access to many Piedmont sites in Cecil and Harford Counties.

Gunpowder Falls State Park

The term *Falls* in Gunpowder Falls State Park does not describe a waterfall but is the name of the stream that falls, or tumbles down, the hard rock of the Piedmont Fall Line, which is actually a zone. On the northeast side of the US 1 bridge is a parking lot from which you can access several interesting trails on both sides of the river.

The rock on both sides of the river upstream and downstream of the bridge is the metamorphosed mafic and ultramafic igneous rock of the Baltimore Mafic Complex, once called the Baltimore Gabbro. This part of the complex is called the Perry Hall Gneiss of the Bel Air Belt, and it runs northeast past Bel Air all the way to the Pennsylvania line in a 2- to 3-mile-wide band straddling US 1. Downstream of the bridge on Lost Pond Trail on the northeast side of the stream are good outcrops and cliffs, but upstream of the bridge the better exposures are on Big Gunpowder Trail on the southwest side of the river. About 1 mile downstream on Lost Pond Trail are potholes in solid rock, formed by strong currents swirling sand and gravel during times of heavy discharge, probably from glacial ice melting at the end of the last Ice Age. At Lost Pond, now drained, you can see the silt that once fell out of its still waters.

Strong swirling water carrying abrasive sand and gravel drilled these potholes into the hard Perry Hall Gneiss at Gunpowder Falls State Park.

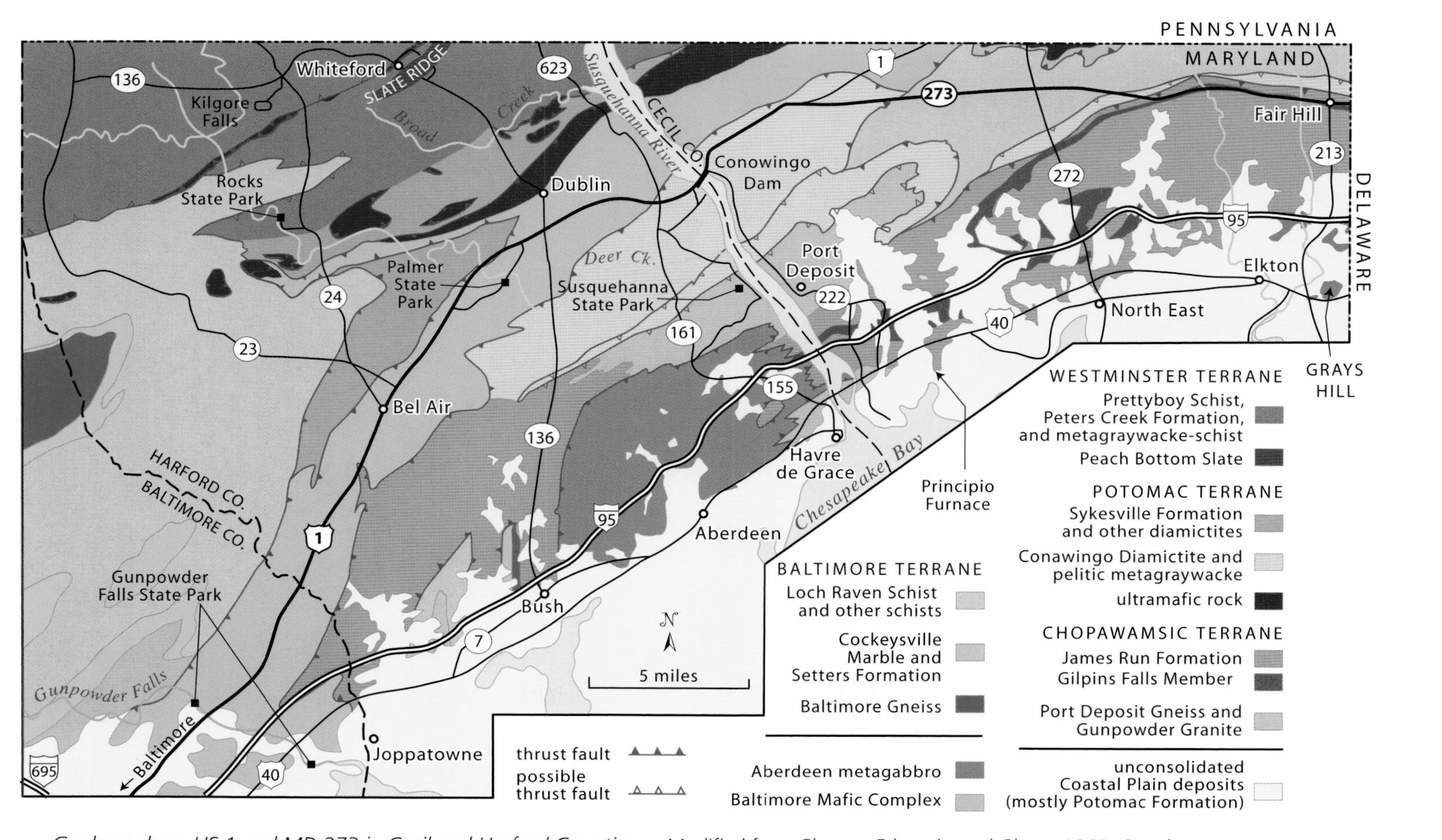

Geology along US 1 and MD 273 in Cecil and Harford Counties. —Modified from Cleaves, Edwards, and Glaser, 1968; Crowley, Reinhardt, and Cleaves, 1976; Southwick and Owens, 1968; Higgins and Conant, 1986; Gates, Muller, and Valentino, 1991

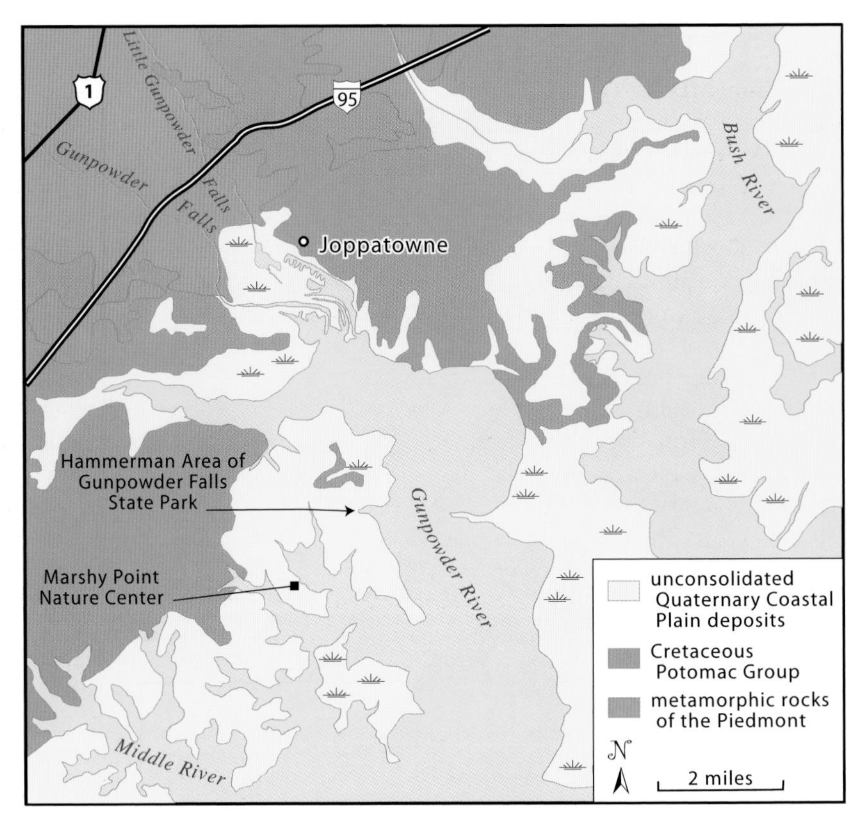

The coastline of Chesapeake Bay in colonial times was near the boundary between the Cretaceous and Quaternary sediments. The deposition of sediments into the bay increased exponentially when settlers cleared large areas for agriculture. —Modified from Cleaves, Edwards, and Glaser, 1968

Upstream of the bridge on Sweathouse Trail at Sweathouse Branch is the fault line where the Baltimore Mafic Complex is thrust over the Baltimore Terrane. It is difficult to determine the exact location, but the creek may have eroded its course along part of this fractured surface between the two terranes.

Joppatowne (colonial Joppa Town), near the marshy mouth of Little Gunpowder Falls, was at the head of a deep harbor in colonial times. Erosion of topsoil, exposed by land clearing and plowing, filled the harbor with sediments, rendering it too shallow for ships. The Hammerman Area of Gunpowder Falls State Park and Marshy Point Nature Center are on low-lying land that did not appear on colonial maps. These two parks occupy the largest area of public-access natural wetlands in the Baltimore area.

Rocks State Park

From the US 1 Bel Air Bypass, you can take MD 24 west to Rocks State Park, where King and Queen Seat, an impressive pinnacle, stands 190 feet above Deer Creek and occupies one side of the creek's water gap. Steep, quarter-mile trails lead to the pinnacle from MD 24 or from the road to the park office, or you can drive up to Rock Ridge Picnic Area and take a more level approach. These towers exist here because they are made of highly resistant, quartz-pebble metaconglomerate and quartzite. This narrow band of hard metamorphic rock is identified on the 1968 *Geologic Map of Harford County* as part of the Wissahickon Formation, but later maps associate it with the Cardiff Metaconglomerate of the Westminster Terrane. Down in Deer Creek below the pinnacle, you will see waterfalls where the stream is making its way over the resistant rock. If you drive up St. Claire Bridge Road a short way, you can see folded rock of this formation.

At the Falling Branch Area of Rocks State Park, you can see cliff-sized exposures of the Prettyboy Schist at Kilgore Falls, Maryland's second highest free-falling falls at 17 feet. The formation is well over 2 miles thick, comprising a large segment of the Westminster Terrane. To reach Kilgore Falls, continue north on MD 24 to Five Forks and then take Clermont Mill Road. Turn onto Falling Branch Road (dirt) to the parking area. There is no sign for the falls, but there is only one trail. A couple hundred yards down the trail you come to a Y. The left branch leads to the area below the falls, and the right to the rocks above the falls. Go left, cross the stream, and look for the schist. The parent rock of schist, like slate, is clay, but the heat and pressure of metamorphism have been greater than they were for slate. The schist foliation is wavy rather than parallel.

Kilgore Falls in the Prettyboy Schist might have originated at a fault about 2 miles downstream and eroded its way to here.

You can see waves and folds in the Prettyboy Schist near Kilgore Falls.

Palmer State Park

US 1 follows a band of the Baltimore Mafic Complex through much of Harford County. Good exposures of the iron- and magnesium-rich rocks are present at undeveloped Palmer State Park. To get to the park, turn off US 1 onto Trappe Road and then go south on Forge Hill Road. There are no signs, but soon after you enter the wooded area you can see the deeply cut valley of Deer Creek. There are multiple roadside exposures of this greenish rock on both sides of the Deer Creek valley.

In Palmer State Park, you can see the igneous rocks of the Baltimore Mafic Complex, formed from magma that oozed up through an ancient ocean floor that was later consumed by tectonic convergence.

Side Trip on MD 136 to Slate Ridge

In the 6 miles between US 1 near Dublin and Whiteford, MD 136 crosses from the Baltimore Mafic Complex, over the Potomac Terrane and its ultramafic intrusions, and onto the Westminster Terrane. But you cannot discern on the landscape that you are making these momentous geologic passages. The metamorphic rock has been weathering into soil in the humid, coastal climate for many years and forms rolling hills and valleys that are remarkably similar from one terrane to another.

Just northeast of the bridge over Broad Creek, MD 136 enters the Peters Creek Formation of the Westminster Terrane, originally sediments in the deep offshore basin of the rifted margin of the craton in late Proterozoic time. The Westminster Terrane was thrust far onto the continent during the Taconic mountain building event and thus lies inland of the other terranes.

At Whiteford, MD 136 crosses over the 10-mile-long, northeast-trending Slate Ridge, a distinct ridge composed of Peach Bottom Slate of the Westminster Terrane. Slate Ridge is the center of the Peach Bottom syncline. The slate is the most resistant rock in the syncline, so the middle of the structural trough is now a ridge. Some of the clays deposited in the Westminster ocean basin were later metamorphosed by tectonic pressures into the parallel, sheetlike planes of slate. Among the uses of this slate have been roofing, flagstone, blackboards, and pool tables. A historical road sign at the junction with MD 165 informs us that the Peach Bottom is the oldest commercial slate in the United States, first used in 1734. In 1850 at the London Crystal Palace Exhibition, it was judged the best in the world. As you pass over the ridge on MD 136, look for the old quarry on top—now filled with water, a private swimming area. West of the ridge where MD 136 turns due west, you can stop and look back at Slate Ridge. If you head northeast on Old Pylesville Road toward Cardiff and then turn east onto Slate Ridge Road, you can see old quarry workings and tailings of slate. As you ascend the ridge, you will see various cuts near the road.

The metamorphic foliation, or planar texture, of slate is along flat, parallel planes. The integrity of its surface makes it useful for many applications.

Peach Bottom Slate of the Westminster Terrane forms 10-mile-long Slate Ridge near Whiteford. View from northwest of Whiteford on MD 136.

Side Trip on MD 623 to Broad Creek

From US 1 about 1.5 miles west of Conowingo Dam, you can turn north on MD 623 and drive about 4 miles to Broad Creek, to the boundary between the Westminster and Potomac Terranes. The terrane boundary is formed by the Pleasant Grove Fault Zone, which is in Broad Creek here. North of the bridge you can park and look at exposures of the Cambrian Peters Creek Formation of the Westminster Terrane—thinly interlayered schist and quartzite with quartz stringers. This metamorphic foliation resulted from the heat and pressure of burial and tectonic convergence, probably mostly during the Taconic mountain building event.

South of the bridge is the Cambrian Sykesville Formation in the overthrust section of Potomac Terrane. This rock is called metadiamictite—*meta* indicating metamorphosed, and *diamictite* indicating a sediment in which pebbles and cobbles float in a sandy or muddy matrix. You can see these pebbles (perhaps from the Morgan Run Formation) in the schist if you look closely. The rock was originally deposited in the fore-arc basin of the volcanic island arc, where debris flows brought in a poorly sorted mixture of sediments. About a half mile south of the creek is the contact between the Potomac Terrane and the overthrust Baltimore Mafic Complex of mafic igneous intrusive rock.

Susquehanna State Park

On US 1 about 1.5 miles west of Conowingo Dam, you can take MD 161 (Darlington Road) south to Wilkinson Road and Susquehanna State Park. A better approach to the park for seeing geology, however, is to take Stafford Road from MD 161. Just before Stafford Road reaches the Susquehanna River, it parallels Deer Creek, and along here you pass onto the Port Deposit Gneiss of the Chopawamsic Terrane, originally granite intrusives that were the magmatic source for the James Run volcanic rocks. This rock was long ago metamorphosed into gneiss but has been quarried for building stone under the name Port Deposit Granite.

At the mouth of Deer Creek, if you turn and look up the hill, you will see a large stone stream that resembles those formed in western Maryland by freeze and thaw during cold, Ice Age climates. Notice the alluvial deposit on the other side of Deer Creek—sediment carried by the creek and dropped where its current was slowed by its intersection with the Susquehanna River. Along Stafford Road on down to the park's historic area, you can see cliffs and exposures of the Port Deposit Gneiss. Out in the river are islands of the same rock.

Take a hike on Susquehanna Ridge Trail for great views of the river when the leaves are down. The trail eventually passes by the head of the stone stream.

Rock Run Road along Rock Run in the state park lives up to its name. You can see good exposures of the Port Deposit Gneiss along the road and in the stream.

Conowingo Dam and Conowingo Diamictite

About 1.5 million tons of sediment are carried by the Susquehanna River each year. Completed in 1928, Conowingo Dam has been stopping the current—impounding about 150 billion gallons of water—and causing much of the river's often polluted

On the slope above the mouth of Deer Creek, you can see a stone stream of angular boulders, indicative of the freeze-and-thaw process rather than alluvial deposition.

Port Deposit Gneiss of the Chopawamsic Terrane is well exposed in Rock Run at Susquehanna State Park.

sediment to fall out of suspension at the bottom of Conowingo Reservoir. Today the reservoir holds almost 200 million tons of this toxic mud, and the U.S. Geological Survey has warned that when the pond reaches its capacity of about 225 million tons, sediment will begin passing over the dam and increase the load that hits Chesapeake Bay by 250 percent, causing an ecological disaster. In 1972, the massive flood of Tropical Storm Agnes scoured eight years worth of sediment from the reservoir, a large pulse of pollution to the bay.

To see rock at the base of the dam, turn south on Shuresville Road just west and uphill from Conowingo Dam. After about a half mile, if you take the first left turn at an acute angle, you can drive down to the river directly below the dam and see Conowingo Diamictite of the Potomac Terrane. Here you can also see the gorge that the Susquehanna has cut in the hard Piedmont rock—similar to that of the Potomac River. In a roadcut on US 1 just west of Octoraro Creek and on MD 273 just east of its junction with US 1 are more exposures of the Cambrian Conowingo Diamictite.

Port Deposit

Just east of Conowingo Dam, you can take the first right onto MD 222 and drive south along the river to Port Deposit. Here you can get a great view of the river and the 200-foot-high bluffs above the town. North of town is a marker for Smith's Falls—not a waterfall but a rapids through the hard, unnavigable, rocky river bottom of the Fall Line, named by John Smith who reached the upstream navigation limit when he sailed here in 1606. Opposite the falls are the old Port Deposit Quarries, where the hard rock was first quarried in 1817. Many buildings in Port Deposit, as well as several in major eastern cities, were built of this gneiss. When used as a building stone, it is called Port Deposit Granite.

Principio Furnace

Principio Furnace is east of the mouth of the Susquehanna River off MD 7. You can stop and look at the 1837 blast furnace but cannot trespass. The furnace location was geologically determined in the 1720s. A nearby stream dropping over the Fall Line provided water power. The flat Coastal Plain was good for the iron works, and the nearby mouth of Principio Creek was a good shipping harbor. Ore was brought in from the Patapsco River and elsewhere.

Mouth of Susquehanna River

From MD 7, US 40, or I-95, you can drive to the grounds of the U.S. Veterans Hospital—bear right after the main entrance—and see the mouth of the Susquehanna River. These grounds—as well as Havre de Grace across the river and North East and Elkton farther east—are underlain by the Talbot gravels and sands of Quaternary age, deposited in river deltas when Chesapeake Bay water levels were higher. The Susquehanna discharges on average about 3 billion pounds of sediment per year, but this quantity can increase by a factor of ten to twenty during a major flood.

At Concord Point, on the western shore of the mouth of the Susquehanna, you can see the oldest continuously operating lighthouse in the United States—since 1827. It is made of Port Deposit Gneiss quarried a short distance upstream. Garrett Island in the Susquehanna River is Port Deposit Gneiss.

Eastern Cecil County

Just west of Fair Hill on MD 273 is Rock Presbyterian Church, where you can see excellent exposures of the James Run Formation of the Chopawamsic Terrane. These basaltic lavas oozed out from a volcanic island chain beneath the cold sea and cooled rapidly into pillow forms. A moldable skin quickly formed on the outer surface of the lava, and as more lava flowed underneath, it ballooned into the pillow shape. Please respect the beautiful church property.

At Fair Hill Natural Resources Management Area, you can exit at the horse track, cross the bridge over MD 273, and drive down to the covered bridge, where you can see that Big Elk Creek has incised into an unnamed gneiss of the Potomac Terrane. The creek flows to Chesapeake Bay, where it becomes tidal Elk River.

A little over 1 mile west of the Delaware line, MD 273 crosses MD 316, which runs along the drainage divide for Chesapeake and Delaware Bays.

At Rock Presbyterian Church you can see the James Run Formation of the Chopawamsic Terrane. These rocks were once lavas that oozed out under the sea, much as Hawaiian eruptions do today.

Grays Hill near Elkton

The major east-west highway US 40 follows the level Coastal Plain through eastern Maryland, but just west of the Delaware border it passes south of Grays Hill, an outlier of Piedmont igneous gabbro that stands over 160 feet above the Potomac Group sediments of the Coastal Plain. The rock at Grays Hill is poorly exposed but is probably related to the gabbro that forms nearby Iron Hill in Delaware. This gabbro intruded rocks of the volcanic island arc in Silurian time, probably during crustal thinning that occurred after the Taconic mountain building event.

US 15

Point of Rocks—Emmitsburg (the Frederick Valley)

36 MILES

US 15 runs north-south along the western edge of the Gettysburg Basin of Triassic age, a unique trough in which the rock is about 300 million years younger than that lying to the east and west. As the African Plate diverged from the North American Plate during the disassembly of Pangea about 200 million years ago, the tension of pulling apart caused fracture and downslippage of huge blocks of crust. The Triassic Border Fault runs from Connecticut to Alabama, and it has been compared to a giant trapdoor, with the hinge to the east. Downslippage of about a half mile occurred on the eastern limb of the South Mountain anticlinorium, the east slope of Catoctin Mountain. As the basin subsided, sediments from the adjacent highlands eroded into it. The 500-million-year-old limestones and sandstones that once formed some of the upper layers of the anticlinorium contributed cobbles and pebbles to the formation of 200-million-year-old conglomerates in the basin.

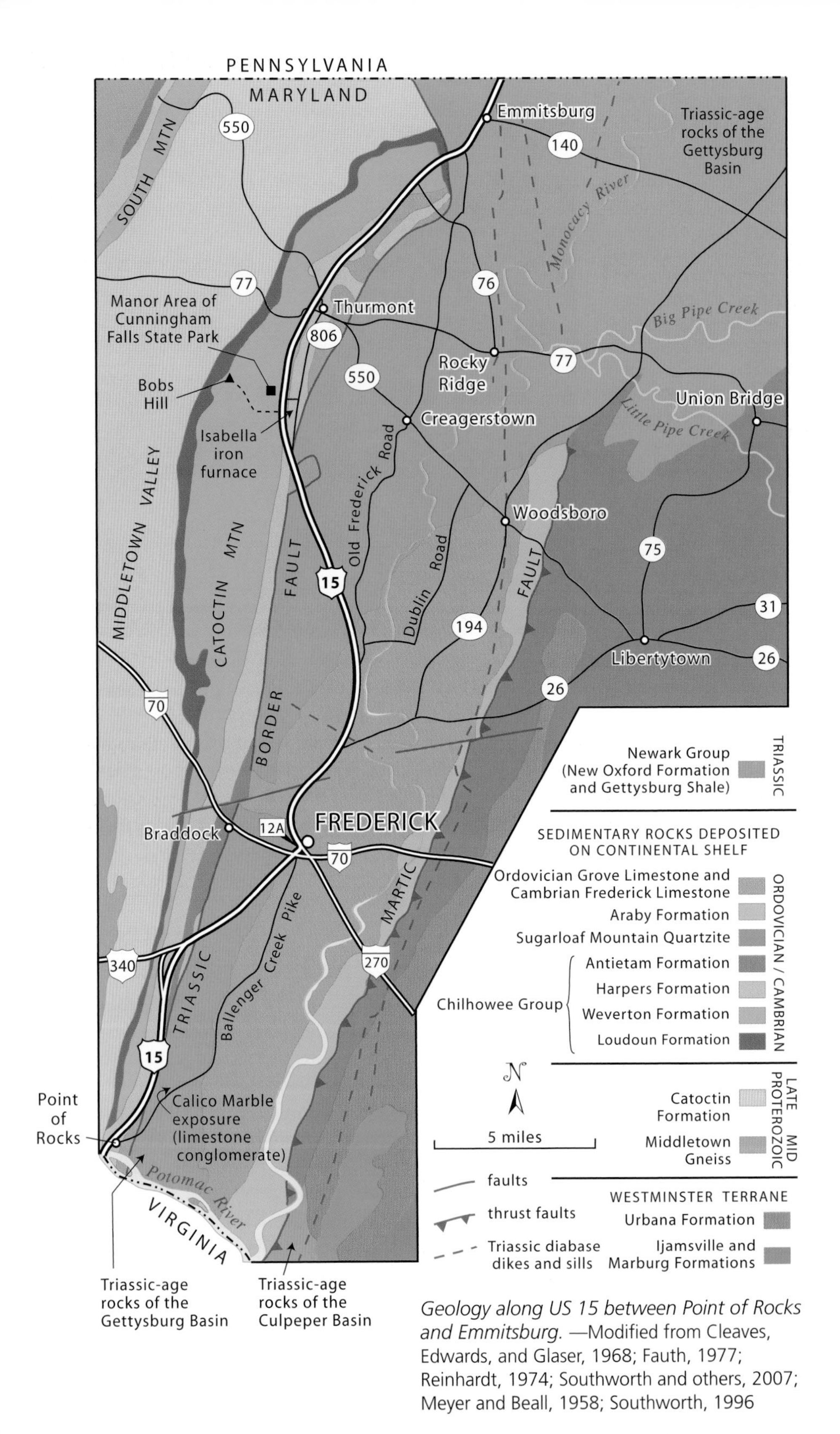

Geology along US 15 between Point of Rocks and Emmitsburg. —Modified from Cleaves, Edwards, and Glaser, 1968; Fauth, 1977; Reinhardt, 1974; Southworth and others, 2007; Meyer and Beall, 1958; Southworth, 1996

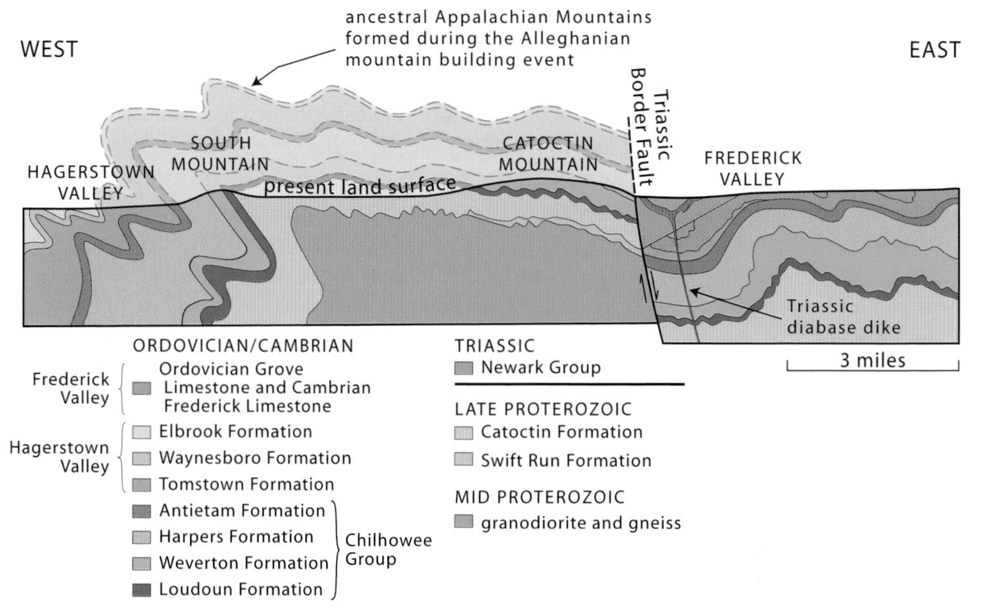

The South Mountain anticlinorium before and after the normal faulting of the Triassic period. The Frederick Limestone exposed along the Monocacy River may once have been part of the mountains. —Modified from Cleaves, Edwards, and Glaser, 1968; Cloos, 1947

Other sediments washed down as alluvial muds and sands, which in the hot, arid climate of the time turned red due to oxidation of iron within the rock. Collectively, these lake and floodplain sediments are called the Newark Group. You can easily tell when you are in the basin because you see red rock and red soil. Also, if you look west, you see Catoctin Mountain jutting up from the relatively flat basin. These are fault-block mountains like the Grand Tetons in Wyoming, but not as young or as high.

US 15 is not the best route for seeing rock exposures, and the traffic will get most of your attention, but you can take side trips and parallel roads and use US 15 as a connector.

Calico (Potomac) Marble

At Point of Rocks you can take Point of Rocks Road (MD 464) east, drive about 100 yards, and turn left onto Ballenger Creek Pike to get to one of the most accessible exposures of the Calico or Potomac Marble. It is not a true marble but the limestone conglomerate of the Gettysburg Shale of the Newark Group—a stone used for columns in the interior of the National Capitol. The exposures are at the entrance to Doubs Substation, where the power line crosses the road. This conglomerate is a sedimentary rock resembling concrete. Here pebbles and cobbles eroded down from the fault scarp to the west when the mountains were much higher and then were cemented in a reddish sand matrix. You can see the rock cycle continuing as weathering has freed some of the gravels. If you go east on Mountville Road, you can see exposures of this rock in pastures.

Old Frederick Road and Dublin Road

If you want a less traveled and more geologically scenic route through the Triassic basin, turn onto Old Frederick Road just a few miles north of Frederick. It rejoins US 15 about 15 miles north, just south of Emmitsburg. Less than 1 mile north of your turnoff, turn east onto Devilbiss Bridge Road, go about 1 mile, and turn north on the old bridge road just west of the bridge. This leads you to a small park under the bridge and to the main stream that drains the Triassic basin, the Monocacy River. If you look upstream to the west of the river you can see a small ridge made of the moderately resistant quartz conglomerate member of the New Oxford Formation of Triassic age.

If you walk downstream along the river for several hundred feet and look across, you can see cliffs and outcrops of the laminated layers of the Cambrian Frederick Limestone. If you are driving over the bridge, you can get a quick look down at this exposure. Here you are east of the deep portion of the basin that downfaulted and then filled with Triassic sediments, and the Frederick Limestone is thought to correlate with limestones in Washington County more than 20 miles to the west. The limestone layer may have been continuous across the high arch of the then alpine-sized South Mountain anticlinorium of Pangea times.

To see a different limestone formation, proceed east over the bridge and turn north on Dublin Road. You will see pinnacle outcrops of the Ordovician Grove Limestone in pastures, and you can look west and see the quartz conglomerate ridge of the New Oxford Formation of the Newark Group. North of Links Bridge Road, you drive up on a significant ridge of this conglomerate—not as high as Catoctin Mountain, but well above the surrounding lowlands.

This laminated Frederick Limestone on the Monocacy River exhibits an eastward dip similar to that of the easternmost, overturned limestones of Hagerstown Valley about 20 miles to the west and may have been physically continuous with them as limbs of the once lofty, alpine-scale South Mountain anticlinorium.

You can turn west off Dublin Road onto Woodboro-Creagerstown Road (MD 550) to return to Old Frederick Road at Creagerstown. Between Creagerstown and Emmitsburg, you can see small roadcuts in the red shale of the Gettysburg Shale of Triassic age and red-tinted soil in fields.

Isabella Iron Furnace (Catoctin Furnace)

From MD 806 off US 15 a couple of miles south of Thurmont, you can park and visit the reconstructed Isabella iron furnace. The site of the iron furnace had everything it needed nearby: forests for charcoal production, limestone as a flux for removing silicon impurities, a stream for water power for the bellows and for washing out the ore, and iron deposits. The iron oxide limonite occurred in several nearby locations intermingled with clay. Furnaces operated here for over one hundred years, and in early days supplied cannons and cannonballs for Washington's Continental Army.

The colonial-era Isabella iron furnace consumed trees and rocks to produce iron, as the Industrial Revolution began altering the environment.

Cunningham Falls State Park

From the Manor Area of Cunningham Falls State Park you can take a strenuous 1.5 mile, 1,100-foot-elevation-gain hike up to the rocky summit of Bobs Hill in the Maryland Heights Member of the Weverton Formation—the highly resistant, ledge-maker quartzite. Far below you can see the Triassic basin of Frederick Valley, and on a clear day, you can see similarly resistant quartzite-topped Sugarloaf Mountain to the southeast, poking through its less resistant Westminster Terrane tectonic window. The Weverton Formation appears again some 8 miles west on the parallel ridge that is South Mountain. These are the eroded limbs of the slightly overturned South Mountain anticlinorium, and at one time they would have arched thousands of feet higher in a continuous formation.

Part of the trail is on old nineteenth-century logging roads—some related to the charcoal production required for the Isabella iron furnace. Common in the Catoctins, these roads are now sunken due to the creeklike erosion that has occurred on their unvegetated surfaces.

From the upper end of the Manor Area campground you can take an easy hike of about a quarter of a mile to Isabella's Rock. Here you can clearly see the east dip in outcrops of the Weverton Formation of the South Mountain Anticlinorium. See **MD 491: Smithsburg—Cascade** for photos of this east limb and the comparably east-dipping west limb at Buzzard Knob.

Thurmont Area

Between the Manor Area and Thurmont, US 15 crosses a couple of the many fan-shaped Quaternary mountain wash deposits that lie at the eastern base of Catoctin Mountain. These are poorly sorted gravity and stream deposits derived from the quartzite and metabasalt of Catoctin Mountain.

Near Thurmont US 15 goes directly along and over the Triassic Border Fault. You cannot see the fault, but you will be aware that Catoctin Mountain immediately to the west climbs upward steeply from the relatively flat Triassic basin. Catoctin Mountain Park is accessible from US 15 near Thurmont and is discussed in **Maryland 77: Thurmont—Smithsburg.**

Rocky Ridge

MD 77 east of Thurmont crosses the aptly named Rocky Ridge, made of a diabase dike that runs over 30 miles south into Montgomery County. After sediments had been deposited in the Triassic basin, mafic magma oozed up through the faults and fractures caused by the continued stretching of the earth's crust. The magma solidified underground in dikes that reflected the mostly parallel configuration of the faults. As weathering lowered the landscape over millions of years, the more resistant diabase dikes emerged as ridges above the more rapidly weathering, red sedimentary rock.

The Maryland Midland Railroad cuts through a diabase dike at Rocky Ridge. Photo taken from the MD 76 overpass.

If you turn north onto MD 76 at Rocky Ridge and park near the railroad overpass, you can look down into the railroad cut and see an exposure of the diabase dike, 200 feet wide here. You can see diabase boulders at the volunteer fire company and nearby Mount Tabor Park. If you drive north on MD 76 you can follow the ridge, encounter more boulders, and see to the west how much lower the sedimentary rock has eroded compared to the hard igneous rock that had originally intruded it. As soon as you see red rock or soil, you know that you have passed from igneous back to sedimentary.

Side Trip on MD 550: Thurmont to Libertytown

On MD 550 between Thurmont and Woodsboro, you cross the relatively flat Triassic basin, and you can look back at Catoctin Mountain. Note the red soils in roadcuts and excavations—the nonmarine Triassic deposits. At Creagerstown Park just west of the Monocacy River, you can get a close look at the red deposits and the river.

About 1 mile west of Woodsboro is a small ridge of quartz pebble conglomerate of the New Oxford Formation of Triassic age, containing hard material derived from the adjacent mountains. Immediately west of Woodsboro, you will see a large quarry in the Ordovician Grove Limestone, a rock correlated in geologic age not with the Triassic basin but with Valley and Ridge.

If you follow old MD 550 into Woodsboro, you will go over a ridge that is the same diabase dike that runs through Rocky Ridge to the north. On new MD 550, you can see a bit of the diabase in the roadcut.

East of Woodsboro on MD 550 is another quarry, this one in the Frederick Limestone, and another ridge, this one in the resistant Araby Formation, and then, at a little over 1 mile southeast of Woodsboro, you pass over the Martic Fault.

Spheroidally weathered diabase boulders on old MD 550 in Woodsboro are part of the igneous ridge that sits above the less resistant sedimentary rock of the Triassic basin.

Gettysburg Shale and Diabase Dikes

On US 15 near Emmitsburg is an excellent exposure of the red shale and siltstone of the Triassic Gettysburg Shale just north of the MD 140 exit in the northbound lane. Dinosaur footprints have been found near here. The red coloration in the shales, siltstones, and sandstones is characteristic of the rapid, nonmarine, lake and river-floodplain deposition that occurred in the Triassic basin. Because deposition was not under the sea, iron was oxidized to impart the red color. The rocks dip west, conforming to the direction down which the fault slippage was occurring, as deposition into the trough from adjacent mountains almost kept pace with the downward slippage. None of the Triassic rock is folded, not yet at least, because there has been no mountain building event since its deposition 200 million years ago.

East of Emmitsburg you can exit US 15 onto MD 140 and drive east a couple of miles over two ridges made of resistant igneous diabase, intruded into the Triassic basin during the divergence that divided Pangea.

Maryland 28 and C&O Canal
Point of Rocks—Blockhouse Point
24 MILES

MD 28 covers about the same geology as I-270 but affords more places to stop and look. In addition, it provides easy access to the Chesapeake & Ohio Canal (C&O Canal) and the Potomac River. This road guide will discuss what you can see from the road and the towpath.

Point of Rocks

Near the MD 28 junction with US 15, you can turn off MD 28 and drive down to the C&O Canal Towpath. From the towpath under the US 15 bridge you can see a railroad tunnel cut through the 600-million-year-old metabasalt of the Catoctin Formation—the tunnel necessary because the Potomac River has cut a water gap in Catoctin Mountain. You can walk up the towpath a few hundred feet and see an excellent exposure of the former lava flow, with large quartz veins.

At Point of Rocks, a tunnel for the railroad had to be dug through the hard metabasalt of the Catoctin Formation because in the narrow water gap, there is no floodplain upon which to lay the track.

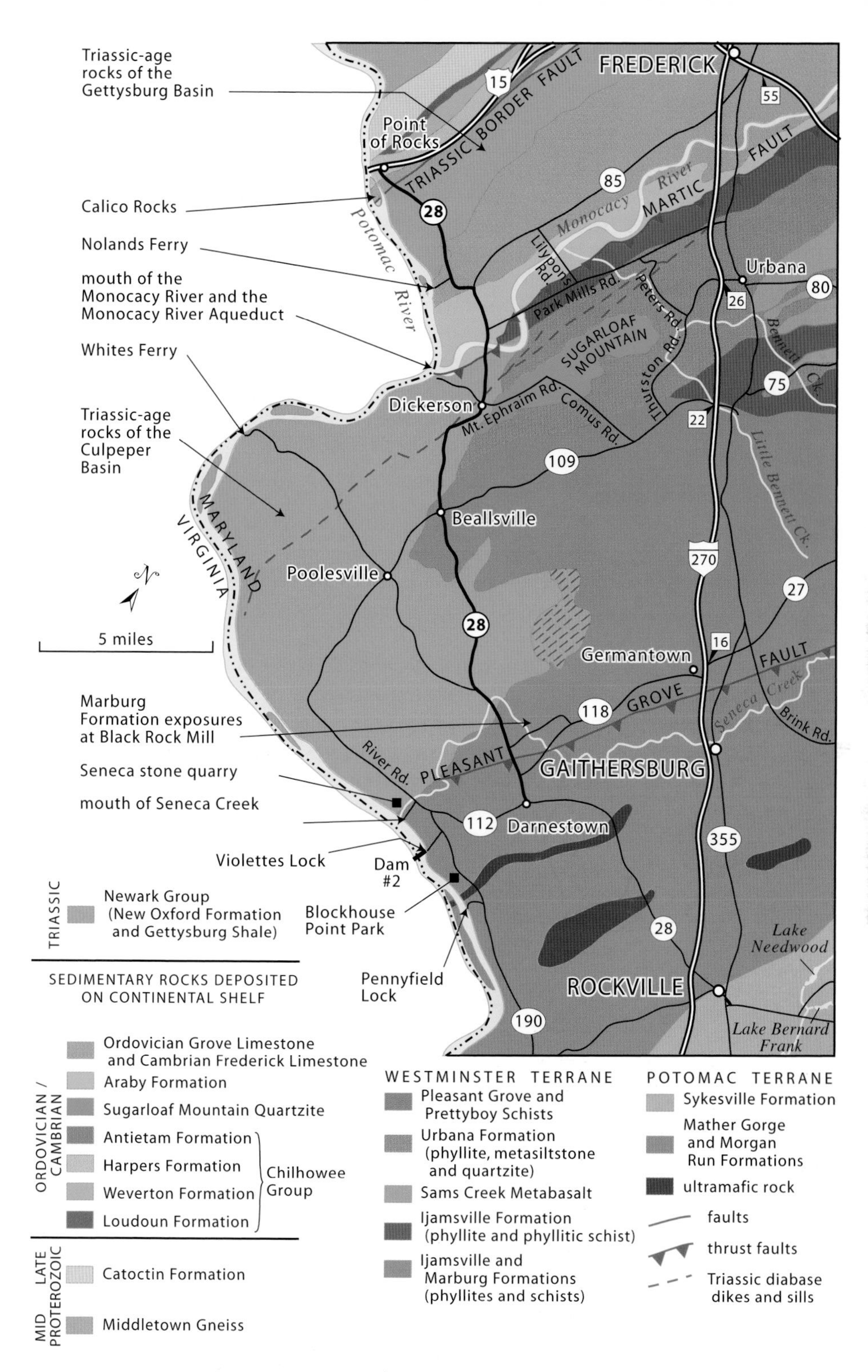

Geology along MD 28 and the C&O Canal between Point of Rocks and Blockhouse Point. —Modified from Cleaves, Edwards, and Glaser, 1968; Reinhardt, 1974; Southworth and others, 2001; Southworth and others, 2007; Southworth and others, 2006; Southworth, 1996; Kunk and others, 2004; Cloos and others, 1964

Triassic Border Fault

Just a few miles east of Point of Rocks on MD 28, you pass over the Triassic Border Fault (not visible on the surface) and onto sedimentary rocks about 200 million years old. A sinking trough formed to the east of the fault when tensional forces pulled the earth's crust apart as Africa diverged from North America. Presumably, downfaulted sections of the Catoctin metabasalt and Weverton Formation lie buried under thousands of feet of sediment that were deposited by streams and debris flows running down from nearby mountains west of the fault. In Triassic time, the mountains standing west of the basin were still young, high, and rapidly eroding, resembling a range like the Rockies. Eroded fragments of those ancient mountain slopes can be found today in the rocks of the basin.

Potomac Marble

About 3 miles east of Point of Rocks on either side of MD 28 are roadcuts of the limestone conglomerate of the Gettysburg Shale of the Newark Group. Here gray limestone pebbles of different sizes and shapes are cemented in a red, silty, sandy matrix. The formation looks a bit like ordinary concrete, but the story of its origin is unique.

About 800 to 500 million years ago, a thick wedge of sandstones, shales, and limestones built up off the ancient eastern shoreline of North America. The rocks of this wedge eventually were pushed up by tectonic collisions to form mountains. Then, as divergence created the Triassic basin, vast quantities of limestone in the form of gravel and pebbles were eroded from substantial outcrops of Cambrian limestone exposed in the highlands to the west and east and were deposited, along with sand and silt, in alluvial fans, floodplains, and streambeds in the basin. This sediment eventually compacted and cemented in water that was rich in calcium carbonate (one of the compounds in cement). The limestone pebbles are the same Cambrian age as the limestones of the present Frederick and Hagerstown Valleys, while the conglomerate is Triassic age, 300 million years younger. Quarry workers

Variation in size and angularity of these conglomerate pebbles indicates differing conditions of deposition—the more angular have generally spent less time in moving water.

near here cut this stone, called Potomac Marble, for the columns in the National Statuary Hall in the Capitol Building. It is not, of course, a true marble, which is a metamorphosed limestone.

You can also see the rock at Calico Rocks near milepost (MP) 47 of the C&O Canal. At MP 47.1 you can cross the walkway and the railroad tracks (stop, look, and listen!) and take a relatively safe and lengthy inspection of this unique limestone conglomerate and breccia. If walking along a railroad track spooks you, you can go just downstream from the walkway and find several other excellent outcrops that you can examine by simply walking across the unwatered canal bed and climbing up a few feet. Calico Rocks is yet another name for the conglomerate formation.

Other places that you can see the conglomerate are in fields north of Mountville Road between Adamstown and US 15 (if they are not consumed by development), on Ballenger Creek Pike north of Point of Rocks near the substation, and east along US 40, in the highly developed "Golden Mile" in Frederick.

Old Limestone and Young Potomac River Gravel

About 4 miles east of Point of Rocks on MD 28 at New Design Road in a farm field (in 2007) across from an electric power substation is an outcrop of the Frederick Limestone. Part of the Cambrian offshore wedge and of the same general age as the limestone of the Hagerstown Valley to the west, these thin-bedded, muddy limestones weather to form soils excellent for cultivation. This limestone forms part of the floor of the Frederick Valley. Some sections of the limestone remained elevated, that is, still visible on the valley floor, while others slid downward and were buried under Triassic-age deposits. Some pebbles of the Triassic limestone conglomerate (Potomac Marble) of the Newark Group have been identified as being derived from the Cambrian Frederick Limestone.

The outcrop of limestone appears to be a minor sinkhole. Contained within the hole are rounded quartzite and quartz stones of cobble size. Rounded stones have been tumbled in moving water, and these may have been carried here by the ancestral Potomac River during the wetter climates of Pleistocene Ice Ages, as well as during post–Ice Age times about 10,000 years ago. The particular stones in this sinkhole may have been carried here by hand or machine from the surrounding fields to facilitate plowing and planting. Remarkably, these rounded cobbles are about 200 feet above the present level of the Potomac River and about 1 mile away from it. The Potomac River may have changed course, may have carried much more water, and certainly has cut downward into the landscape since these rounded stones were deposited here. A Maryland Geological Survey map indicates that this entire field and the hilltop above it are covered with these river terrace gravels.

Nolands Ferry

Upstream and downstream from Nolands Ferry, accessible from New Design Road off MD 28, the C&O Canal Towpath crosses a broad, low floodplain. When the river overtops its banks and spreads across the flats where the trees are growing, its current slows and the river drops quantities of mud and silt, building up the floodplain even more. Near MP 44 below Nolands Ferry, the towpath crosses the Martic Fault—not visible here under the alluvial floodplain.

Monocacy River Bridge

The Monocacy River drains most of the Frederick Valley and the Triassic basin. Just to the west of the MD 28 bridge over the river, you can see a roadcut of the Araby Formation, consisting of slightly metamorphosed siltstone and phyllite. This rock fractured but did not slide downward during the Pangea breakup, and it is over 300 million years older than rocks to the west in the Gettysburg Basin and rocks to the east in the Culpeper Basin, a southeastern section of the Triassic basin.

You can explore the floodplain of the Monocacy River with a little walking. Some of the river banks are 5 to 10 feet above the normal flow, and they were deposited there by muddy floodwaters that crested the banks, slowed in velocity, and dropped mud and sand. You can find pieces of glassy slag in muddy flats here, left over from the operation of an iron furnace once located nearby.

The Monocacy roughly follows the Martic Fault here, and east of the bridge you can see the stream-deposited red rock characteristic of the New Oxford Formation of Triassic age—also known as the Seneca Red Sandstone—which fills much of the Culpeper Basin.

Sugarloaf Mountain

From MD 28 you can see to the north an isolated mountain standing well above the relatively flat surrounding area. This is Sugarloaf Mountain, quartzite of Cambrian age that was overthrust by the Westminster Terrane but has since been exposed by erosion. See **I-270: Frederick—Rockville** for further discussion. You can reach it from Mount Ephraim Road north out of Dickerson or from MD 109 north from Beallsville. You can also see the mountain from the towpath between MP 44 and MP 42.

Monocacy Aqueduct

Resistant to the ravages of floods is the beautiful Monocacy Aqueduct, just above MP 42 of the towpath and also accessible from MD 28 on Mouth of Monocacy Road. This water-bridge carries the C&O Canal over the Monocacy River. It was made of the white Sugarloaf Mountain Quartzite—originally a beach sand—quarried from the southern base of the mountain 4 miles to the north.

From the aqueduct you can watch the Monocacy—the river that drains the downfaulted basin—empty into the Potomac and increase its flow considerably. You can also see the 90-degree turn the Potomac makes to the southwest. The river bends because it encounters the hard, interbedded siltstones and sandstones of the red New Oxford Formation. Also known as the Seneca Red Sandstone, it was deposited in Triassic time in the Culpeper Basin. If you want to go downstream another mile or so, you can see cliffs from MP 42 to Lock 27 at MP 41.5. Then between MP 41.5 and MP 41, the cliffs become much more open and lie just opposite the adjacent, watered canal. On the bluffs above sits the large Dickerson power plant.

The beautiful Sugarloaf Mountain Quartzite of the Monocacy Aqueduct originated as a white beach sand.

Invisible Igneous Dike

About 1 mile south of Dickerson or about 2 miles northwest of Beallsville, at the intersection of MD 28 and Martinsburg Road, MD 28 makes a ninety-degree turn on a ridge underlain by an igneous dike. When Pangea broke up and pieces of crust fractured, hot molten rock called magma oozed up from below through some of the fractures. As it neared the surface, the magma cooled and solidified into an extremely hard rock called diabase. Although the rock is not visible at this ridge, diabase dikes form linear ridges throughout many areas of the Triassic basin. This particular dike extends for over 35 miles to Rocky Ridge in Frederick County, where the diabase is visible.

Slate

A roadcut on MD 28 just west of Beallsville, accessible by foot but hazardous due to traffic, features slate—metamorphosed shale—tilted to a vertical dip or angle. Both tilting and metamorphism were due to the great pressures exerted during one or more of the mountain building episodes. This exposure is phyllitic slate of the Cambrian Ijamsville Formation of the Westminster Terrane.

Seneca Creek State Park

At Darnestown you can pick up MD 112 (Seneca Road) off MD 28 and drive to Seneca Creek State Park and the C&O Canal Towpath. West of Seneca Creek may be found the ruins of the Seneca Stone Cutting Mill, where red sandstone of the New Oxford Formation from various quarries was cut by steel blades, at the rate of

Even in ruins, the red sandstone Seneca Stone Cutting Mill is an impressive structure.

1 inch per hour, into large blocks. This stone, known as the Seneca Red Sandstone, was used in numerous C&O Canal structures and in the Smithsonian Institution Castle on the Mall. The sandstone was originally deposited in the Culpeper Basin of Triassic age by streams that flowed into the low-lying trough.

From the mouth of Seneca Creek, near MP 23, you can walk upstream on the towpath to MP 24 and see cliffs and old quarries of the red sandstone on the other side of the canal—more visible during winter than summer. These rock layers are not folded because they have never been caught in the mountain building of a tectonic convergence, but they do dip slightly to the west at the angle of the downfaulted Triassic basin, which was formed by tectonic *divergence*.

From the mouth of Seneca Creek, you can look downstream and see, about 1 mile away, the hills on either side of the river that mark the transition from Triassic basin to metamorphic Piedmont rock.

Violettes Lock and Blockhouse Point

About 1 mile downstream from Seneca Creek on the towpath, near MP 22, accessible by car via Violettes Lock Road off MD 190 (River Road), are Lock 23 and Feeder Dam No. 2, which supplied water for the canal. Here is the eastern edge of the Culpeper Basin, where the red stone of Triassic age lies in direct contact with the highly erosion-resistant Mather Gorge metagraywacke and schist of Late Proterozoic–Early Cambrian-age, an unconformity representing a gap of about 330 million years of geologic time. This hard rock is part of the Potomac Terrane. If you are approaching on foot or bicycle from upstream, you can see the hills of the Mather Gorge Formation rising to the north and south of the river like a water gap, while upstream from this point, the land along the river is relatively flat.

You can see many exposures of red sandstone from the towpath in the mile above the mouth of Seneca Creek. This one reflects the westward dip of the hinged downfault.

The original dam of stone-filled cribs was built here so that it could be anchored and supported by the hard rocky islands and outcrops. Upstream of the dam the Potomac River is relatively straight, shallow, and wide—flowing through the siltstone and sandstone; but downstream the river cuts a gorge with rapids where it meets the hard metamorphosed rock. About 1 mile below Violettes Lock, near MP 21, are the cliffs of Blockhouse Point. About 150 feet high and made of a hard phyllite of the Mather Gorge Formation, these cliffs had to be blasted to make room for the canal, and you can still see numerous vertical drill holes. The many folds and fractures in the rock here result from the great forces of tectonic convergence.

If you bike from Violettes Lock at MP 22 to MP 0 at the Coastal Plain, you can get some idea of the great thickness of the rocks of the Potomac Terrane. Most of this terrane is composed of material eroded into the ancient ocean trench from the

Thick, hard metamorphic rock of the Mather Gorge Formation had to be blasted to make room for the canal near MP 21 at Blockhouse Point.

nearby volcanic island arc. The volcanic mountains must have been of considerable size to have eroded enough sediment to make a terrane over a dozen miles thick. The river and towpath passage through the terrane is longer because the river meanders.

Pennyfield Lock

You can reach this lock, near MP 20, on Pennyfield Lock Road off River Road. You can see more cliffs of the Mather Gorge phyllite just below Pennyfield Lock or 1 mile downstream, near MP 19. These are the best sheer cliffs between here and Great Falls, near MP 14. Between MP 20 and 19, you can also see piles of sediment in the canal where streams have carried eroded particles and dropped them when their currents were slowed by the still water of the canal.

Black Rock Mill

On MD 28, 2 miles northwest of Darnestown, turn north on Black Rock Road, and at Seneca Creek you can visit Black Rock Mill, a former grain and lumber mill. Inside the mill are displays depicting historic flood levels of Seneca Creek. Here the creek is funneled into a relatively narrow valley by the resistant interbedded phyllite and metasiltstone of the Marburg Formation of the Westminster Terrane, making it an ideal location to channelize water. The millrace, across the road from the mill, was hand cut into this stone. The impressive Black Rock Mill was constructed in 1815 of schist/phyllite walls with dovetailed cornerstones of Seneca Red Sandstone. Across the bridge on the north bank of the creek is a cliff exposure of the Marburg Formation

Black Rock Mill was constructed in 1815 of stone quarried nearby.

Maryland 75 and Maryland 31
New Market—Westminster
25 MILES

At the New Market exit on I-70, you can travel north on MD 75 and MD 31 over rocks of the Westminster Terrane, including the Sams Creek Metabasalt, a rock that cooled from lava in late Precambrian time. It was later metamorphosed into metabasalt, a dark greenish rock called greenstone. Not much of this greenstone is visible from MD 75, but you can see it south of Union Bridge and also on MD 31 near Sams Creek.

A few miles south of Libertytown is the Pinnacle, standing 60 feet above the road. The Pinnacle is made of fine white quartzite, a highly resistant rock that may have originated as a beach deposit in Cambrian time, seemingly part of the Ijamsville Formation of the Westminster Terrane. The quartzite, as well as the Sams Creek Metabasalt, may be the same age and have the same geologic history as the metabasalt and quartzite located farther to the west in Catoctin and South Mountains. There are no "No Trespassing" signs here, and you can try climbing up for a close look at the quartzite.

MD 75 north of Libertytown goes over typical Piedmont rolling terrain of various metamorphic phyllites of the Westminster Terrane. A few miles south of Union Bridge you can see exposures of the Sams Creek Metabasalt by the road. Union Bridge is located in a valley because it lies on late Proterozoic Wakefield Marble of the Westminster Terrane, which wears down faster than the adjacent metabasalt. Rain, which is acidic, can dissolve marble over time. A huge quarry and cement plant at Union Bridge mines this marble. The quarry is closed to the public but is partly visible from the road.

The hard quartzite of the Pinnacle along MD 75 south of Libertytown projects above the surrounding landscape.

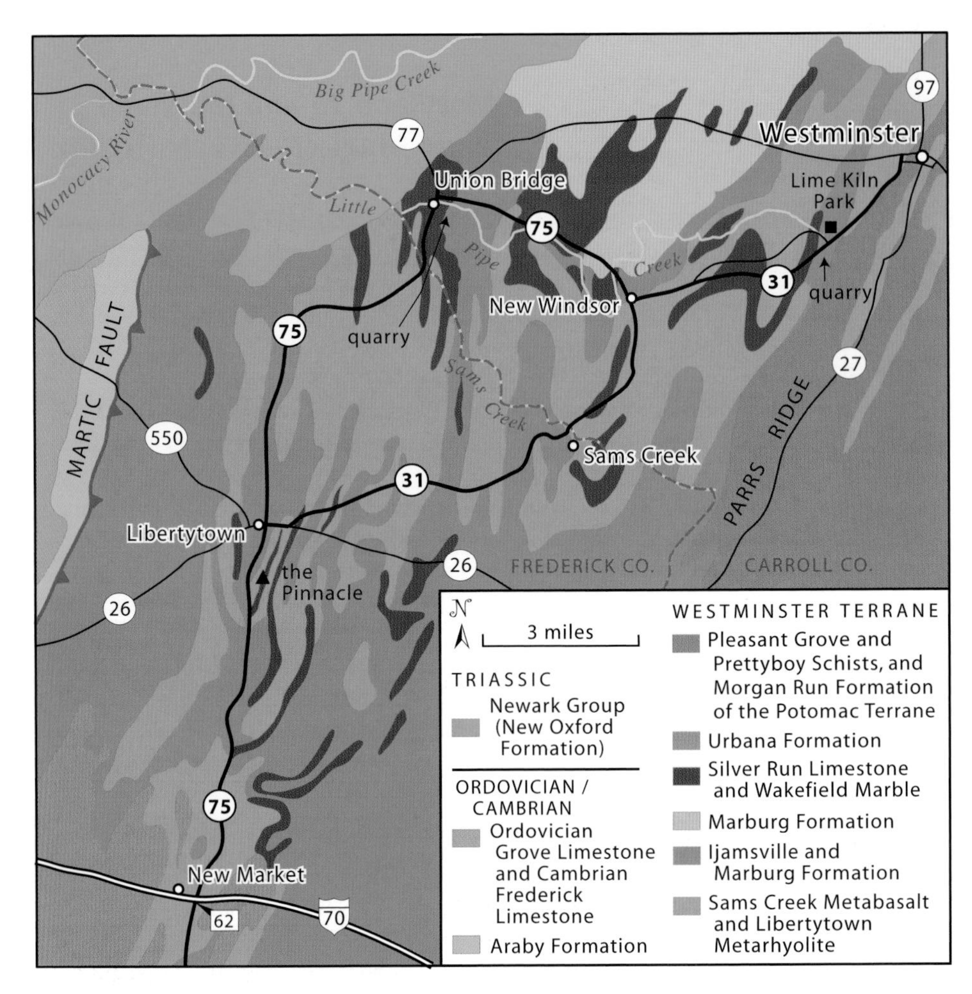

Geology along MD 75 and MD 31 between New Market and Westminster.
—Modified from Cleaves, Edwards, and Glaser, 1968; Meyer and Beall, 1958; Southworth, 1996; Southworth and others, 2006

You can see the floodplain and the meandering Little Pipe Creek at a park in Union Bridge and along MD 75 between Union Bridge and New Windsor. This creek drains westward toward the downfaulted Triassic basin and the Monocacy River.

Sams Creek Metabasalt on MD 31

You can see exposures of the Sams Creek Metabasalt along MD 31 between Libertytown and Sams Creek, and near Sams Creek. This metabasalt is similar in age and appearance to its relative to the west, the metabasalt of the Catoctin Formation. They both formed between 800 and 600 million years ago when the ancient North American craton was pulling apart. As the crust stretched and fractured, molten rock oozed up through the fractures and flowed onto the surface as lava. These

Little Pipe Creek meanders across a flat area of the Piedmont on its northwesterly drainage toward Monocacy River in the Triassic basin.

flows occurred intermittently over millions of years and contained huge volumes of magma that poured onto the surface. Later geologic events—offshore deposition above the basalt, plate collision and compression, and thrusting of different beds of rock over one another during episodes of mountain building—served to interbed the basalt (metamorphosed to metabasalt) with metamorphosed limestone (marble) and metamorphosed shale (phyllite and schist). As these tilted and contorted layers were exposed to millions of years of erosion, the beautiful ridges and valleys were formed because the adjacent rock types eroded at different rates.

Just north of Sams Creek, to the west you can see an old quarry located in a valley floored by marble. From either MD 31 or Old New Windsor Road between New Windsor and Westminster, you can see hills and ridges of metabasalt and valleys of marble. The marble is not exposed on the surface, except in a large quarry near Medford (closed to the public). You can stop and look at a large roadcut in the Sams Creek Metabasalt near Medford on MD 31.

Fenby Farm Quarry and Lime Kiln Park

About a couple of miles southwest of Westminster on MD 31 is a sign for Fenby Farm Quarry and Lime Kiln Park. Here you can walk into the old quarry and see marble interbedded with phyllite. These are both metamorphic rocks, but before tectonic forces changed them through heat and pressure, the marble was limestone and the phyllite was shale. Walk across from the parking lot and take a close look at the well-preserved kiln. It is constructed of rocks from the surrounding area: marble, metabasalt with marble, and purple phyllite.

Parrs Ridge

From Westminster to Mount Airy or even to Germantown, you can follow Parrs Ridge, the highest ground of the Piedmont, on MD 27, also known as Ridge Road. This highly developed area has heavy traffic at times and very few rock exposures,

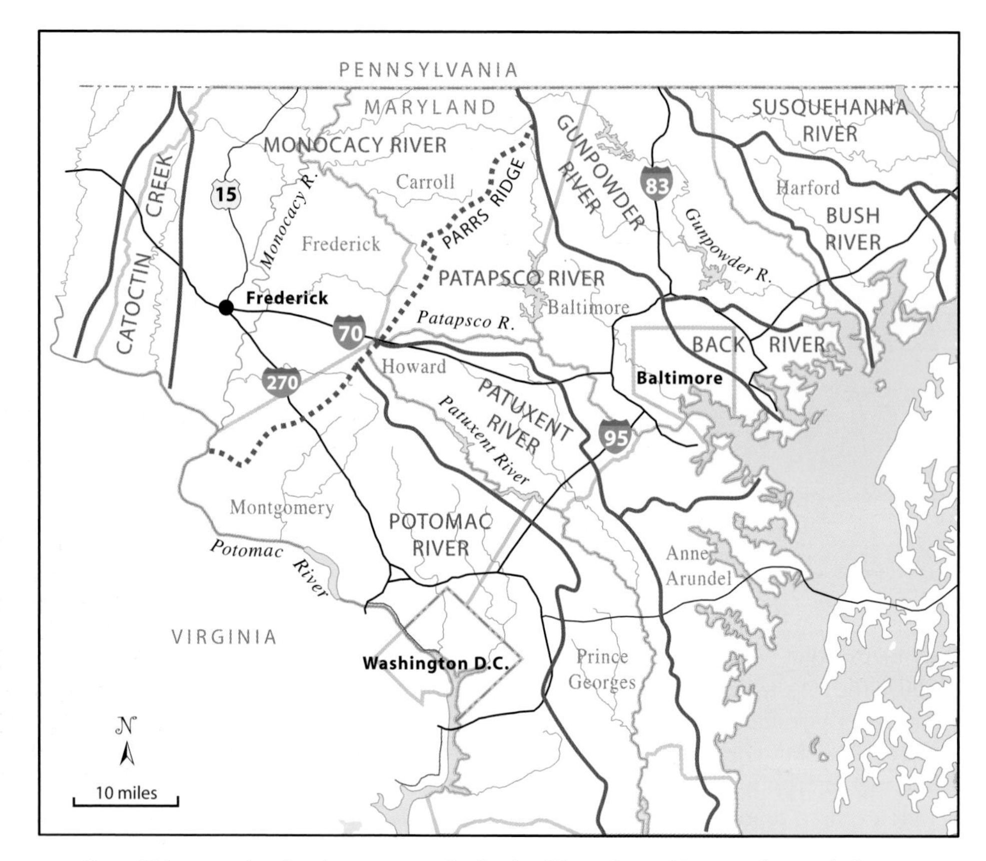

Parrs Ridge, made of resistant quartzite in the Westminster Terrane, forms drainage divides in the Piedmont. —Modified from Vokes, 1961

but on a clear day in a few places along the road you can get good distant views. To the west you can see the terrain sloping downward toward the Triassic basin and, farther west, Catoctin Mountain. To the east you can see the land sloping downward to the Fall Line and Chesapeake Bay. Ridge Road generally follows the drainage divide for the westward-flowing tributaries of Monocacy River, which empties into the Potomac and the eastward-flowing tributaries of the Patuxent, Patapsco, and Gunpowder Rivers, which empty into Chesapeake Bay. Parrs Ridge is composed of erosion-resistant quartzites that form the upper part of the Marburg Formation of the Proterozoic Westminster Terrane.

Hashawha/Bear Branch Nature Center

To visit a natural Piedmont area in the resistant quartzites of the Marburg Formation of the Westminster Terrane, go north from Westminster on MD 97 about 4 miles to Hashawha/Bear Branch Nature Center. In rolling, undeveloped hills, the area has over 300 acres of forests, fields, and mowed areas with trails—a great place to hike. This hilly band of hard quartzite runs for almost 10 miles from here northeast into Pennsylvania.

Dug Hill Ridge and Prettyboy Schist

A significant feature northeast of Westminster is northeast-trending Dug Hill Ridge, composed of a band of resistant quartzite interbedded in the Prettyboy Schist of the Westminster Terrane. Another band of quartzite-determined hills runs northeast of Manchester several miles, topped by Schalk Road No. 1.

Another good place to see the Prettyboy Schist is at the uppermost tract of Gunpowder Falls State Park, downstream from Prettyboy Dam, which impounds the upper reaches of Gunpowder Falls. You can reach it either on backroads from

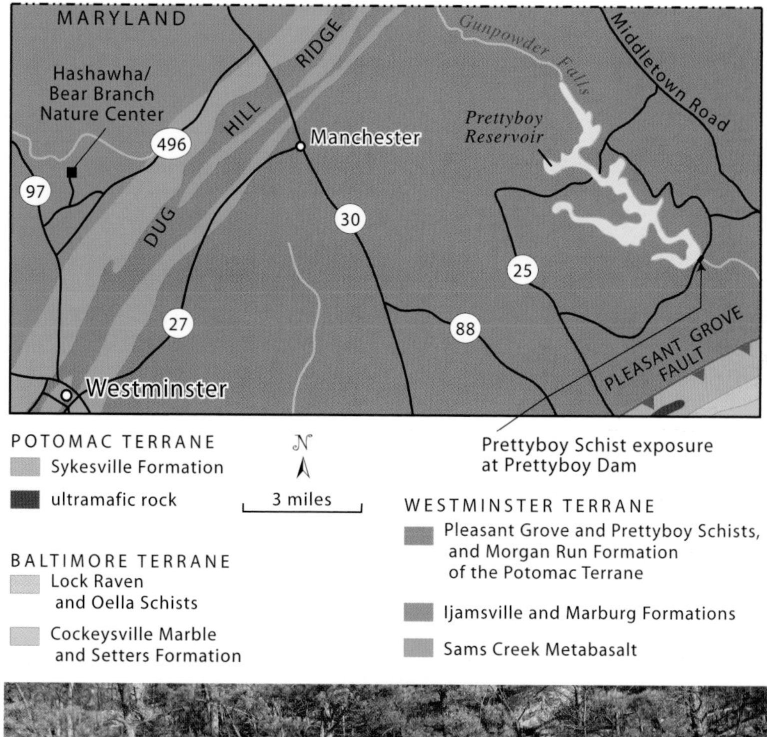

Geology of the Manchester area. —Modified from Cleaves, Edwards, and Glaser, 1968; Crowley, Reinhardt, and Cleaves, 1976; Meyer and Beall, 1958

Prettyboy Schist of the Westminster Terrane near Prettyboy Dam.

Manchester or from exit 31 off I-83. The dam was built in a valley eroded into Prettyboy Schist of the Westminster Terrane. Take the steps down to Gunpowder Falls right below the dam for a view of the dam braced against natural rock.

Washington, D.C., and the Capital Beltway Area

About half of the area inside the Capital Beltway is metamorphic rock of the Piedmont Province, and the other half is unconsolidated sediments of the Coastal Plain Province. Both provinces feature hilly terrain, and if you are driving the beltway, you will not be able to discern when you pass from one to the other. The District is ringed with hills, and the beltway passes over them. However, if you drive into the city from the northwest on, say, Connecticut or Wisconsin Avenue, you will be able to tell when you pass from the Piedmont slopes down onto the flat Coastal Plain, where downtown and the Mall are located.

The Washington National Cathedral, at the intersection of Massachusetts and Wisconsin, affords one of the best views of the District. Ride the elevator up to the seventh level for a view to the east of the Piedmont sloping down to the Coastal Plain of downtown, and to the west of Sugarloaf Mountain.

Since colonial times, the city has grown outward in a roughly concentric manner from the head of navigation at the Fall Line, located at the mouth of Rock Creek on the Potomac River at Georgetown. Upstream from this point are Potomac Gorge and Great Falls, geologic sites not to be missed. The Potomac Terrane of the Piedmont Province is well exposed along waterways in the D.C. area.

In early Cretaceous time, about 140 million years ago, the Potomac Group of Coastal Plain sediments were beginning to be deposited on the continent edge, which had been subsiding for tens of millions of years as the African Plate diverged from the North American Plate. The Potomac sediments are roughly divided into sand-gravel and silt-clay sequences, both containing sediments eroded from the uplifting Appalachian Mountains to the west. Plant, animal, and dinosaur fossils found in these deposits indicate that the sediments were not marine but fluvial, laid down by meandering streams and rivers. They stretched in an eastward-thickening

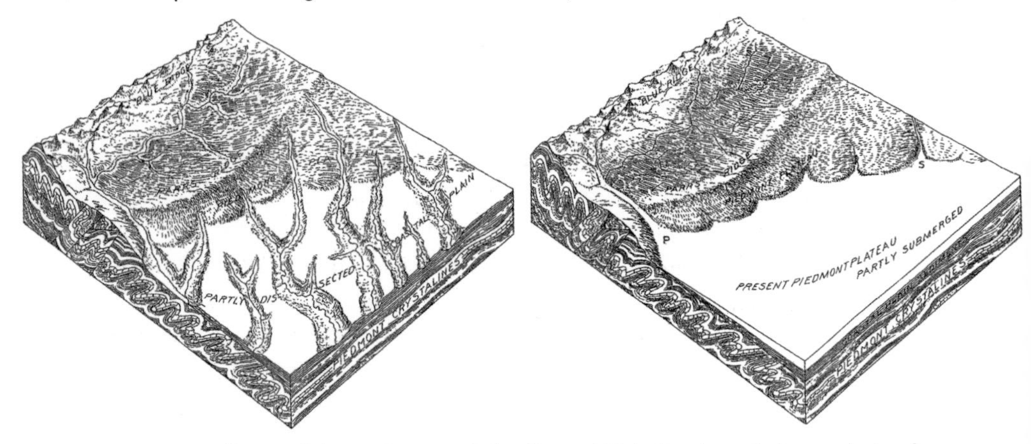

Streams dissected the sediments of the Coastal Plain Province during periods of low sea level. —Modified from Abbe, 1899

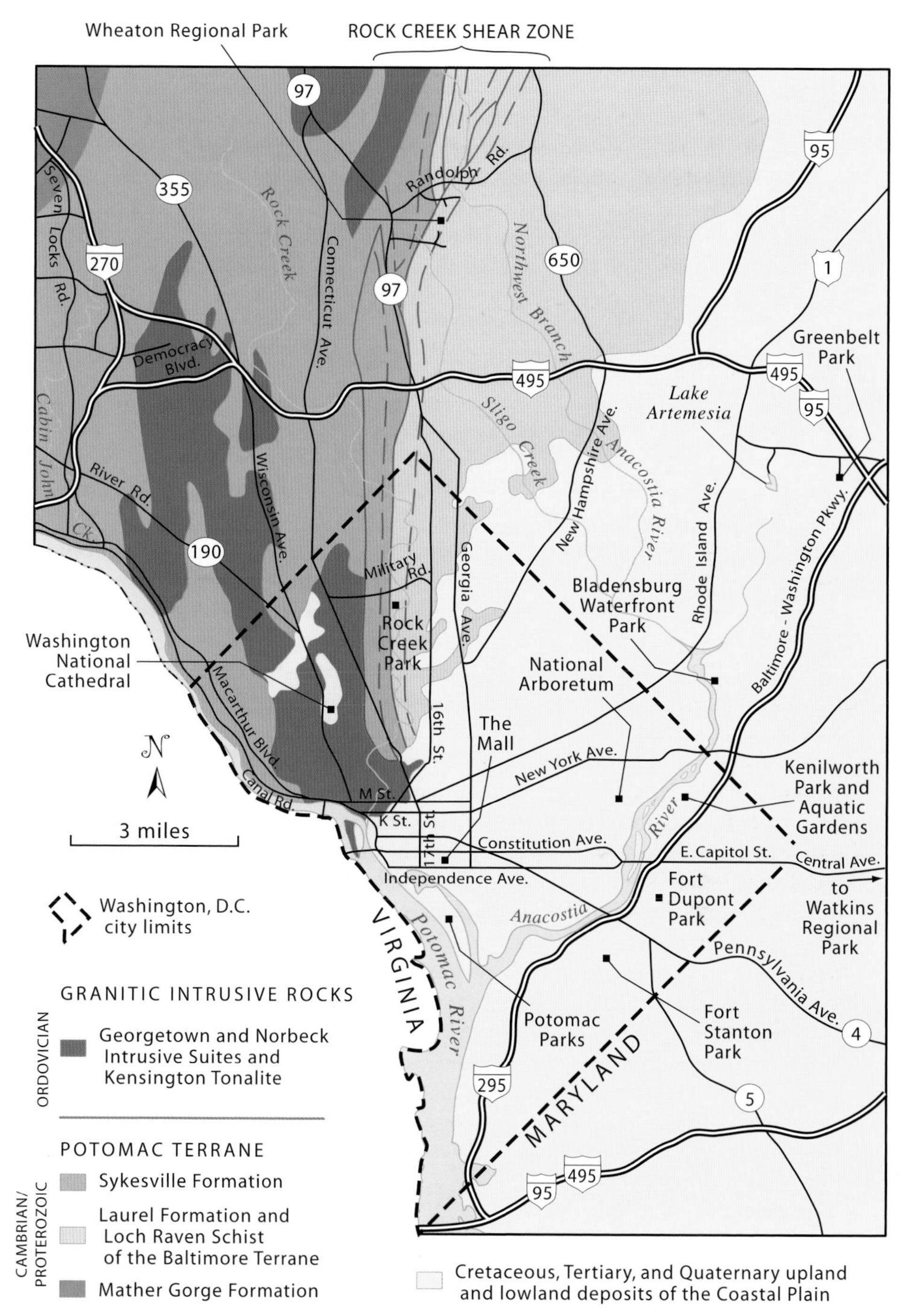

Geology of Washington D.C. and the Capital Beltway area. —Modified from Cleaves, Edwards, and Glaser, 1968; Kunk and others, 2004; Southworth and others, 2001; Cloos and others, 1964; Southworth and others, 2006

wedge far onto the subsiding continental shelf. In D.C. and in Prince Georges County south of D.C., the Potomac Group is up to 1,000 feet thick.

In the Coastal Plain Province, the thick, uncemented, once-flat alluvial deposits of Cretaceous age have long been dissected by streams, especially during periods of low sea level. The dissection created many hilly areas, such as the Anacostia Hills, Landover Hills, and those of Greenbelt Park. On the surface these hills are indistinguishable from the hills of the Piedmont to the west, but Piedmont hills contain solid rock, while the Coastal Plain hills are unconsolidated sediments. Deep wells have revealed that Piedmont rock lies hundreds of feet under the hilly Coastal Plain sediments.

PIEDMONT SITES

Chesapeake & Ohio Canal National Historical Park: Great Falls—Georgetown

The Chesapeake & Ohio (C&O) Canal National Historic Park has a free public-access walking and biking trail on the canal towpath that runs 184.5 miles from Georgetown, D.C., to Cumberland, Maryland. In the first 15 miles (Georgetown to Great Falls), you can drive on roads parallel to the towpath, park at the numerous access points, and bike or walk to see many rock formations. To access these parallel roads from Great Falls, take MacArthur Boulevard to Clara Barton Parkway and follow it to the D.C. line, where it becomes Canal Road down to Georgetown.

The towpath is measured with the original locations of mileposts or mile markers from milepost 0 at the canal entrance at the mouth of Rock Creek. MP or MM are common abbreviations for these posts, and they are helpful reference points for geologic features because the roads, towpath, and river are roughly parallel from Great Falls into the city.

The access points with parking areas include MP 14.3 Great Falls Tavern Visitor Center; MP 12.3, opposite Anglers Inn at Cropley on MacArthur; MP 10.4–10.9, Carderock Area, with four large parking lots, on Clara Barton; MP 8.8 (Lock 10) on Clara Barton; MP 7.0 (Lock 7) on Clara Barton; MP 5.6 (Lock 6) on Clara Barton; MP 4.2, Chain Bridge, with parking just north of the bridge, on Canal Road; and MP 3.1, the Boathouse at Fletchers Cove, on Canal Road—difficult entry and restricted exit. Be aware that Clara Barton and Canal Road change travel direction according to morning and evening rush hours, and some access points may not be available at certain times.

Great Falls and Mather Gorge

If you want to make just one stop to see the most (for your money), pay the entry fee and go to Great Falls, Maryland—preferably off season. From I-495 (Capital Beltway) exit 41, take Clara Barton Parkway west onto MacArthur Boulevard to the park. From I-270 in Rockville, exit onto Falls Road (MD 189), and follow it south to the park. The parking area is on a bedrock terrace incised by the ancestral Potomac River about 30,000 years ago, an ancient riverbed from a time when the river and surrounding landscape were at a less eroded, higher elevation.

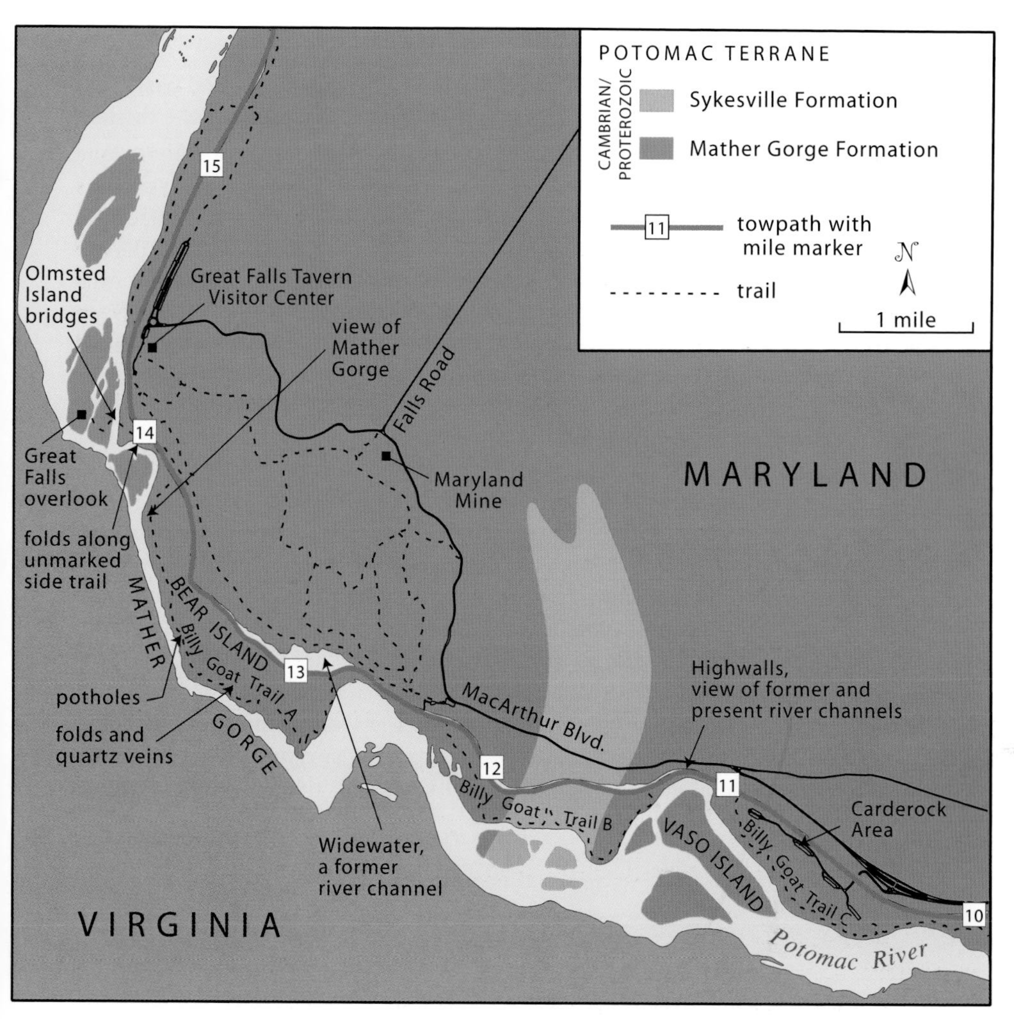

The many trails of the Great Falls section of the C&O Canal National Historic Park give you access to many spectacular geologic sites. —Modified from National Park Service, 2008; Southworth and others, 2001

The park has a visitor center, trails over rocky islands, and trails through the uplands above the river, including one to an old gold mine. Gold was found in quartz veins in many nearby areas. Many of the old mines are today on private property, but from the Great Falls Visitor Center you can acquire a trail map that shows trails to the remains of mine diggings. The Maryland Mine (1867–1939) was located near the intersection of Falls Road and MacArthur Boulevard, and you can see some of the old structures as you begin the descent to the park. While gold has excited many prospectors, it should be noted that here one ton of hard quartz had to be crushed to extract half an ounce of gold.

The Potomac River at Great Falls and Mather Gorge flows over and past rocks of the Mather Gorge Formation of the Potomac Terrane—at perhaps the thickest section of this metagraywacke formation along the Potomac River. Sediments of the terrane were deposited in an oceanic trench, a fore-arc basin, at the edge of a volcanic island arc generated above a subduction zone. The best and most easily accessible view of Great Falls and the head of Mather Gorge is from the overlook at the end of Olmsted Island Bridges Trail, a 0.2-mile, wheelchair-accessible trail over bridges and boardwalks. The island is part of another bedrock terrace of the ancestral river. The bridge to the island crosses over a few modern side channels that have cut down into the rock. In the middle of the viewing platform for Great Falls and in many rock exposures nearby, you can see folded quartz veins that reveal the deformation and metamorphic activity associated with the subducting oceanic trench.

About 2 million years ago, before there were falls and a gorge, the highly resistant metagraywacke of the Mather Gorge Formation upstream from today's falls created rapids that probably spread the river into several, broad channels at an elevation of over 50 feet above the present river. Ocean level was much higher than it is today, and thus the Potomac would have flowed more slowly through this section. The river bottoms of those and more recent times are today the surfaces of not only Olmsted Island but also Widewater, Bear Island, MacArthur Boulevard, and other rocky terraces along the gorge. When the Pleistocene Ice Ages began about 1.6 million years ago, ocean level began dropping as more and more ice was frozen into continental glaciers. About 35,000 years ago, river currents increased

Great Falls of the Potomac, from the viewing platform on Olmsted Island.

Some outcrops of the highly folded and metamorphosed Mather Gorge Formation challenge the climbing skills of young geologists—on the unmarked trail below Olmsted Island, MP 13.85.

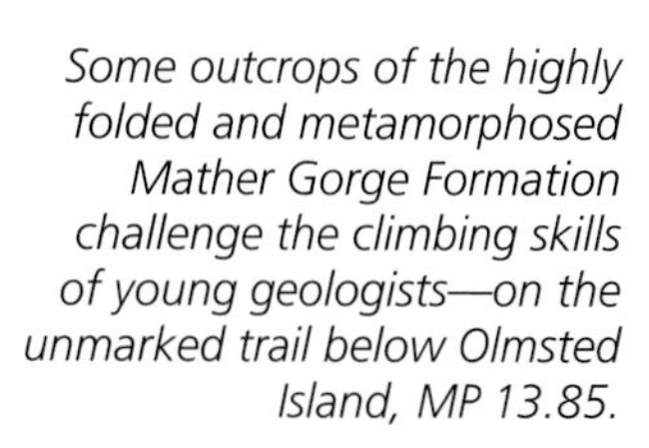

here and began downcutting into the rock, creating the 50-foot cliffs of Mather Gorge. By the end of the last Ice Age, about 10,000 years ago, rocky terraces like Bear Island were vegetated.

Today over the 15 miles from Great Falls to the Coastal Plain at Georgetown (a section called Potomac Gorge), the river drops from elevation 140 feet to 10 feet, with a 60-foot drop in a half mile at Great Falls. Of the seventy-four lift locks on the 184.5-mile C&O Canal, a total lift of 610 feet, the highest concentration is the twenty between Georgetown and Great Falls, where the total lift is about 130 feet.

Bear Island: Billy Goat Trail, Section A

The best place to see rocks, the river, and the narrow 2-mile section of Potomac Gorge called Mather Gorge is probably Bear Island, on Billy Goat Trail, Section A. The upstream trailhead (at MP 13.8) is located about a half mile down the towpath from the Great Falls visitor center. The trail maps warn that this strenuous, physically demanding trail involves climbing and scrambling over angled rocks and large boulders. The warning is true.

If you are approaching from Great Falls, you can first take an unmarked trail down to a side channel just below Olmsted Island and see good exposures of folded rock. Then, on Billy Goat A you can walk 0.1 mile on the north end of Bear Island and get a great view of Mather Gorge without any dangerous climbing. At this point you can turn and go back if you want to avoid rock scrambling.

The wild, rocky confines of Mather Gorge demonstrate the erosive power of nonstop water current.

If the gorge looks unusual for the American east, it is. Bear Island is one of the bedrock terraces that the Potomac River cut across about 35,000 years ago. The length of Mather Gorge along Bear Island is well over 1 mile.

There are three trail markers on Billy Goat A, designated TM. Between TM1 and TM2 are numerous potholes in the rock some 50 feet above the present river. These were formed when the river was running at the level of this terrace, before cutting the gorge. Water swirling in eddies with suspended sand and gravel bored these many pothole depressions, possibly during high-water discharge when the glaciers melted at the beginning of interglacial times. They are proof positive that the river flowed at this level.

At TM2 you can scamper over, through, and down some rock to see less dramatic views of the gorge but more dramatic exposures of rock folds and quartz veins. The many wavelike folds you see in the rock attest to the compressive tectonic forces at work in the convergent boundary that generated the ocean trench. Heat from deep burial and pressure from plate convergence were the forces responsible for such contorted shapes, probably mostly during the Taconic mountain building event.

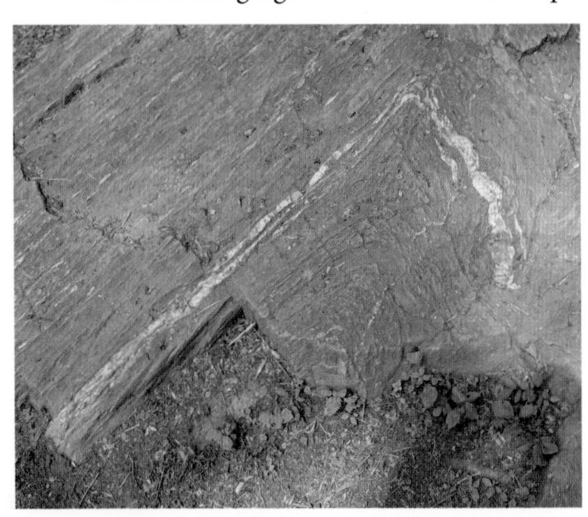

A symmetrically folded quartz vein in the Mather Gorge Formation.

You can also see quartz veins here, most of them folded. Quartz veins form when hot fluids from deep magma bodies dissolve silica and flow upward through fractures toward the surface. When the solutions in these hydrothermal vents approach the surface, they cool and solidify into glassy-looking quartz. The same convergence that produced the oceanic trench drove ocean crust so deep that it melted into magma and then was pressed upward to generate the Chopawamsic volcanic island arc. Dissolved in the magma-heated solutions was not only the silicon dioxide of quartz but also gold! (Note that "gold" is usually spelled "gold!") Several quartz veins within a few miles of here have been mined for gold!

Widewater

Widewater, from MP 12.62 to 13.45, is an abandoned meander of the ancient Potomac and lies well over 50 feet above present-day river level. The canal builders found that this ancient, elevated channel in bedrock of the Mather Gorge Formation worked well for part of their route. You can see folds in the rocks across the canal from the towpath.

The metagraywacke of the Mather Gorge Formation of Great Falls, Bear Island, and Mather Gorge (Billy Goat A) is exposed along the towpath down to just below MP 12, where it has eroded to show, in a tectonic window, the rock beneath the Plummers Island Thrust Fault. The rock of the Mather Gorge Formation was thrust over the younger diamictite of the Sykesville Formation along this fault. The Sykesville is exposed along the canal for about a half mile, and then the Mather Gorge resumes. At Highwalls, where another former river channel in rock was used for the canal, just above MP 11, you can look down far below and see today's river channel at Vaso Island, itself the remnant of a former terrace, as are most of the

Widewater, a wide stretch along the C&O Canal, was once a meander of the ancestral Potomac River. It is floored by solid rock and well above areas that might be flooded by the present river.

rocky islands in the 12 miles downstream from here. This is a good place to see where flowing water has cut into solid rock over thousands of years and established lower-elevation channels.

Billy Goat Trail, Sections B and C

Billy Goat B is a 1.4-mile-long trail of moderate difficulty that requires some minimal rock scrambling. You can see exposures of the Sykesville Formation in the middle of your hike and the Mather Gorge Formation in the more upstream portion. These ancient formations are covered with Quaternary alluvial (floodplain) deposits over much of this tract between the towpath and the river. Out in the river you can see that you are still in the rocky terrain of Potomac Gorge. All of the 1.6-mile-long Billy Goat C trail is on the Mather Gorge Formation and the overlying, recent floodplain. You can see good exposures and cliffs without walking too far on the upstream section of the trail.

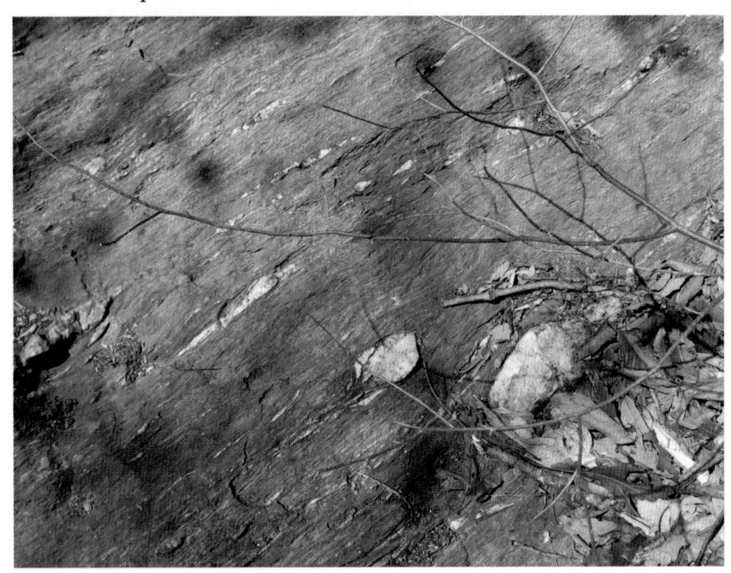

Heat and pressure from tectonic compression and burial have flattened these quartz clasts in the Mather Gorge Formation.

Plummers Island Thrust Fault

The Mather Gorge Formation, hidden beneath much younger Quaternary deposits, continues down to the Plummers Island Thrust Fault (also hidden) near MP 9.5. This fault was the surface along which the older Mather Gorge rocks were slowly shoved westward over the younger Sykesville during the Taconic mountain building event, a tectonic convergence in Ordovician time. The Sykesville mélange, containing a mix of clasts from the overthrusting Mather Gorge, runs all the way down to Georgetown. During Ordovician time it was intruded by plutonic masses of magma such as the Georgetown and Dalecarlia Intrusive Suites.

You can get a close look at the Sykesville Formation just east of Lock 12 at MP 9.3 by going several feet toward the river from the towpath on a dirt path. Here you are

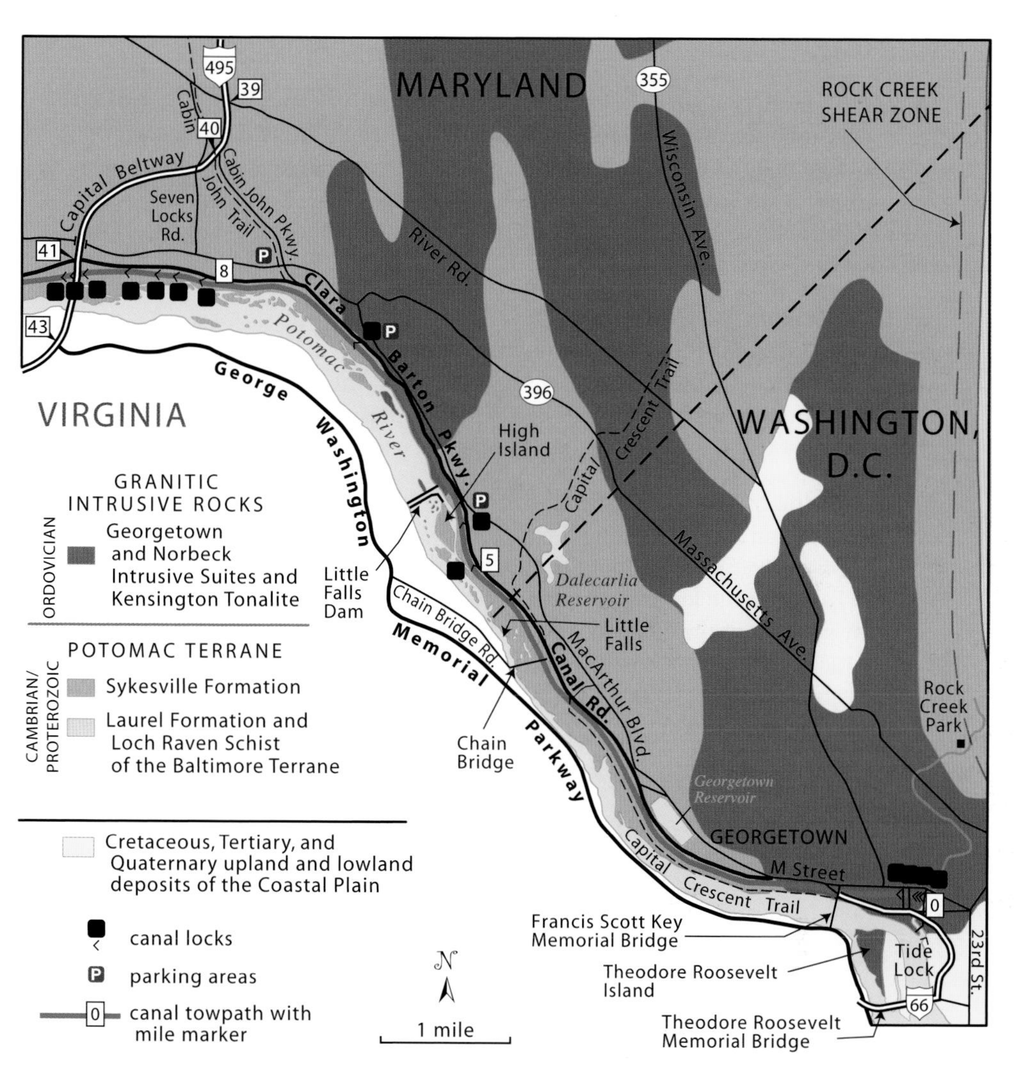

Geology along the C&O Canal between Georgetown at MP 0 and the seven locks area around MP 9. —Modified from National Park Service map; Cleaves, Edwards, and Glaser, 1968; Kunk and others, 2004; Southworth and others, 2001

not far from the Plummers Island Thrust Fault. In the outcrops of this metamorphic rock, you can see the planar foliation that resulted from the shearing, flattening stress applied millions of years ago by plate convergence. Nearby and overhead you can hear the continuous rush of traffic on the Capital Beltway, I-495.

Dalecarlia Intrusive Suite

The quartz veins near MP 8 are evidence that a hot, intrusive igneous body located nearby was heating groundwater to the point where it would dissolve silica. Imagine water hot enough to dissolve sand. After the hot solution flowed through fractures and cooled, it precipitated as quartz veins. At MP 6.5, near the cable ferry to

Sycamore Island, you can see a good outcrop of the Ordovician granitic rock of the Dalecarlia Intrusive Suite. Granite is a rock that crystallizes slowly underground from a large magma body, and the heat from this particular granite is probably responsible for the quartz veins near MP 8. Molten rock that solidifies underground is called intrusive igneous rock, and here the granite of the Dalecarlia Intrusive Suite intruded the Sykesville Formation.

Little Falls Pumping Station

At MP 5.8 Little Falls Pumping Station draws water from a dammed section of the river to supply about a fourth of Washington's water, the remainder coming from above the dam just upstream from Great Falls. Just below the pumping station is a short path to a kayak run in a channel formed by rocky High Island.

A Tour of the Sykesville Formation

At MP 4.5 you can take a surviving concrete road once associated with a hydroelectric plant down to river level and out through boulders to a platform with a great view of Little Falls and the lower Potomac Gorge. This road is lined on either side with excellent samples of the Sykesville Formation, many containing clasts that reflect the chaotic mixing that occurred in the deep-trench mélange environment.

Just above Chain Bridge on Canal Road is a good exposure of the Sykesville Formation, accessible from the towpath via a ramp onto the bridge at MP 4.2. At the Boathouse at Fletchers Cove, MP 3.2, you can easily switch over to the paved Capital Crescent Trail if you wish. This trail descends slightly onto a younger and lower terrace, or former river bottom.

Fragments of the Mather Gorge Formation were mixed into the Sykesville Formation when deep ocean-trench compression and thrusting occurred.

Georgetown Area

From the intersection of Foxhall and Canal Roads, which is visible from the towpath at about MP 1.6, you can see the igneous rock of the Ordovician Georgetown Intrusive Suite beside Canal Road where it runs east toward Key Bridge at Georgetown. This igneous pluton intruded the Sykesville Formation.

Passing through Georgetown on the towpath, you may encounter foot traffic, but continue to Lock 1 and find a large slab of Sykesville rock with a bronze plaque on it. This boulder clearly shows the rock fragments derived from the older Mather Gorge Formation imbedded in a sandy matrix. Continue on to the paved bike path adjacent to Rock Creek Parkway and then turn toward the river and Thompson Boat Center. You are now on the Coastal Plain geologic province. Cross the paved area in front of the boathouse and find the Tide Lock, MP 0, and the confluence of Rock Creek and the Potomac—all beneath the Watergate Hotel of Nixon-era infamy.

Across the Potomac you can see Theodore Roosevelt Island. With Piedmont rocks at its upstream end and unconsolidated sediments downstream, it marks the boundary of the Piedmont and Coastal Plain Provinces.

Theodore Roosevelt Island National Memorial

The 88-acre Theodore Roosevelt Island was formed by deposition of stream sediment around bedrock due to the slowing of the current. Sediment carried by the Potomac River and by Rock Creek, which enters the Potomac east of the island, falls out of

Piedmont rock—diamictite of the Sykesville Formation—marks the upstream end of Theodore Roosevelt Island. The downstream end is swampy, unconsolidated deposits of the Coastal Plain. The Watergate Hotel is in the distance.

suspension where the currents meet and slow, and also where rapid currents from Potomac Gorge are slowed by incoming tides. The north end of the island is right on the Fall Line and has a core of Piedmont bedrock. A couple of miles of trails cover this island, some through marshes and muddy areas. You can see the Piedmont rocks from the north ends of the park's Upland Trail and Swamp Trail. Across the Potomac you can see the mouth of Rock Creek on the Fall Line, just upstream from the Watergate Hotel. The parking area for Theodore Roosevelt Island is accessible only from the northbound lanes of George Washington Memorial Parkway in Virginia, across the Roosevelt Bridge from D.C.

Cabin John Regional Park

The 9-mile Cabin John Trail follows Cabin John Creek from Montrose Road to MacArthur Boulevard and the Potomac River over metamorphic rock of the Potomac Terrane. Most of Cabin John Creek upstream from Locust Grove Nature Center at Democracy Boulevard is in the Mather Gorge Formation, while most sections downstream from there are in the Sykesville Formation. You can access most sections of the trail from Seven Locks Road. You can get a good sense of the descent of the creek down the Fall Line from Rockville to the Potomac by just following Seven Locks Road from Montrose Road down to MacArthur Boulevard.

Fast-moving water from street runoff can erode deep gullies, such as this one near the northern end of Cabin John Trail.

To reach the northern trail terminus, turn east onto Goya Drive from Seven Locks Road and follow it to its end, where there is room for a few cars to park. The creek and woods here are within sight and hearing of the mightily congested junction of I-270 and the Beltway. At the very beginning of the trail, you can see gullying that has resulted from rain and meltwater running rapidly over streets and through concrete pipe to discharge into a natural channel.

From Tuckerman Lane, which crosses the creek, you can walk up along the creek toward the campground to see examples of schist of the Mather Gorge Formation. To see floodplains, natural bank erosion, and more exposures of the Mather Gorge, continue east on Tuckerman Lane and turn right (south) on Westlake Drive. Turn in at Shirley Povich Field and drive past baseball fields to the last parking lot, where you can follow the blue blaze down to Diamond Circle.

Locust Grove Nature Center, off Democracy Boulevard, also provides access to the Cabin John Trail. Where Bradley Boulevard crosses the creek, you can park and walk north 50 to 100 yards and see excellent exposures of the Sykesville Formation.

The southern terminus of the trail is at a park and parking lot off MacArthur Boulevard just west of the Cabin John Bridge. From the parking lot you can walk down into the deep ravine cut by the creek and see exposures of the Georgetown mafic igneous rock that intruded the Mather Gorge and Sykesville Formations in Ordovician time.

The impressive Cabin John Bridge, now part of MacArthur Boulevard, was constructed in the years just before the Civil War. Today civil engineers consider it the finest existing stone arch in the United States. Maryland rocks used in the span include gneiss of the Sykesville Formation quarried nearby, Manassas (Seneca Red) Sandstone, and Port Deposit Gneiss (formerly called granite).

Rock Creek Park

Rock Creek Park, the fourth oldest national park, connects with Rock Creek Regional Park, a Montgomery County park. The parks occupy land on both sides of Rock Creek from Lake Needwood to Georgetown. A 23-mile-long, nonmotorized path runs the length of the parks. The creek descends over metamorphic and igneous Piedmont rock and flows into the Potomac River at the Fall Line in Georgetown.

Beach Drive through Rock Creek Park is the only nonurbanized, preserved natural passage from the Beltway to downtown. On the way you can see stream and rock formations from your car. However, between the D.C. line and Military Road, Beach Drive is closed from 7 a.m. Saturday to 7 p.m. Sunday and on holidays.

The Rock Creek Fault and Rock Creek Shear Zone run from just south of Rockville to just north of Georgetown and pass roughly along the western boundary of the park. The fault and zone separate the Sykesville Formation and its suites of plutonic intrusions on the west from the Laurel Formation on the east, both of the Potomac Terrane. Most of the rock visible along Rock Creek from the District line to Connecticut Avenue is diamictite of the Laurel Formation of the Potomac Terrane. The Laurel is like the Sykesville in that it was originally a sedimentary

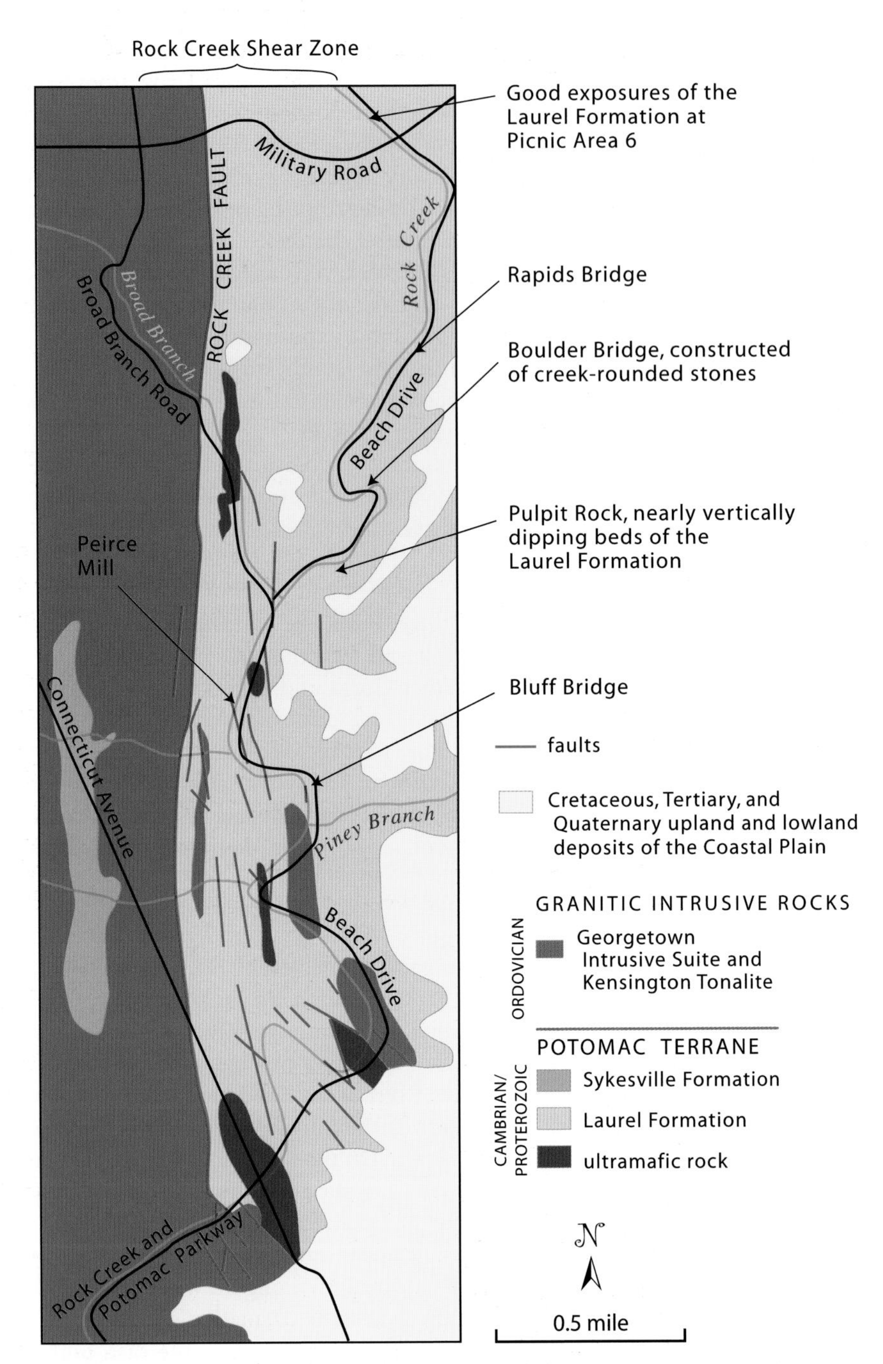

Rock Creek Park, which straddles the Rock Creek Shear Zone, contains Potomac Terrane rock intruded by Ordovician plutons, as well as some areas of Cretaceous Coastal Plain sediments. —Modified from Southworth and others, 2006

This diamictite of the Laurel Formation, exposed near Picnic Area 6 in Rock Creek Park, contains pebbles that have been associated with the Loch Raven Schist. The pebbles were mixed with sands and muds during underwater landslides. The pebbles were elongated when all of the rock was metamorphosed during Taconic mountain building or during shearing associated with the Rock Creek Shear Zone.

mélange, Cambrian or older in age, containing granitic plutons that intruded during Ordovician time. Picnic Area 6, just north of Military Road, is a good place to stop and walk down to the creek and see Laurel Formation rocks. On Broad Branch Road you can see Kensington Tonalite (a type of granitic rock) exposed along Broad Branch in the Rock Creek Shear Zone, but there is no good place to stop. The Rock Creek Shear Zone, an area wider than the fault, was subjected to pressures of slippage. The fault and zone here are not thrust faults like the Plummers Island, Pleasant Grove, or Martic Faults of the Piedmont. Instead, they are strike-slip faults in which the movement, or slip, is lateral rather than vertical. A well-known example of this type is the San Andreas Fault in California. During intrusion of the Kensington Tonalite in Ordovician time, the Taconic convergence caused lateral slippage to the left—from the viewpoint of the observer looking across the fault.

The rounded stones of Boulder Bridge have been tumbled in the waters of Rock Creek. Laurel Formation bedrock lies on the far bank.

Later, during the very early tectonic pressures of the Pennsylvanian Alleghanian mountain building event, the fault was reactivated but this time with right-lateral slippage. Rocks in this high-pressure shear zone were deformed and sheared.

Rapids Bridge, a footbridge with parking nearby, is a good place to stop along Beach Drive and have a close look at the creek. Also, look here for the Laurel diamictite. Boulder Bridge, made of creek-rounded rocks, is an interesting structure, but there is no place to park close to it. Below the bridge is Pulpit Rock of the Laurel Formation, with near-vertically dipping beds that reflect the forces of converging plates.

At Peirce Mill you can see a mill built in the 1820s with beautiful "blue granite," a stone quarried nearby, probably from one of the Ordovician intrusive rocks. The mill used water-powered millstones that were 4.5 feet across and weighed about 2,400 pounds to grind wheat into flour—geology and hydrology at work. Below the mill, southward from the adjacent National Zoo, there are no pulloffs or parking areas along Beach Drive, but Laurel Formation rocks are still visible along the road, with a good exposure below Bluff Bridge, just above Piney Branch.

At the Q and P Street overpasses, take note because you are close to the Coastal Plain here, and in colonial times Rock Creek was navigable to P Street. Clearing of land caused siltation of these lower stretches well before 1800. When you emerge from the Rock Creek valley and can see the Potomac River, you have passed from the Piedmont over the Fall Line to the Coastal Plain.

Capital Crescent Trail

The Capital Crescent Trail, once a railroad and now an 11-mile paved trail for nonmotorized uses, climbs gently over 300 feet from the Coastal Plain at Georgetown to the Piedmont at Silver Spring. Most of the upper end, mileposts 0 to 5.0, passes through developed areas. From MP 7.0 down to the old railroad bridge over Canal Road and the C&O Canal, just below MP 7.5, there are many good exposures of the Sykesville Formation. You can stop at the fence in this section and look down at Little Falls. Here, if you want to climb down and explore, you can see large, clifflike exposures in what looks like an old quarry. Below here, along the paved trail, is a large exposure. From the railroad bridge down to Georgetown, the trail is level and parallels the C&O Canal Towpath. Near Foxhall Road, MP 9.5, the trail descends to a former river terrace, one that is younger and lower in elevation than the one through which the higher canal is partially excavated at this point. At the Boathouse at Fletchers Cove, near MP 8.0, you can gain easy access to the trail and usually find plenty of parking. There is also an exposure of the Sykesville Formation on the road down to the underpass.

Sligo Creek Parkway

The Sligo Creek Parkway runs along Sligo Creek from University Boulevard (MD 193) in Wheaton down to New Hampshire Avenue (MD 650) in Takoma Park. The slope is gradual, and there are many exposures of the metamorphic schist and gneiss of the Laurel Formation in the creek and along the road, more on the lower end. The Fall Line is about 1 mile downstream from the southern end of the parkway. Sligo Creek joins the Northwest Branch of the Anacostia River in Chillum Park on the Coastal Plain. A paved trail runs the length of Sligo Creek from Wheaton Park, Shorefield Road Entrance (off Georgia Avenue—MD 97) to the creek mouth at Northwest Branch, where it joins the Northwest Branch Trail. The parkway will provide you with some views and access, but the streamside trail will get you much closer to the rocks and the water.

Wheaton Regional Park and Northwest Branch Trail

In Wheaton, from Georgia Avenue (MD 97), follow Randolph Road to the east and then Glenallan Avenue to the Brookside Nature Center parking lot, where a half-mile park trail will connect you to the trailhead for Northwest Branch Trail. You can also park on Kemp Mill Road below the intersection with Glenallan and be right at the trailhead. The trail follows the meandering Northwest Branch of the Anacostia River, which in places exhibits the classic erosional cutbank on the outside of a bend and the depositional bar on the inside. A little over 2 miles down from the trailhead is an excellent outcrop of highly folded schist of the Laurel Formation.

You can join the trail about 7 miles downstream at Adelphi Mill on Riggs Road (MD 212) in Langley Park. About a half mile downstream from Adelphi Mill, under some power lines, you come to the Coastal Plain, and you can sense the change in terrain. The stream banks no longer exhibit rock faces but are now composed of the unconsolidated sands and gravels of the Coastal Plain. This border of geologic provinces marks an unconformity, or gap in the rock record, of about 400 million years. This time gap in the rock record was created in Cretaceous time when gravels and sands eroding from the young, tall Appalachian and Piedmont Mountains were laid down onto the subsiding older rocks. The trail continues to Bladensburg, the colonial head of navigation on the Anacostia River. Runoff from farm fields silted up the river, making it too shallow to navigate.

COASTAL PLAIN SITES

The Mall and West and East Potomac Parks

When the site of the city of Washington was proposed, Tiber Creek, a tributary of the Potomac River, drained 2,500 acres—over 40 percent of the District area—and emerged onto the Coastal Plain lowlands near the western end of Capitol Hill, in the area that is today the Mall. Between present-day First Street and Seventeenth Street, the tidal Tiber Creek overflowed into marshes that were several hundred feet wide. Also known as Goose Creek, the Tiber met the Potomac at a bluff called Braddock's Rock, near Seventeenth Street and Constitution Avenue

In 1815, with the White House and the Capitol in place, the Washington City Canal was built to channel the flow of Tiber Creek. The canal ran along the north edge of the Mall (Constitution Avenue today) from the Potomac River at Seventeenth Street to Sixth Street, where it turned south for half the width of the Mall and there turned again toward the Capitol and then ran up to Third Street in front of the Capitol, where it turned south again and ran down to the Anacostia

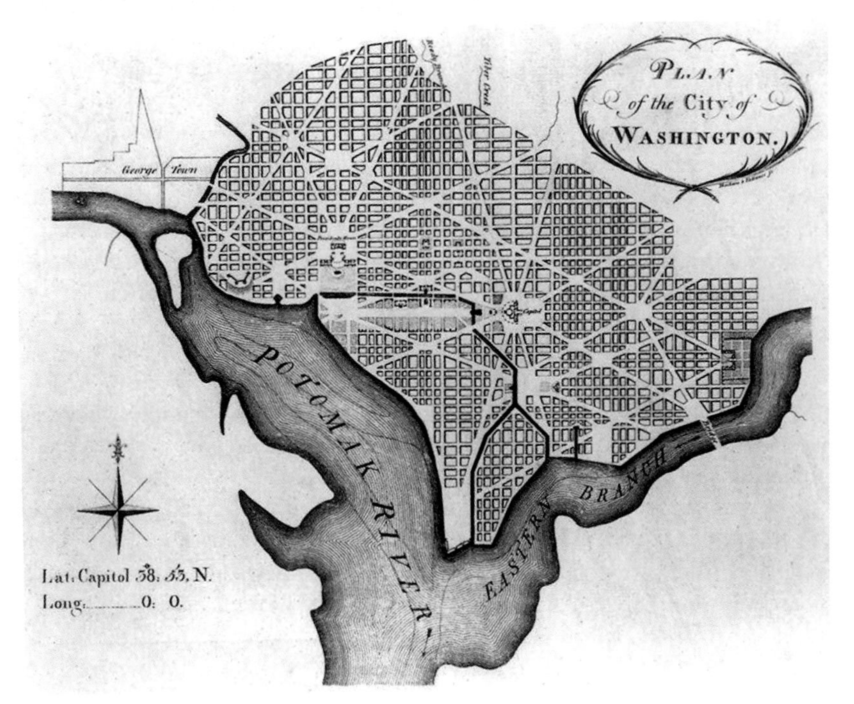

1792 map of Washington, D.C., with plan of the Tiber Canal, which was opened in 1815. —Ellicott, 1792

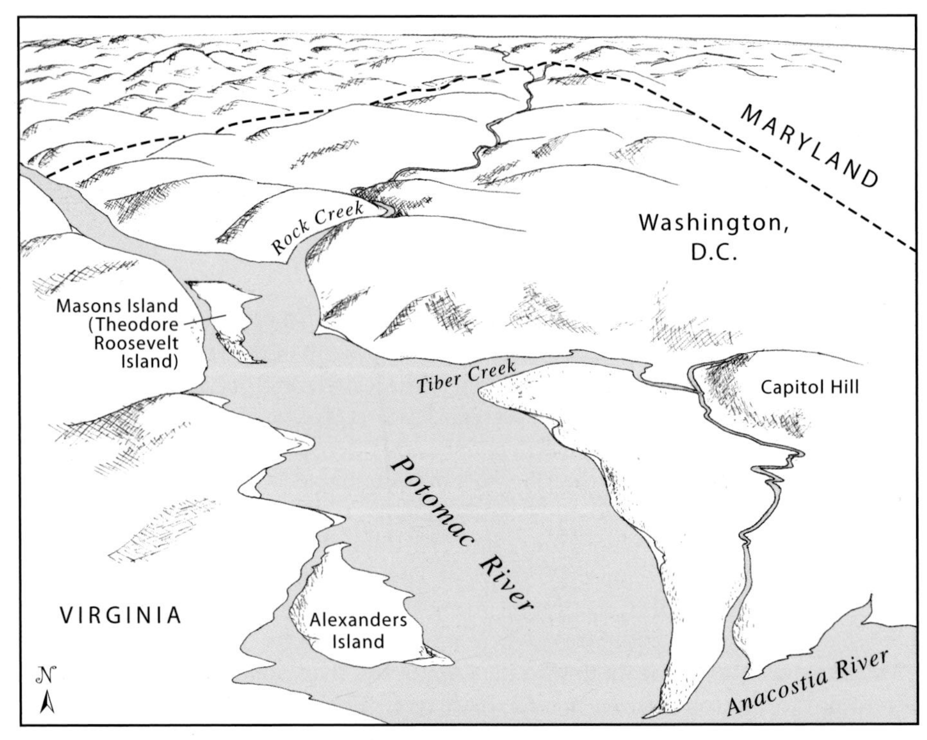

Predevelopment D.C. was a meeting of Piedmont highlands, marshy Coastal Plain, and tidal rivers and creeks. —Matt Moran sketch

River, then called Eastern Branch. This Tiber Canal was designed to control spring floods and provide a thoroughfare for passengers and freight, but by the 1860s it had become a mosquito-infested open sewer. In 1872 it was covered over and the outlet of the original Tiber Creek redirected to the remaining canal passage that connected to the Anacostia River. Thus, the sewage drainage was diverted from the Mall area between the Capitol and the Washington Monument, and the covered section became B Street, later renamed Constitution Avenue.

The Potomac River still flowed close to the Washington Monument and the White House, and in 1881, after severe flooding threatened this area, Congress appropriated funds to dredge the river channel and reclaim the marshy tidal flats by the riverbanks. Over a twenty-year period, dredging and dumping created more than 600 acres of land above flood and high-tide levels: West and East Potomac Parks. Washington Channel behind East Potomac Park created a new harbor, and the Tidal Basin was equipped with gates that collected water at high tide and released it at low tide to flush the harbor.

If you drive along the Potomac River in either East or West Potomac Park (Ohio Drive from Lincoln Memorial) or if you drive down to Hains Point to see where the Anacostia joins the wide Potomac, remember that the land you are on is not the kind of island that is often formed where the currents of two rivers meet, decrease in velocity, and drop sediment out of suspension. This land is sediment dredged from the river bottom by the U.S. Army Corps of Engineers in Holocene time.

The tallest rock in D.C. is the 555-foot, 5⅛-inch Washington Monument, with its 897 steps. It is constructed out of marble quarried in Maryland. The land in the foreground used to be part of the Potomac River.

National Museum of Natural History

The National Museum of Natural History, on the Mall about halfway between the Washington Monument and the Capitol, is located right where the wide, tidal Tiber Creek used to run. If you want the informational equivalent of several college-level geology courses, you could spend a couple of tuition-free weeks here viewing and absorbing the excellent exhibits in Dinosaur Hall on the first floor and the Janet Annenberg Hooker Hall of Geology, Gems and Minerals on the second floor. Sections on these two floors include Fossils, Historical Geology, Ice Age and Dinosaur Skeletons (with a *T. rex*!), Meteorites (the world's largest collection), Plate Tectonics with Volcanoes and Earthquakes, Rocks Gallery, Building Stones, Ores and Mining, and the Minerals and Gems Gallery, which includes the National Gem Collection.

You can touch one of the oldest rocks ever found—a 3.96-billion-year-old piece of gneiss from Northwest Territories, Canada. It was metamorphosed from granite in the earliest continental crust. You can see an incredibly colorful and large collection of rocks and minerals. You can see a section of rock layers from Raton, New Mexico, that contains the light clay layer with quartz and iridium that is the Cretaceous-Tertiary (K-T) boundary, material thought to be deposited from the skies when a meteorite struck the earth 66 million years ago, an event that many geologists think began the demise of the dinosaurs. You can see samples of the sandstone used for building the White House and the Capitol. You can see a 107-pound, flawless sphere of polished, clear quartz. Overall, it is an amazing place.

Kenilworth Park and Aquatic Gardens

D.C.'s last tidal marsh is the partially restored, seminatural, 78-acre Kenilworth Marsh. Off the Eastern Avenue exit of DC Route 295 (Kenilworth Avenue) you go south on the service road to Douglas Street and then west a couple of blocks to Anacostia Avenue. If you are coming from the south on 295, you will need to make a U-turn after the exit. This area contains numerous ponds of flowering water plants, an aquatic garden started as a hobby in the nineteenth century by a U.S. Treasury Department clerk. A boardwalk leads through marshes and open water, and the River Trail goes through restored freshwater marshes to the polluted, tidal Anacostia River. Here you are across from the hills of the National Arboretum and 7 miles upstream from the confluence of the Anacostia with the Potomac. In spite of the distance, this river still rises and ebbs by about 3 feet with the tide. At one time it was navigable not only to here but upstream to Bladensburg.

Bladensburg Waterfront Park and the Anacostia River

The story of the Anacostia River, D.C.'s forgotten river, is tragic. Four hundred years after John Smith called it a "crystal" river, the Anacostia is considered one of the most polluted rivers in the nation. Draining most of D.C. and two adjacent Maryland counties where over 800,000 people reside, the river receives about 1.5 billion gallons of untreated wastewater per year. One-quarter to one-half of the drainage basin is covered with impervious roofs and pavement. Rain that would soak into an undeveloped, vegetated landscape runs rapidly from these hard surfaces and causes flooding that not only gathers sediment and toxic pollutants along the way but also overwhelms the leaky, one-hundred-year-old sewage systems.

The Bladensburg Waterfront Park, site of an eighteenth-century port, is off MD 450 (Annapolis Road) and Baltimore Avenue on the floodplain of the Anacostia River. Before colonization, Native Americans lived along this river, which teemed with fish and was bordered by over 2,000 acres of tidal wetlands. Today there are

During colonial times, Bladensburg was a major port, with the channel 40 feet deep. Today the Anacostia River is 4 feet to 4 inches in depth, and the former port is a sandbar.

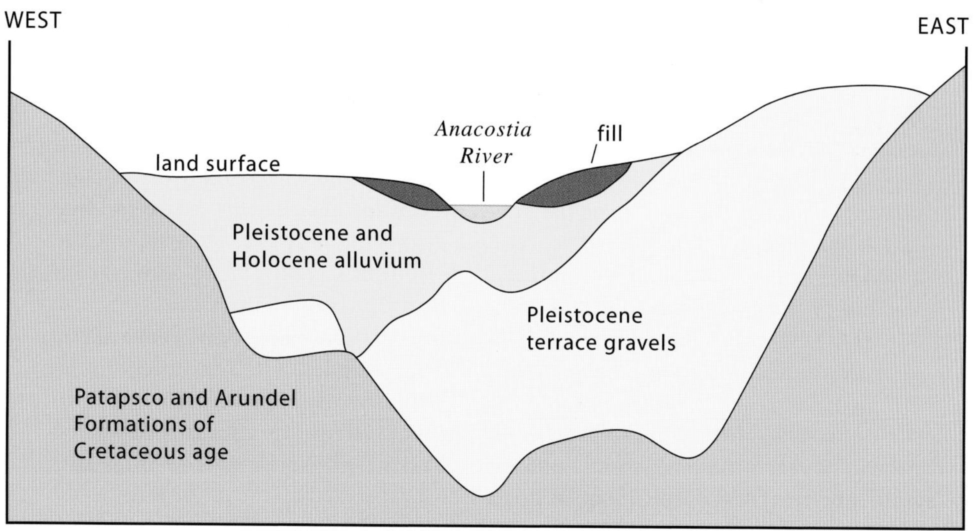

The Anacostia River flows over unconsolidated sediments of the Coastal Plain. —Modified from Tenbus, 2003

fewer than 200 acres. By the early 1740s, Bladensburg was one of the major seaports of the colonies, shipping tobacco and cotton to England, but by the mid-1800s, huge amounts of muddy runoff from large areas of plowed fields had filled the river channel with silt. With river depth reduced from 40 feet to less than 10 feet, seagoing vessels could no longer navigate here. Ironically, the port was rendered useless by uncontrolled erosion from fields of the very crops that had made the port so profitable.

During the first half of the twentieth century, it was not bare crop fields but urbanization and the increase in impervious surfaces that caused problems: pollution, sedimentation, and flooding. Big floods in 1933, 1942, and 1953 covered roads in Bladensburg with up to 8 feet of water. In the 1950s the U.S. Army Corps of Engineers constructed 3.5 miles of levees on both sides of the river. In the park you can see these levees or flood walls. At the park and in its visitor center, plaques tell the story of the history of this river.

Recent coring by the U.S. Geological Survey into the river bottom of the Anacostia revealed a depositional history of Cretaceous nonmarine sands and clays, Pleistocene terrace gravels (from high-energy glacial meltwater discharge), Holocene alluvium, and recent urban silt and clay—including one sample dated by the presence of shards of a broken windshield. (About 20,000 tons of trash wash into the Anacostia annually.) These findings represent a revealing archaeological dig into the climatic and human history of the Anacostia watershed.

Another plaque at the park describes Dinosaur Alley, a corridor roughly along US 1 between D.C. and Baltimore where about a dozen categories of dinosaur bones have been found. Teeth from a dinosaur eventually named *Astrodon johnstoni* were found in a hand-dug iron-ore pit north of Bladensburg near Muirkirk in 1858. Later to become Maryland's official state dinosaur, this sauropod was more than 30 feet tall and over 50 feet long.

Lake Artemesia Park and Greenbelt Park

A little farther up in the drainage basin of the Anacostia River lie a couple of parks. To reach Lake Artemesia Park from Capital Beltway exit 23, drive south on Kenilworth Avenue (MD 201) and turn west on Greenbelt Road. The park is located in Quaternary sand and gravel floodplain deposits of Paint Branch and Indian Creeks, tributaries of the Northeast Branch of the Anacostia River. A huge quarry was located here to mine the gravel needed to construct the Metro rail line visible just to the west of the park. Because quarrying near the construction site saved millions of dollars, Metro agreed to develop this park. The 38-acre lake on the quarry site and several hiking and biking trails are the result.

The 1,100-acre Greenbelt Park, operated by the National Park Service, has good drives and almost 10 miles of foot trails over wooded hills and valleys in the Potomac Group of the Coastal Plain. Some of these hills have surprisingly steep slopes. In streambeds you can see the sands and gravels of this alluvially deposited group. To reach the park from I-495 exit 23, take Kenilworth Avenue (MD 201) south and turn east onto Greenbelt Road.

National Arboretum

The National Arboretum occupies over 400 acres on rolling hills in one of the higher areas of D.C. It is in the Cretaceous alluvial plain sediments of the Potomac Group and bordered on the east by the marshy Anacostia River. An outdoor museum of trees and plants, it is accessible on 10 miles of roads and footpaths in a Coastal Plain upland. The park entrance is off New York Avenue (US 50) just inside the District line.

Behind the Administration Building, in the National Bonsai and Penjing Museum, you can see a variety of the miniature, artistically trained bonsai ornamental trees, some of them hundreds of years old. But you can also see some very interesting rocks. On exhibit are numerous viewing stones or scholar's stones—naturally shaped, unworked stones that in miniature resemble natural landforms such as mountains, waterfalls, islands, or whole terrains. The art of finding and presenting these stones, called *suiseki* in Japanese, is centuries old in Japan and China, and the stones often accompany bonsai trees.

Fort Circle Trail

In Anacostia, D.C., the Fort Circle Trail connects the sites of five forts and fortifications built during the Civil War to prevent the Confederate Army from taking the Anacostia highlands, from which they could have bombarded Washington. The forts, which no longer exist, were built on hills of the gravelly and sandy Potomac Group. The trail passes through parks and neighborhoods. Your best bet is to drive on different routes through ravines and along hills and catch glimpses of D.C. far below, to the west. From Fort Stanton Park, you can get a spectacular view of Washington and understand why the Union Army wanted to defend this high ground—very high land for a province known as the Coastal Plain.

Looking north from Fort Stanton, you can clearly see how high some of the Coastal Plain can be compared to the former lowland swamp that is now D.C. Note the hills of the Piedmont north of D.C.

Watkins Regional Park

Watkins Regional Park is on very high, hilly ground in the sandy Aquia Formation, a marine formation laid down when sea levels were very high in Tertiary time. The park has a nature center and ten trails that pass through forests, uplands, wetlands, and meadows, and onto areas of panoramic view. Take Central Avenue (MD 214) east and turn south onto Watkins Park Drive to get to this park.

Baltimore and the Baltimore Beltway Area

In general, the geology of Baltimore resembles that of Washington, D.C., in that about half of the area inside the beltway of each city is metamorphic, crystalline rock of the Piedmont Province and the other half is unconsolidated sediments of the Coastal Plain Province. This resemblance is mirrored in human history because each city grew outward in a roughly concentric manner from the head of navigation at the Fall Line on a major tidal river. Like D.C., hills ring Baltimore—Piedmont hills to the north and west and Coastal Plain hills to the south and east.

Much of western Baltimore City and southern Baltimore County is a large expanse of the Baltimore Mafic Complex, which is thought to have intruded beneath the Chopawamsic island arc and possibly into the other terranes of the Piedmont in Cambrian time. The terranes and intruded mafic magmas were later thrust upon the continent and metamorphosed during the Taconic mountain building event in Ordovician time.

Through central and western Baltimore County are found folded, overturned nappes, or domes, of the Baltimore Terrane. When fragmented sections of crust (massifs of 1.1-billion-year-old Grenville gneiss, known as Baltimore Gneiss in Maryland) covered with ocean-bed sandstone, shale, and limestone were metamorphosed and folded during the Taconic mountain building event and shoved into and onto the continent, the wavelike folds were overturned to a recumbent position and the tops of the folds thrust faulted horizontally from the overturned limbs. These structures, called nappes, form only during the intense pressures of a tectonic collision. Between D.C. and Pennsylvania are found over a dozen of these nappes, collectively known as the Baltimore Terrane. Over a half dozen of them lie near Baltimore.

The tops of the Domes have been eroded over hundreds of millions of years to reveal the 1.1-billion-year-old Baltimore Gneiss, the ancient Grenville basement rock, at the core. Today the gneiss of the domes is surrounded in outward succession by surrounding bands of quartzite of the Setters Formation, Cockeysville Marble, and Loch Raven Schist—together known as the Glenarm Group—giving most domes a somewhat oval bull's-eye appearance on a geologic map. The different rates of erosion of the gneiss and of the surrounding formations are responsible for many of the rolling hills and valleys of the Baltimore area northwest of I-95—the hard, resistant gneiss forming hills, and the more easily eroded marble forming valleys.

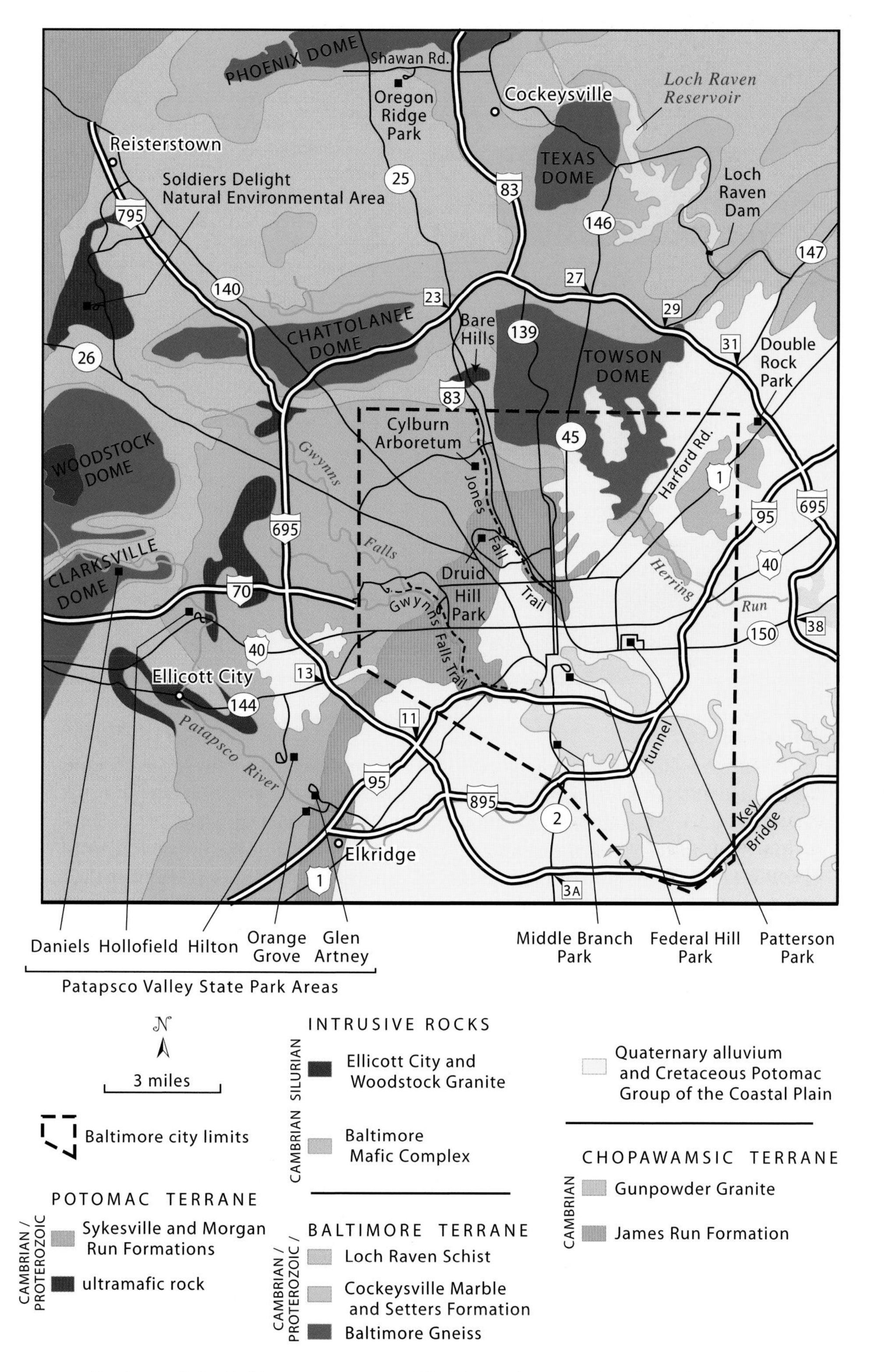

Geology of the Baltimore area. —Modified from Cleaves, Edwards, and Glaser, 1968; Crowley, Reinhardt, and Cleaves, 1976

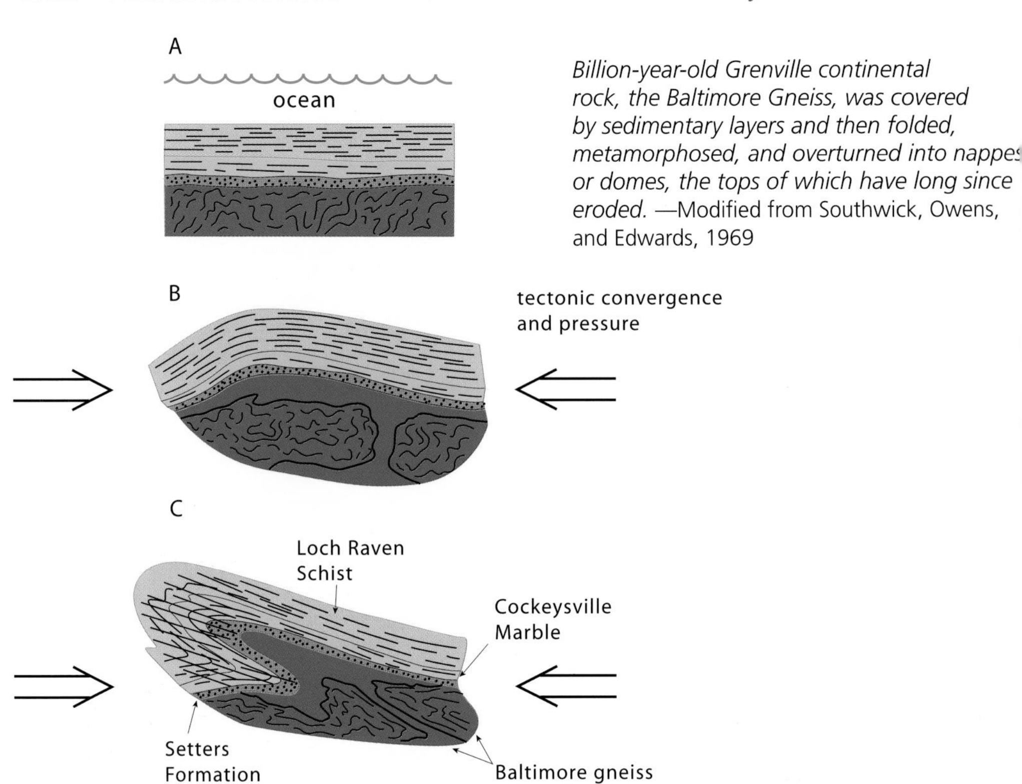

Billion-year-old Grenville continental rock, the Baltimore Gneiss, was covered by sedimentary layers and then folded, metamorphosed, and overturned into nappes or domes, the tops of which have long since eroded. —Modified from Southwick, Owens, and Edwards, 1969

In a band adjacent to the Coastal Plain deposits is the Chopawamsic Terrane, here called the James Run Formation, the remnant of the volcanic island arc. It developed during Cambrian time above a midocean subducting plate.

In the Coastal Plain of east and southeast Baltimore, unconsolidated deposits lie on deeply buried, faulted Piedmont rock. Unevenness in this basement surface may be partly responsible for some of the Coastal Plain hills, such as Federal Hill south of Inner Harbor and Patterson Park in East Baltimore, but a more probable cause is stream dissection of these thick deposits. The unconsolidated sediments, which eroded from highlands to the west, are about 400 million years younger than the Piedmont rocks on which they lie. Streams such as the Patapsco River, Gwynns Falls, and Jones Falls tumble over rapids in the Piedmont rock and then slow at the Coastal Plain where they meet the tidal portion of the Patapsco River, which in the harbor area is more estuary than river. Federal Hill, which John Smith called "a great red bank of clay flanking a natural harbour basin," affords an excellent view of coastal and upland Baltimore.

I-695: Baltimore Beltway

You cannot, of course, see much rock from the beltway, but you can get a general sense of the many different formations responsible for the continual ups and downs in this heavily traveled highway. The ancient soils and rock have been covered and obscured by continuous urban development. In order to have a common reference for this circular road, we begin from the I-70 interchange and head north on the inner loop in a clockwise direction.

First, you travel gradually upslope over the crystalline schists of the Laurel Belt of the Baltimore Mafic Complex, the metamorphosed igneous rocks that intruded either beneath the island arc or into ocean sediments of the other terranes. These Laurel Belt formations, known as the Hollofield and the Mount Washington, are not visible in this highly developed area.

North of the Northwest Expressway, I-795, the Baltimore Beltway climbs more abruptly past the MD 140 exit (Reistertown Road) as it ascends the Grenville basement rock known as the Baltimore Gneiss. The beltway is now on the Chattolanee Dome of the Baltimore Terrane. At the Park Heights overpass, the beltway descends the eastern portion of the dome and, just east of the Greenspring Avenue exit 22, the slope steepens as you cross from the upthrown block of a fault to the downthrown block and leave the hard gneiss basement rock of the dome.

Continuing downslope past the I-83 South exit, the beltway hits bottom at the Joppa Road overpass, which is in the Cockeysville Marble (metamorphosed limestone). This rock weathers rapidly, forming valleys. Jones Falls has downcut into the Cockeysville of Green Spring Valley between two dome cores.

West of the I-83 North exit, the beltway crosses another fault from the downthrown to the upthrown block. Between the Charles Street exit and the York Road exit, the beltway travels upslope onto the sands and gravels of the Potomac Group, a Cretaceous Coastal Plain deposit located far to the west of most other Coastal Plain sediments—perhaps an area where an ancient sea backflooded between domes during a high stand of sea level. At York Road the beltway heads downhill again with varying changes in steepness past the Dulaney Valley Road exit and the Providence Road exit to the valley bottom at the Cromwell Bridge Road exit. The valley bottom is occupied by the less resistant Cockeysville Marble, which weathers more rapidly than the surrounding gneisses. The natural acidity of precipitation dissolves the marble over time.

Between exits 29 and 30, the beltway climbs up and over gneiss of the Towson Dome. At exit 30 (the Perring Parkway exit) the beltway crosses onto Coastal Plain deposits of the Patuxent Formation of the Potomac Group.

The beltway continues south across various formations of the Cretaceous-age Potomac Group of the Coastal Plain. There are ups and downs on the way, but generally the elevation drops as the beltway approaches Francis Scott Key Bridge over the Patapsco River, here a tidal arm of Chesapeake Bay. East of the beltway on the north side of the river is low-lying land that did not appear on colonial maps. The erosion of topsoil after European colonization caused the Patapsco and Back Rivers to carry more sediment, which they deposited at their mouths. North Point State Park and Fort Howard Park occupy this new land.

South of the Key Bridge, the beltway is still on the Coastal Plain deposits of the Potomac Group. Looking west you can see tank farms and harbor facilities of the Port of Baltimore. The beltway continues across formations of the Potomac Group between the Key Bridge and exit 12A. West of exit 12A, the beltway heads onto rocks of the Piedmont, specifically for less than 1 mile over amphibolite of the James Run Formation of the Chopawamsic Terrane, remnants of the ancient volcanic island arc, and then between exits 12B and 13 onto Baltimore Mafic Complex amphibolites. The climb here onto the Piedmont is steady and noticeable—past Wilkens Avenue, Frederick Road, and Edmonson Avenue to US 40. This Piedmont rock, from exits 12B to 13, is the Mount Washington amphibolite and then, just south of the I-70 interchange, the Hollofield ultramafic rock.

If you drive the Baltimore Beltway counterclockwise, on the outer loop south and east of the US 40 exit you will see before you the long, dramatic descent down the Fall Line to Southwestern Boulevard.

PIEDMONT SITES

Patapsco Valley State Park

From its source on Parrs Ridge just south of Mount Airy, the Patapsco River flows more than 50 miles to Baltimore harbor, where it becomes an arm of Chesapeake Bay. In its last 15 miles to the Fall Line, from Woodstock to tidewater at Elkridge (a port that once rivaled Baltimore until it silted up), the Patapsco meanders through a gorge and drops over 200 feet. During the seventeenth, eighteenth, and early nineteenth centuries, rapids along the river were a source of water power for many mills, forges, and textile manufacturers.

Along this stretch and farther upstream is Patapsco Valley State Park. With numerous areas accessible from I-70, I-695, and I-95, the park has dozens of well-maintained trails along the river and on ridges above it, and through hills and valleys of river tributaries. Many of these trails will take you to excellent rock outcrops of the different terranes and igneous intrusions that make up the eastern Piedmont.

Hilton Area

I-695 exit 13 for Frederick Road (MD 144 West) will take you to South Rolling Road (MD 166) and the Hilton Area of Patapsco Valley State Park. Hike down Forest Glen Trail through a stream valley where you can see excellent exposures of Mount Washington amphibolite, a metamorphosed iron- and magnesium-rich igneous rock of the Baltimore Mafic Complex of Cambrian age. Here it weathers to a black soil.

At the railroad arch over this stream, turn north and follow Buzzards Rock Trail to the overlook. Here, far below you can see the incised valley of the Patapsco River and the remnants of the 1907 Bloedes Dam, the world's first internally housed hydroelectric dam, which had turbines located inside the core underwater. In fact, the state park was established in part to maintain forested slopes in order to reduce runoff of silt, which clogged the turbines.

At Buzzards Rock look closely for pegmatite, veins of a light gray to pinkish, coarse-grained granite that intruded the amphibolite. The large mineral crystals in the pegmatite are pink and light gray feldspars, glassy quartz, and black, sheety muscovite. Pegmatite crystallizes from a watery residual magma under high pressure when a slow-cooling granite pluton is almost totally solidified. In this area it forms

From Buzzards Rock of Mount Washington amphibolite, you can see the Patapsco River cascading over Bloedes Dam, which once generated electricity for manufacturing.

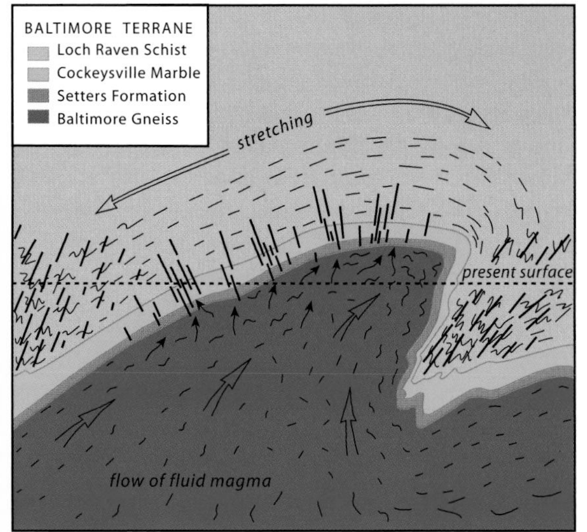

Tension fractures (heavier lines perpendicular to layers and stretching) *in the folded nappes of Baltimore Gneiss were conduits for fluid magmas that intruded the surrounding country rock and cooled, crystallizing slowly into pegmatites.* —Modified from Cloos and others, 1964

tabular intrusive bodies that have oozed through tension fractures in the intensely folded rock layers on the outer curves of Clarksville Dome, a large-scale overturned fold, or nappe. The large, visible size of the minerals requires slow cooling of the magma and watery, low viscosity, which allows atoms to move easily toward growing crystals and add to their size. Look for the highly reflective surfaces of the pink and light gray feldspar crystals.

If you continue on the Buzzards Rock Trail past Hilltop Road, you will come to Ilchester Rocks, another overlook with more pegmatite. From here you can look down to the former site of a nineteenth-century mill town. You can follow the trail down to the river and Bloedes Dam, and return to Forest Glen Trail via a paved trail along the river.

Avalon, Orange Grove, and Glen Artney Areas

The Avalon, Orange Grove, and Glen Artney Areas of the Patapsco Valley State Park are located very close to and even under major interstate routes southwest of Baltimore. From I-695, take exit 10 onto Washington Boulevard South (US 1Alt S) or exit 12A onto Southwestern Boulevard South (US 1 S), or from I-95, take I-195 East to the Washington Boulevard exit. Signs for the park entrance are just west of the I-195 overpass on Washington Boulevard (US 1, here). As you approach the park entry gate, notice the unconsolidated dirt banks of the Patapsco River, which flows through the western edge of the Coastal Plain here. Between the entrance and the park areas, you will pass onto Piedmont rock. The Chopawamsic Terrane (rock of the James Run island arc), is exposed for only about 1 mile, and then, at Lost Lake or across the river about a half mile upstream from the only vehicular bridge over the Patapsco here, the Baltimore Mafic Complex is exposed.

Upon entering the park, the first stone you will see is the 1835 Thomas Viaduct, at 704 feet long the world's oldest multispan, stone arch railroad bridge. Built of granite quarried upstream in the Patapsco Valley (presumably the Ellicott City Granite), the bridge is still in use today and has never been closed. As you drive under I-95 on River Road, you may be able to see rock of the James Run Formation where a stream tumbles down from the north, but this is not a safe place to stop.

From the 1760s until the great flood of 1868 (with a flood crest of 40 feet above normal), Avalon was occupied by various kinds of foundries and iron works, all utilizing the nearby water power of the river. If you follow the signs to the Lower Glen Artney Area on the north side of the river, at Lost Lake, or when you drive under the Thomas Viaduct, you are very close to the contact between the Chopawamsic Terrane and Baltimore Mafic Complex. If you drive up Soapstone Branch, you can

In 1835, when the Thomas Viaduct was completed over the Patapsco River, many people believed that this first curved, multiarched, stone railroad bridge in the United States would collapse, but to this day it has never been closed. View to the south.

see exposures of James Run amphibolite opposite the stream. This rock was once part of a volcanic arc sitting out in the ocean, much like today's Aleutian Islands off the coast of Alaska. Tectonic convergence brought it here and metamorphosed it.

On the south side of the river, on the road to the Orange Grove Area, you can see cliffs of the Baltimore Mafic Complex, but this is not a safe place to stop. Your best bet for seeing rocks up close is to drive to Orange Grove, the site of a water-powered flour mill during the last half of the nineteenth century, and take the Cascade Falls Trail. You will need to walk uphill only a few hundred yards to see the falls and rock outcrops of the salt-and-peppery looking amphibolite of the Baltimore Mafic Complex and intruded pegmatite. If you investigate closely here, you can find some remarkably large crystals of quartz, pink feldspar, and mica in the granite pegmatite.

The swinging bridge at Orange Grove will give you a good perspective of the Patapsco River and its steep-sided valley. The predecessor of this bridge was carried away in the massive floodwaters of Hurricane Agnes in June 1972.

On Cascade Falls Trail, you can find rock of the Baltimore Mafic Complex (left) *and pegmatite* (right), *the rock that intruded it.*

Ellicott City

The Patapsco River passes through Ellicott City, an often flooded former mill town that still has many early granite buildings on its main street. The stone was quarried in the Ellicott City Granite, which is 420 million years old. You can see outcrops of

The Ellicott City Granite is exposed beside an Ellicott City sidewalk.

the granite along the sidewalks here. It intruded during the last stages of the Taconic mountain building event when, according to one interpretation, greatly thickened crust melted rock and generated the intrusion.

Hollofield Area

If you have time for only one stop in Patapsco Valley State Park, go to the overlook at the Hollofield Area off US 40 just west of where the highway crosses the river. This view, particularly in winter when the leaves are off the trees, will give you a good sense of the steepness of the Patapsco River valley. Don't miss the impressive stone masonry in the fireplace of the picnic shelter near the parking lot.

Here you can also take the approximately half-mile round-trip hike down to Union Dam and see the river and its clay and sand floodplain. The trail down through the woods offers many excellent outcrops of the dark amphibolite of the Baltimore Mafic Complex, which several hundred yards west of the park entrance has been thrust faulted onto the limb of the Clarksville Dome (or nappe) of the Baltimore Terrane—an area of intense tectonic compression.

You also can see excellent exposures of the amphibolite on the River Ridge Trail, which leads down another quarter-mile to the river. To find the trail, walk around the entry gate for the campground and down the paved road about a quarter-mile. There is a cliff-sized exposure along the railroad tracks (watch for the train).

The CCC-built fireplace at the Hollofield Area of Patapsco Valley State Park is an artful collection and assemblage of nearby rocks.

Daniels Area

The Daniels Area of Patapsco Valley State Park affords an excellent opportunity to see the tremendously contorted folding that occurred when tectonic forces created the Clarksville Dome. To reach this area, take Daniels Road off Old Frederick Road, just west of where it crosses the Patapsco River. Look carefully for the street sign for Daniels Road—there is no park sign for this geologically interesting area. At the river you will see one of the many Patapsco River dams. This one furnished four hundred horsepower for canvas and denim mills here and hydroelectric power for Baltimore.

Just below the dam at a broad, excavated area that looks like a former quarry, you might be able to see freshly fractured pieces of quartzite and schist of the Setters Formation, which originated from deposits laid down offshore in high-energy ocean waters on the ancient basement rock of Baltimore Gneiss. The Setters Formation is on the limb of the Clarksville Dome, one of the dozen nappes of the region. Look for exposures of the large-crystal pegmatite intrusions, which oozed through fractures in the gneiss and the overlying Setters Formation during the tectonic compressions

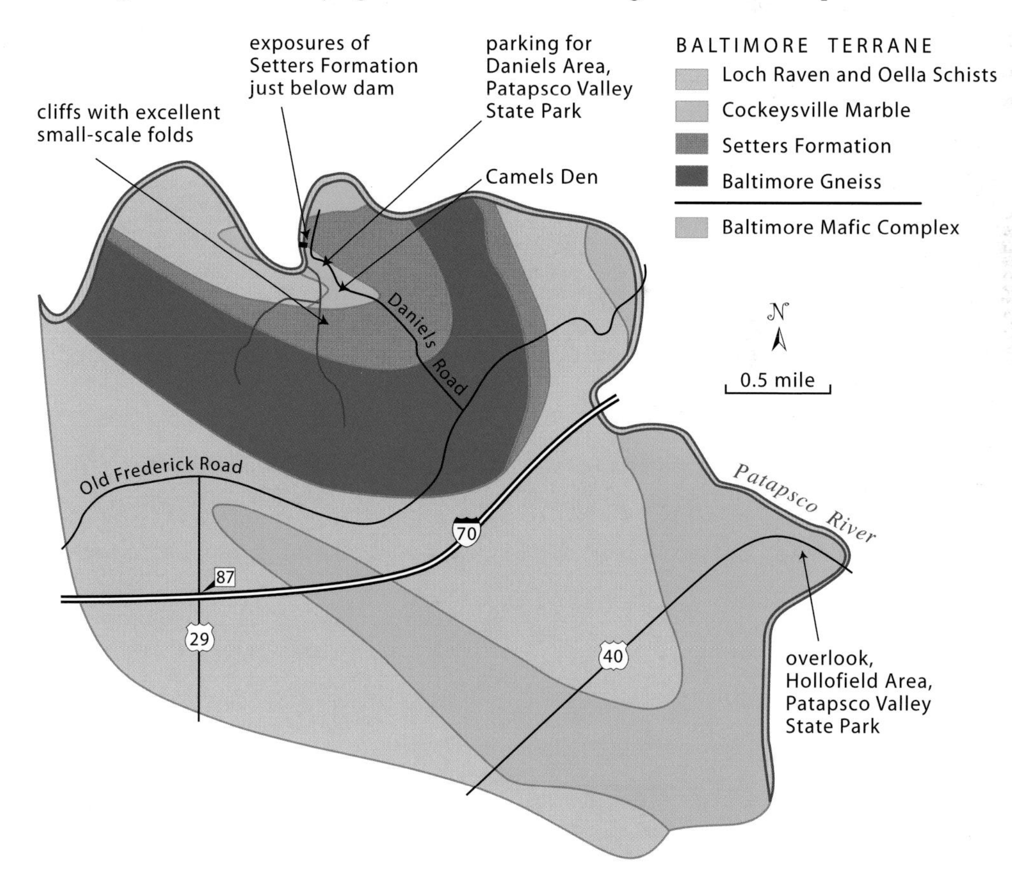

Daniels Area of Patapsco Valley State Park is a good place to explore the limb of the Clarksville Dome of the Baltimore Terrane. —Modified from Edwards, 1993

of the Taconic mountain building event in Ordovician time. Just below the open area is an excellent example of root wedging in rock just below the church.

Above the dam is a parking area and a sign for Camels Den Trail. Camels Den is a cave in the Cockeysville Marble, a carbonate rock which, like the limestones of many other caverns, is easily dissolved by groundwater. After the initial sign, the trail can be a bit difficult to follow. You can either cross the tributary creek and walk up the dirt road to where it joins Daniels Road, or you can walk up Daniels Road to where it meets the dirt road, at an "Authorized Vehicles Only" sign, only 0.1 mile either way. From where the two roads meet, you can see a trail with wooden beams as steps. Walk up about 100 feet and then down another set of steps to the cave.

With a little walking and exploring you can find some spectacular folds in schist and gneiss of the Setters Formation. Cross the small stream at Camels Den and turn left upslope a short distance, and you will see a trail off to the left and a wooden footbridge. Follow this and you will come to a stream crossing where a few iron bars are cemented into the rock. Cross carefully, go up the trail, and then look for the folds in the rock above the trail.

When you return over the footbridge to the main trail, you can turn left and hike uphill on switchbacks to the top of the ridge and the Loch Raven Schist, originally muddy sediments deposited on the Baltimore Terrane massifs. Here you can find some very thick quartz intrusions along the ridgetop and, during winter, a view of the river valley.

Camels Den is a cave weathered in a sliver of Cockeysville Marble in the contorted rocks of the Clarksville Dome.

Small-scale folds in the Setters Formation are indicative of the forces that were required to create the Clarksville Dome, a giant overturned fold or nappe.

Gwynns Falls Park

Gwynns Falls drops over 500 feet in about 20 miles from Reistertown to the tidal Middle Branch of the Patapsco River at Baltimore harbor, thus falling through the Fall Zone of metamorphic Piedmont rock to Coastal Plain deposits near sea level. You can see the intruding Baltimore Mafic Complex and the James Run volcanic rock of the Chopawamsic Terrane at Gwynns Falls Park along the stream's lower reaches in urban Baltimore.

To reach the park and Gwynns Falls Trail, follow I-70 East to its end and turn north on Security Boulevard. After a few hundred yards, turn right on North Forest Park Avenue and follow it to a right-hand turn onto Windsor Mill Road. You will first come to Leakin Park. Next you can turn into Carrie Murray Nature Center for information and maps of the park and trails. A half mile farther on Windsor Mill Road you cross Gwynns Falls and come to a parking area for trail segment T3. Here you can hike downstream on the wide gravel trail and see excellent exposures of the Mount Washington amphibolite, a metamorphosed rock of the Baltimore Mafic Complex. Down below, you can see the urban floodplain of the stream.

About 1 mile downstream from the Windsor Mill parking lot, the trail is paved, and several access points provide parking for walking or biking: Leon Day Park, T4, off Edmundson Avenue; Frederick Avenue, T5; trailhead at the Carroll Park Golf Course, T6; and Carroll Park off Washington Boulevard. Of particular interest is the half mile or so downstream from Leon Day Park. Above the trail are cliffs and below the trail are waterfalls. The rock here is gneiss from the volcanic James Run Formation of the Chopawamsic Terrane. In Carroll Park, the Piedmont rock meets Coastal Plain sediments down the slope from the Carroll Mansion, but you cannot see the contact because of the grassy surface of the park.

Foliations and fractures caused by metamorphism are well exhibited in the James Run Formation along Gwynns Falls.

Bare Hills and Soldiers Delight Natural Environmental Area

There are two large masses of serpentinite in Maryland, one at Bare Hills and the other at Soldiers Delight. This green ultramafic rock, part of the fore-arc basin deposits of the Potomac Terrane, is called the Soldiers Delight Ultramafic Complex and includes material from ancient oceanic crust and perhaps from as deep as the mantle. Serpentinite is low in silicon and high in magnesium and iron. The component mineral serpentine produces dry, nutrient-poor soil that blocks most plants' ability to absorb calcium, and only certain plants have adapted to grow in it. Because of rapid weathering, the soil layer is thin, and being low in clay it does not hold water well. In addition, some areas are high in toxic heavy metals, such as chromium and nickel.

The green serpentinite at Bare Hills along Falls Road may have come from ocean crust or possibly from as deep as the mantle.

At Bare Hills, along Falls Road (MD 25) just north and south of its intersection with Old Pimlico Road, you can see exposures of serpentinite, which was once mined for chromium here. The rock is recognizable because it is intricately fractured.

Soldiers Delight Natural Environmental Area has foot trails crossing 2,000 acres of rolling terrain in serpentine barrens. To reach Soldiers Delight, take Deer Park Road north off Liberty Road (MD 26).

Cylburn Arboretum and Jones Falls

Jones Falls flows through Baltimore over the hard rock of the Piedmont. Interstate 83 parallels Jones Falls through the city, but several parks provide access to rock exposures. Cylburn Arboretum has wonderful grounds and gardens, and on the Woodland Trail you can see fractured and mottled Hollofield ultramafic rock of the Baltimore Mafic Complex, which intruded Piedmont terranes in Cambrian time. The trail leads down into the valley of Jones Falls.

The Jones Falls Trail, an asphalt sidewalk between Druid Hill and Penn Station, runs along Jones Falls as it descends through gneisses of the James Run Formation of the Chopawamsic Terrane. You can walk or bike the trail, or drive parallel to it on Falls Road. From the Wyman Park Drive Bridge, you can look down for a good view of Jones Falls. If you descend on the switchbacks of the trail to the platform at Round Falls (actually a dam), you can see good exposures of the Jones Falls Gneiss, here associated with the James Run Formation of the Chopawamsic Terrane. Farther down the trail, under the Twenty-ninth Street Bridge, you can see some excellent exposures of gneiss of the James Run Formation.

Gneiss of the James Run Formation, called the Jones Falls Gneiss here, is well exposed along Jones Falls Trail under the Twenty-ninth Street bridge.

Loch Raven Reservoir

Loch Raven Reservoir impounds Gunpowder Falls. The roads around the reservoir afford good exposures of rocks of the Baltimore Gneiss, the 1.1-billion-year-old basement rock that was buried under carbonates and ocean floor sands and clays and later deformed into nappes, or domes. The rock here is on the limb of the Towson Dome. Cromwell Bridge Road, south of the lower dam, runs through the valley of the less resistant Cockeysville Marble, and 0.3 mile up Loch Raven Road, just below the lower dam, you can park beside an old quarry of marble. The lower dam is made of blocks of this quarried marble. The marble is still quarried at the Texas Quarry in Cockeysville at the north end of Lock Raven Reservoir. This quarry is visible from I-83 and MD 45.

At the upper dam, about a half mile upstream of the lower dam, you can see the Loch Raven Schist, originally one of the ocean sediments laid down on the offshore blocks of Grenville basement rock. Large outcrops of this same formation also appear along the shoreline in the parklike area above the bridge over the reservoir.

Cockeysville Marble, quarried nearby and still visible along Loch Raven Road, was cut into this keystone shape, each having a volume of a perch, which is 24.75 cubic feet, for use in the lower dam at Loch Raven Reservoir.

Loch Raven Schist on the shore of Loch Raven Reservoir above Loch Raven Road bridge.

Oregon Ridge Visitor Center is a good place to familiarize yourself with the rocks of the Baltimore Terrane. The rocks, from left to right, are the Baltimore Gneiss, Cockeysville Marble, and Loch Raven Schist.

Oregon Ridge Park

Oregon Ridge Park, west of I-83 exit 20, provides an excellent example of differential weathering. Shawan Road on the lowland leading to the park is in the Cockeysville Marble, a carbonate rock that reacts chemically with naturally acidic precipitation and weathers more rapidly than most other rock. Oregon Ridge is composed of Loch Raven Schist, which here contains quartz veins and quartzite that is relatively resistant to chemical weathering. Iron was once mined from this formation. Outside the visitor center is a display of three huge boulders of different rocks found in the domes of the Baltimore Terrane: Baltimore Gneiss, Loch Raven Schist, and Cockeysville Marble. No Setters Formation is exposed here.

COASTAL PLAIN SITES

Patterson Park

Located between Baltimore Street and Eastern Avenue in East Baltimore, this public park is on a perched water table—a body of groundwater located above the main water table. Most of Patterson Park is clay, deposited in a Cretaceous swamp. However, the higher western part, under the pagoda and high point once called Hampstead Hill, is sand that was probably dropped in the swamp area by a major flood when the whole area was underwater during Cretaceous time.

Sand is a good aquifer, with excellent permeability—the capacity to transmit fluids. Clay has poor permeability, acting like a barrier to water flow. When sand is located above clay, precipitation soaks readily into the sand and infiltrates downward until it reaches the clay layer, where it pools or moves downslope on the top of the subsurface clay. Where the clay intersects the surface, the water will flow out as a spring. While nineteenth-century grading and filling have altered the natural locations of springs, the lake in the park is still fed by this perched water table.

Double Rock Park

At Double Rock Park in the northeast corner of Baltimore, Stemmers Run cuts through Coastal Plain sediments into hard rock of the Perry Hall Gneiss—an intrusive rock of, perhaps, the Baltimore Mafic Complex. If you take I-695 exit 31 at Parkville and head south on Harford Road (MD 147), you will be traveling on the Cretaceous sands of the Patuxent Formation. You are east of the Fall Line here, on the Coastal Plain. In eastern Baltimore City and Baltimore County, however, the Fall Line boundary is irregular and interrupted, with metamorphic Piedmont rock cropping out in many places.

About a half mile south of exit 31, turn right on Texas Avenue and follow it to Double Rock Park. If you park in the first lot on the right and hike down the blue blaze asphalt trail and even farther down, you will see an interesting series of waterfalls and cascades over the Perry Hall Gneiss. Because the Fall Line—the Cretaceous shoreline—is only a mile or two to the west, the Coastal Plain sediments are not too thick here. The erosive action of the stream has cut down to expose the metamorphic rock below the sediments.

Just off the beltway, this downcutting stream in Double Rock Park reveals Cambrian intrusive gneisses, perhaps of the Baltimore Mafic Complex, that lie very close to Cretaceous Coastal Plain sediments here.

Herring Run Park

Herring Run Park occupies the floodplain of Herring Run between Harford Road and I-895 in eastern Baltimore. You can walk under the Harford Road bridge and on the south end see the bridge concrete anchored in the Loch Raven Schist. This schist and its quartz veins are in one of the metasedimentary limbs of the Towson Dome. The Baltimore Gneiss, which occupies the core of the dome, is located just about a tenth of a mile upstream from here. In northeast Baltimore, Cretaceous Coastal Plain sediments cover metamorphic Piedmont rock except where erosion has revealed the much older rock, as Herring Run has done. Most of Harford Road inside the beltway, for example, passes over Coastal Plain sediments of the Potomac Group.

The Delaware Piedmont

The geologic history of the Delaware Piedmont, as described by the Delaware Geological Survey, is very similar to that described for Maryland, D.C., and Virginia, to the south. However, the names of some formations are different, and the Delaware Geological Survey does not use the more regionally applicable "terrane" classification.

Northwest of the island-arc Wilmington Complex in Delaware and into adjoining southeastern Pennsylvania lie the Wissahickon Formation and a few nappes of Baltimore gneiss and Glenarm Group rocks. While the Maryland Geological Survey has dropped the name *Wissahickon* and renamed different parts of the formation, Pennsylvania and Delaware geologists have tentatively divided it into three subunits: a Wilmington Complex arc-related unit, a Glenarm-related unit, and another related to neither.

Gale Blackmer, of the Pennsylvania Geological Survey, who has done much of the work on the three subunits, has said, "The term Wissahickon Formation has a history which . . . is no less complicated than that of the rocks it is used to define." Categorizing has been difficult because of limited rock exposures and because the rocks have been greatly metamorphosed and deformed, and "questions remain." Investigations into this formation provide a good example of how scientific inquiry works: sometimes data lead to reliable, consensus conclusions, but sometimes data indicate that we can conclude very little, or as geologist LeeAnn Srogi has said of the Wissahickon, "The tectonic significance of our results is murky." Because the Delaware Geological Survey still uses just "Wissahickon Formation" and not the subunits, only "Wissahickon" will be used here. Keep in mind, however, that it might refer to fore-arc (Wilmington Complex–associated) basin origin, Baltimore Gneiss and Glenarm association, or some other as yet undetermined origin.

In capsule form, the Delaware Piedmont geologic history describes a volcanic island arc converging toward and eventually colliding with the ancestral North American continental crust—an event that was part of the larger-scale Taconic mountain building event of Ordovician time. The ocean basin that had separated the continent and island arc closed along an eastward-dipping subduction zone, which created the island-arc volcanoes, a deep ocean trench, and a fore-arc depositional basin. The tectonic convergence ramped the igneous rock of the island arc, named the Wilmington Complex in Delaware (generally correlative to the Chopawamsic Terrane in Maryland), over the thick fore-arc basin sediments, named the Wissahickon in Delaware (parts of which generally correlative to the Potomac Terrane in Maryland and Washington, D.C.), and over the nearshore Baltimore Gneiss massifs that were covered with the Setters Formation and the Cockeysville Marble. During the pressure of this convergence and collision, all of these rocks were intensely folded, faulted, and metamorphosed.

The Wilmington Complex, the rocks of the volcanic arc, is divided into three distinct units: (1) metavolcanic units that were originally extrusive volcanic flows and ejected clasts interlayered with sediments; (2) metaplutonic units that were originally plutons, large intrusive magma bodies, some of felsic (high silicon) content and some of mafic (high iron and magnesium) content; and (3) undeformed plutons of both felsic and mafic composition.

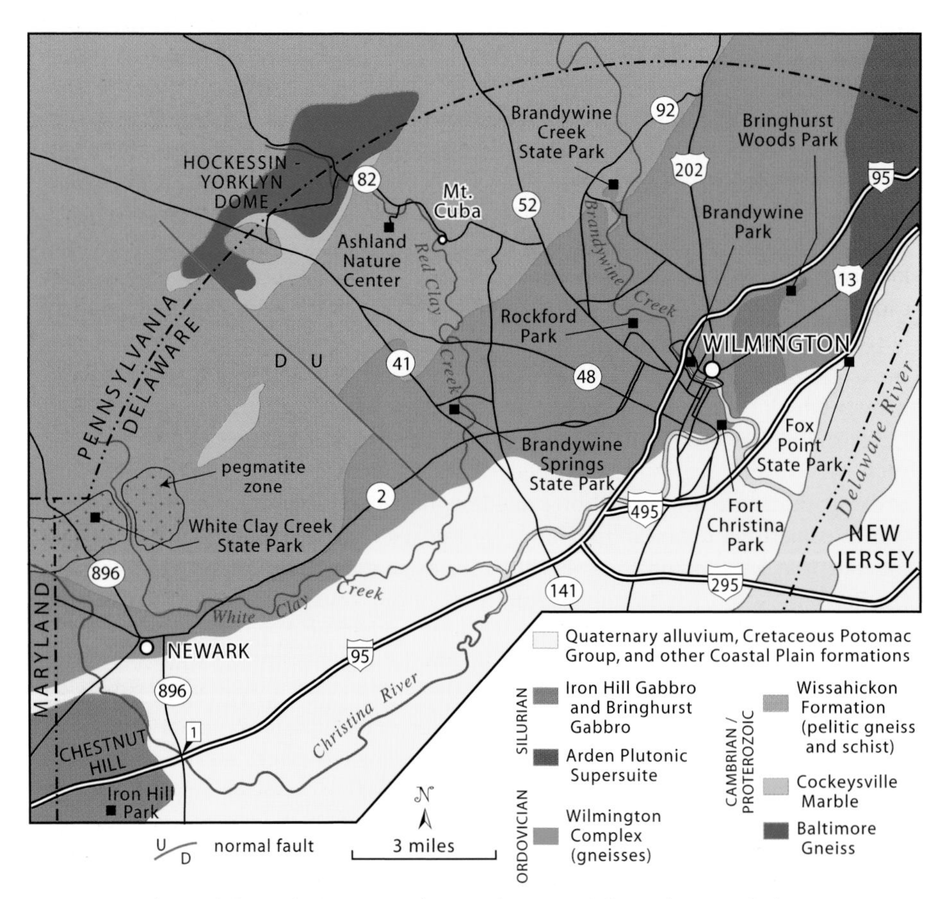

Geology of the Wilmington and Newark areas of the Delaware Piedmont.
—Modified from Schenck, Plank, and Srogi, 2000

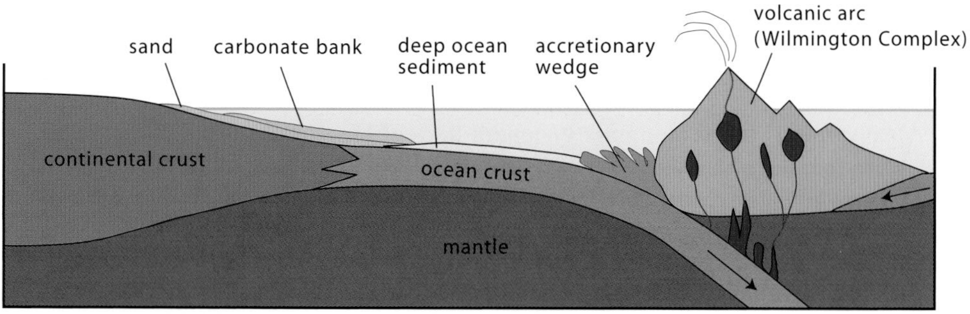

The Wilmington Complex was a volcanic arc that was sutured to the growing continent during the Taconic mountain building event in Ordovician time.
—Modified from Plank and Schenck, 1998

The ages of the metavolcanic and metaplutonic units are in the 488- to 470-million-year range (Ordovician time), while the age of the undeformed plutons is about 434 million years (Silurian time). The Taconic collision of Ordovician time metamorphosed the extrusive and intrusive rock of the island arc. An example of a metaplutonic rock is the Brandywine Blue Gneiss, known in the building stone industry as the Brandywine Blue Granite. An example of the metavolcanic unit is the Rockford Park Gneiss. The undeformed plutons—such as the Silurian Arden, Bringhurst, and Iron Hill—intruded as the earth's crust thinned after the collision was over. The plutons occur in the noncolliding margin of the former volcanic islands. Being on the far side, or ocean side, of the Taconic collision, these plutons are today located along the southeast or Coastal Plain edge of the Wilmington Complex. Not caught in the convergent collision, the plutons were not metamorphosed, so their igneous fabrics were preserved.

A 2000 Delaware report separates the Wilmington Complex from the similarly formed, island-arc James Run Formation in Baltimore. However, the James Run Formation in adjacent Cecil County, Maryland, may be equivalent to the Wilmington Complex. There may have been several arcs or arc fragments in early Paleozoic time off the coast of the ancestral North American continent.

Sediments that became part of the Wissahickon Formation in Delaware eroded a great distance from the slopes of the volcanoes down into the fore-arc basin, near where one plate was subducting another. The sediment moved down the long submarine slopes in a turbid, densely murky undersea current. Mostly sands and clays, such sediments are called distal turbidites, and because the fore-arc basin that contained them was between the island arc and ancient continent, they were metamorphosed, deformed, and thrust onto other formations to the west during the Taconic mountain building event.

Rocks of the Delaware Piedmont are known mostly from exposures along streams and from rocks discovered during excavations, quarrying, and well boring. Parks along the Christina River, Brandywine Creek, White Clay Creek, Red Clay Creek, and a few other streams provide outcrops you can visit. A good place to begin is where the Swedes landed in 1638 along the lower Christina River.

WILMINGTON AREA

Fort Christina Park

Many visitors to the landmark Plymouth Rock of Massachusetts are disappointed when they learn that the rock has been moved from its original location. Much less famous but more clearly documented and still in its original location are The Rocks, where Peter Minuit's expedition of Swedish settlers landed on March 29, 1638. You can see them at Fort Christina Park. To reach it, take exit 1 from I-495 and follow US 13 North (not Business 13). After crossing to the north side of the Christina River, follow the sign that directs you to turn right toward the industrial park. In a couple of blocks you will need to make another right and follow the sign to Fort Christina Park. Walk through the grassy, wall-enclosed park to Old Swedes Landing at the river. Look down and you will see the Fall Line at the edge of a tidal river—a platform of Brandywine Blue Gneiss of the Wilmington Complex. For the weary Swedish seafarers, The Rocks was a natural wharf, easily accessible by gangplank. Across the river lies the Coastal Plain. This gneiss was once the magma chamber below a volcano in a volcanic arc about 480 million years ago.

Beside a concreted edge at Old Swedes Landing in Fort Christina Park, you can see the natural dock provided by the Brandywine Blue Gneiss.

Brandywine Park in Downtown Wilmington

In downtown Wilmington you can see exposures of the Brandywine Blue Gneiss of the Wilmington Complex at Brandywine Park (not Brandywine Creek State Park, which is farther north). Follow the signs to Brandywine Zoo, on the northeast side of Brandywine Creek between Washington Street and I-95. If you walk a paved walkway along Brandywine Creek downstream from the parking lot, under the

Near Brandywine Zoo you can see a wall of blue gneiss built atop an exposure of Brandywine Blue Gneiss.

Washington Street Bridge you can see good exposures of the gneiss. The rock is blue on unweathered surfaces but is black or dark gray on weathered surfaces.

Farther downstream is the Market Street Bridge. Stop in the middle and compare upstream with downstream. Downstream lie the tidal waters of the Coastal Plain, and here you stand at the colonial head of navigation for Wilmington. William Penn, in a letter of 1683, included Brandywine Creek in a list of coastal rivers "any one of which have room to lay up the royal navy of England, there being from four to eight fathom of water." This would be 24 to 48 feet of water.

Looking upstream you will see Brandywine Creek flowing over a rock-strewn course, unnavigable. You can walk upstream on the west side of the creek and see the rocky section more closely. Here along the Brandywine were once located many water-powered mills, and you can still see one surviving millrace. Flour, cotton, and snuff from the mills were transported the short distance down to the navigable waters for shipment across the sea.

The rocks you see along the creek here were formed in a volcanic island arc similar to the Aleutians off the coast of Alaska, created by plate subduction. Then, between about 441 and 432 million years ago, they collided with and were attached to the ancient continent edge, buried at great depth, heated to temperatures over 850 degrees Fahrenheit, folded, and metamorphosed. Hundreds of millions of years of erosion and subsequent uplift have brought them to the surface in present geologic time.

Brandywine Creek is rocky downstream to the Market Street Bridge (in distance), *which was the colonial head of navigation for Wilmington.*

Rockford Park

Rockford Park is along Brandywine Creek a few blocks north of the intersection of Delaware Avenue and Greenhill Avenue. In mostly unmarked trails through the woods here, you can find boulders and exposures of the Rockford Park Gneiss of the Wilmington Complex. When the tower is open, you can climb up for a spectacular view of the hills of the Piedmont to the north and west, and the low-lying, flat areas of the Coastal Plain to the south and east. This is one of the best and most convenient places to see how the terrain changes from one province to another.

Brandywine Creek State Park

Brandywine Creek State Park, near the intersection of DE 92 and DE 100, or just west of US 202, has many good rock exposures. Adjacent to the south (Hawk View) parking lot, you can see numerous boulders of the Wilmington Complex up close, in this area the Montchanin Metagabbro.

The contact between the originally igneous Wilmington Complex and the originally sedimentary rocks of the fore-arc basin, the Wissahickon Formation, is located in the park. An excellent, flat, 1.5-mile-long trail runs along the east bank of Brandywine Creek from the parking lot at Thompson Bridge Road to a smaller parking area off Rockland Road on Millrace Lane. A geologic trail guide says that the contact of the light-colored Wissahickon Formation and the dark Wilmington is 500 feet below the footbridge over Rocky Run, but it is difficult to find. You can,

Gneissic foliation in the Montchanin Metagabbro of the Wilmington Complex, right beside the parking lot at Brandywine Creek State Park.

however, see the bluish Brandywine Blue Gneiss of the Wilmington Complex on the southern, downstream part of the trail—along with a good interpretive sign—and the Wissahickon Formation on the northern, upstream part.

Bringhurst Woods Park

At Bringhurst Woods Park you can see rock of an igneous pluton of Silurian age—a body of hot magma that formed far beneath the surface and intruded the Wilmington Complex during the crustal thinning that followed the Taconic mountain building event. To visit the park from I-95 North, take exit 9 and make a hard right from the exit ramp onto Carr Road. From I-95 South at exit 9, exit onto Marsh Road (DE 3) and turn left and then right onto Carr Road. Drive a short distance to an office, and park roadside near the old gravel lot, now gated off. Below the gravel lot you can explore Shellpot Creek and look for the speckled Bringhurst Gabbro.

At Bringhurst Woods Park, the stream has eroded and separated boulders of the layered, metamorphosed Brandywine Blue Gneiss (right) *and the coarse-grained, unmetamorphosed Bringhurst Gabbro* (left) *that intruded it.*

In and around Shellpot Creek, you can find Bringhurst Gabbro that encloses numerous dark pieces, or xenoliths, of Brandywine Blue Gneiss.

This unmetamorphosed intrusive igneous rock is not folded and layered like the highly metamorphosed Brandywine Blue Gneiss into which it intruded. The gabbro is very coarse-grained, meaning the mineral crystals are fairly large, with grain size averaging about 0.5 inch in diameter.

If you explore the hillside from the asphalt path or from the creek, you may find dark fragments of the blue gneiss that were incorporated by the molten, intruding magma. These are called xenoliths. Also, slow-cooled, large-crystalled pegmatitic gabbroids can be found here. *Pegmatite* usually refers to granitic rock, but here it is applied to the relatively larger grain size found in some mafic gabbro intrusions.

Brandywine Springs State Park

Brandywine Springs State Park, on Red Clay Creek near Faulkland Road, was once the location of a nineteenth-century spa fed by a spring that discharged brown water rich in iron. This public park once was a favorite of mineral collectors seeking

This unusual opening in hard, metamorphic rock of the Wissahickon Formation at Brandywine Springs State Park probably formed as groundwater weathered layers weakened by the tight folding.

Delaware's state mineral, sillimanite, an aluminum-rich metamorphic mineral that forms where temperatures have exceeded 1,100 degrees Fahrenheit. The sillimanite, which is in the Wissahickon Formation, occurs in nodules and in large fibrous clumps from 2 to 3 feet in diameter.

The park is laced with informal trails. If you park in the main lot and go downhill from the easternmost baseball field and follow the gravel road across from the railroad trestle, you can find a cave where a groundwater-eroded mica schist overlies a thick layer of resistant garnet-bearing quartzite. You can also reach this location if you park along Faulkland Road by Red Clay Creek, walk west on the road, and turn south past the maintenance building and into the fire lane. If you walk south along the railroad tracks on the eastern side of the creek, you can see exposures of the garnet-bearing quartzite of the Wissahickon Formation.

Mount Cuba Area and Ashland Nature Center

The Mount Cuba area, west of Hoopes Reservoir, has many interesting exposures of the Wissahickon Formation—rock of the fore-arc basin. If you drive north on DE 82 along Red Clay Creek, just south of Way Road you will see a roadcut in the gneiss featuring recumbent folds—the limbs have been overturned so far that they are horizontal or nearly so. The rock was buried so deeply that the heat and pressure softened it into a plastic state resembling modeling clay. You can park just south of the cut, where the road crosses a small tributary of Red Clay Creek.

At DE 82 and Brackenville Road, near the covered bridge, you can pull into Ashland Nature Center and take some very interesting hikes. The circular kiosk just above the parking lot has trail guides for Floodplain, Sugarbush, and Treetop

Trails at Ashland Nature Center take you to the Wissahickon Formation, which was once so plastic and pliable from heat and pressure that it bent and folded on a small scale like molding clay.

Trails. The latter two climb into the woods, where you can see numerous excellent examples of intensely folded metamorphic rock. The trail guides tell you that the rock is Brandywine Granite with garnet and quartz, but the Delaware Geological Survey maps this area as the Wissahickon Formation, originally a layered, sand and clay turbidite that was metamorphosed into gneiss. The connecting Floodplain Trail gives you a good view of the flat, sandy and clayey deposits laid down by a stream with its repeated flood events over thousands of years.

DE 82 north of Ashland has many rock exposures on its margins, but there are few places to stop. At the junction with Yorklyn Road, you can pull off and look at the ancient basement rock of the North American continent, the 1.1-billion-year-old Baltimore Gneiss.

White streaks in the outcrop at the Yorklyn turnoff from DE 82 are quartz and plagioclase feldspar in 1.1-billion-year-old Baltimore Gneiss of the Hockessin-Yorklyn Dome.

Fox Point State Park

Take exit 4 from I-495 and follow signs to Fox Point State Park to see the wide, tidal Delaware River, which descends from Trenton, past Philadelphia, in Coastal Plain deposits. Just upstream from Trenton near the I-95 bridge, the river meets the Fall Line of Piedmont rock.

NEWARK AREA

White Clay Creek State Park

Just southeast of the three-corners state boundary north of Newark, you can turn off DE 896 into the Walter S. Carpenter Recreation Area of White Clay Creek State Park. Here you can hike to a small rocky knob that is one of Delaware's few rock outcrops not exposed by a modern, downcutting stream. From the parking lot, hike to Millstone Pond, where you will see a large exposure of gneiss of the Wissahickon Formation. In adjacent Cecil County, Maryland, this formation is called pelitic gneiss—a clay-based, deep-ocean sediment that was later metamorphosed by the pressure and heat of deep burial during tectonic collision. If you look closely in the rock, you will find the distinct, large minerals identified with pegmatite, crystallized from slowly cooling fluids that had been superheated by a magma body—a pluton—that had intruded the Wissahickon Formation during its deep burial.

At White Clay Creek State Park you can hike many other trails and enjoy fields and woods of hilly Piedmont terrain. At the far northwest boundary of the park, you can see a stone commemorating the surveying of the arc boundary of Delaware and Pennsylvania. This arc has a 12-mile radius measured from the cupola of the courthouse in New Castle, the colonial capital of Delaware.

Iron Hill Park

Areas generally south of DE 2 lie in the Coastal Plain, but there is an outlier, or pinnacle, of Piedmont rock—Iron Hill. You can see this isolated 340-foot-high hill from I-95, which passes just north of it east of the Maryland border. DE 896 South from I-95 exit 1 will take you to Old Baltimore Pike and Iron Hill Park. Here is a large ore pit where iron was once mined. The Iron Hill Gabbro is an igneous pluton that probably intruded in Silurian time during the crustal thinning that followed the Taconic mountain building event. The surface of the gabbro is coated with iron oxides, which were mined for iron. Iron Hill and geologically related Grays Hill in adjacent Maryland protrude from the Coastal Plain so dramatically that they are visible from the high C&D Canal bridge on MD 213 some 8 miles to the south. The gabbro is also exposed along the banks of the Christina River at Rittenhouse Park, which is off Chestnut Hill Road southwest of DE 896 and north of Iron Hill. Chestnut Hill, northwest of Iron Hill, is also formed by the gabbro but is much lower.

•— Coastal Plain —•

Before we discuss the geologic history of the Coastal Plain Province, we should define some geographical terms. The Delmarva Peninsula is the Coastal Plain area between Chesapeake Bay on the west and the Delaware River and Atlantic Ocean on the east. It is named for the three states that comprise it—Delaware, Maryland, and Virginia. The Maryland portion of the Delmarva Peninsula is also known as the Eastern Shore, a region with its own historical, cultural, and political identity. The mainland side of Chesapeake Bay is called the western shore, a term that is not capitalized. The western shore includes Southern Maryland, a region generally bordered by the Fall Line (followed roughly by I-95) to the northwest, the Potomac River to the southwest, and Chesapeake Bay to the east.

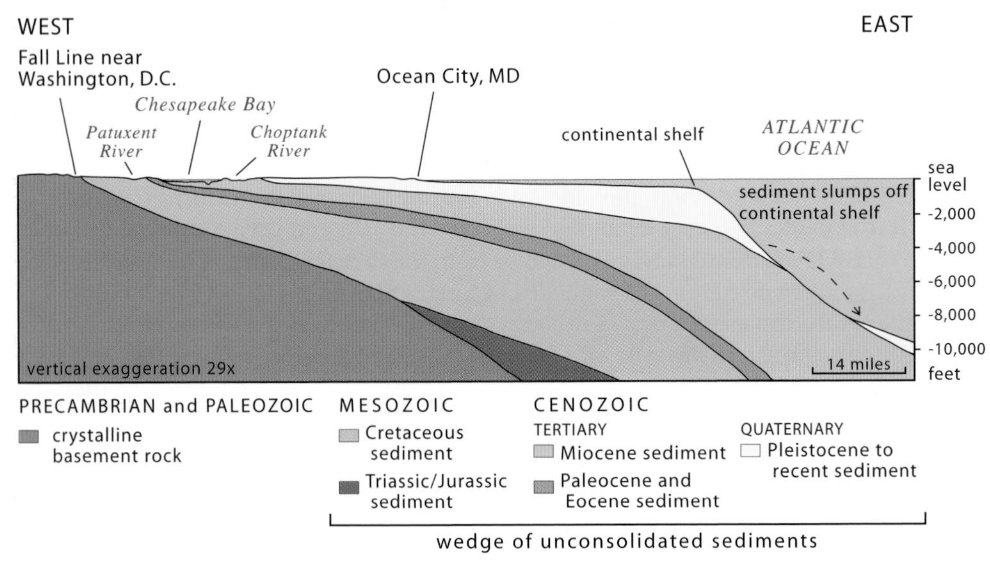

The eroded remains of ancient mountain ranges now lie in the Coastal Plain's thick layers of unconsolidated sediment. —Modified from Cleaves, Edwards, and Glaser, 1968

WEDGE OF UNCONSOLIDATED SEDIMENT

The Coastal Plain differs distinctly from the other provinces because it does not contain solid rock and is relatively flat. The geologic layers of the Coastal Plain were laid down in a giant sedimentary wedge beginning with the breakup of Pangea at the end of Triassic time, about 200 million years ago. The wedge tilts or dips seaward, thickens seaward, and generally has the oldest layers on the bottom and youngest on top. All of the formations are unconsolidated, loose material (like soil) that lacks the carbonates or other cements necessary for them to solidify into sedimentary rock. Even the oldest exposed formations, those of the Cretaceous Potomac Group, are about 400 million years younger than the mostly Cambrian Piedmont formations

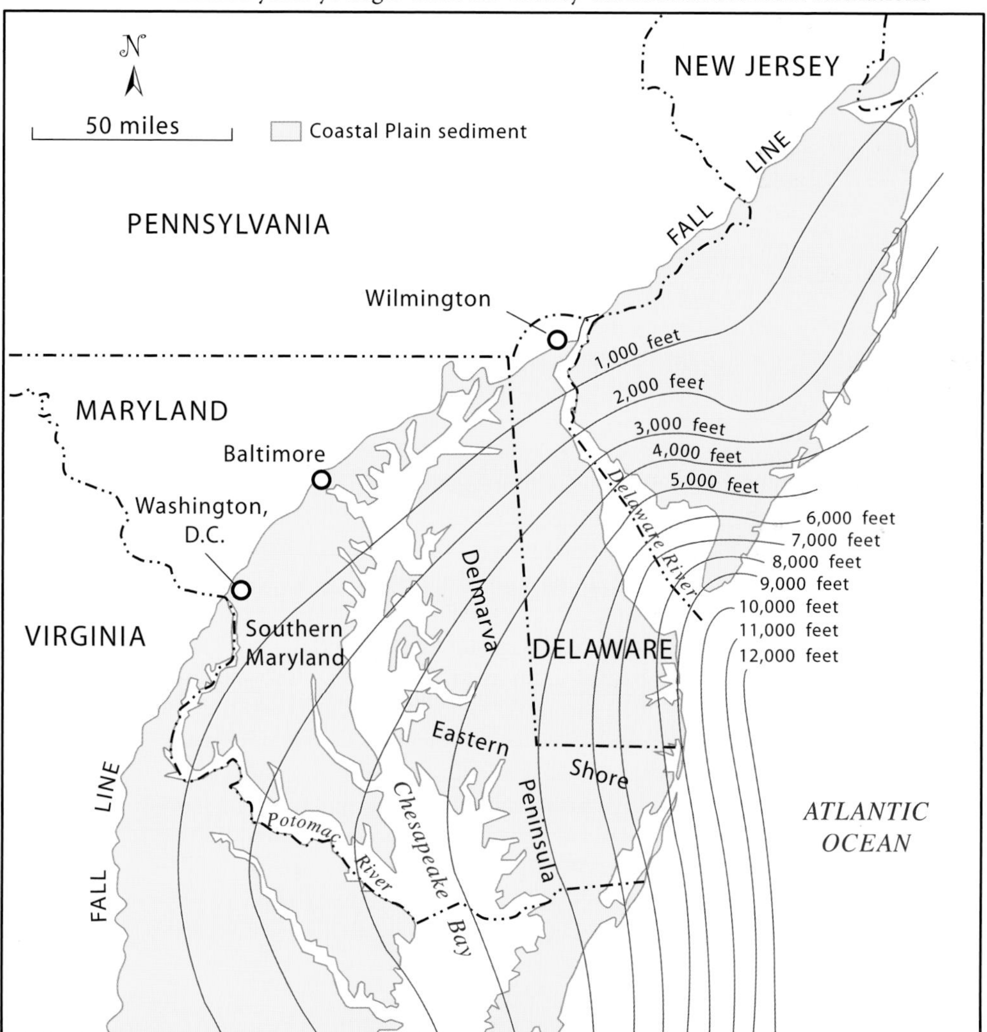

Hard rock to the east of the Fall Line dropped along normal faults and was covered with the thousands of feet of sediment that compose the Coastal Plain. Sediments thicken to the east, forming a wedge. —Modified from Higgins and Conant, 1990; Southwick, Owens, and Edwards, 1969

that abut them. Today, millions of years after deposition, these deposits remain as unconsolidated gravels, sands, silts, and clays—ancient, recycled particles of the ancestral Appalachian Mountains. The wedge is about 1.5 miles thick beneath the present Atlantic shoreline.

The Fall Line is the contact between the soft sediments of the Coastal Plain and the hard rocks of the Piedmont. Regionally it runs from Trenton, New Jersey, to Richmond, Virginia. Because it marks the zone where the deep Coastal Plain rivers issue from rapids tumbling over metamorphic Piedmont rock, the Fall Line was the head of navigation during colonial times. Cities such as Trenton, Philadelphia, Wilmington, Baltimore, Washington, and Richmond were located at these inland port sites.

The Fall Line formed when North America and Africa separated during Triassic time about 200 million years ago. As the crust stretched, numerous fractures created sunken rift basins. In basins below sea level, located beneath present-day Southern Maryland, Delmarva, and offshore, Triassic and Jurassic marine deposits were laid down. These created the sedimentary base for the Coastal Plain depositional wedge as the Atlantic Ocean opened wider and wider.

During most of the early Cretaceous period (145 to 100 million years ago), nonmarine fluvial, floodplain, river-channel, and deltaic deposition occurred along this eastern continental margin. Meandering rivers fanned out onto the continental shelf and deposited large volumes of clay and silt, with some sand and gravel, creating the thick Potomac Group. Uplift in the Appalachians is thought to have supplied this sediment, which buried the older Triassic and Jurassic layers. Today the Potomac Group is exposed at the surface in Maryland along a 5- to 10-mile-wide strip east of the Fall Line on either side of I-95 and is thought to be over 1,000 feet thick in this zone. From the Fall Line, the formations of the Potomac Group gradually tilt downward and thicken as they run beneath younger formations of the huge Coastal Plain depositional wedge. The tilted layers extend from beneath Southern Maryland, Chesapeake Bay, and Delmarva to hundreds of miles out to sea. Deep beneath the present-day coast, the Potomac Group is over 3,000 feet thick and makes up about 75 percent of all Coastal Plain material.

Next, during late Cretaceous time (just before the dinosaurs went extinct around 66 million years ago), the earth was without glaciers during an extended episode of global warming, and a worldwide rise in sea level flooded many coasts (as well as the American Midwest). Marine sandy shelf deposits, such as the Magothy Formation, Matawan Group, and Monmouth Formation, were laid down on top of most of the Potomac Group. The growing offshore continental margin subsided and tilted seaward for the next 30 million years, possibly due to the weight of the offshore sediments piling up. This tilting and downslippage of previously fractured Proterozoic and Paleozoic basement rock created the Salisbury Embayment, which covered Virginia, Maryland, Delaware, and southern New Jersey up to the Fall Line during late Cretaceous and most of Tertiary time.

Through four epochs of Tertiary time, about 65 to 5 million years ago, layer upon layer of sediment accumulated on the subsiding Salisbury Embayment floor

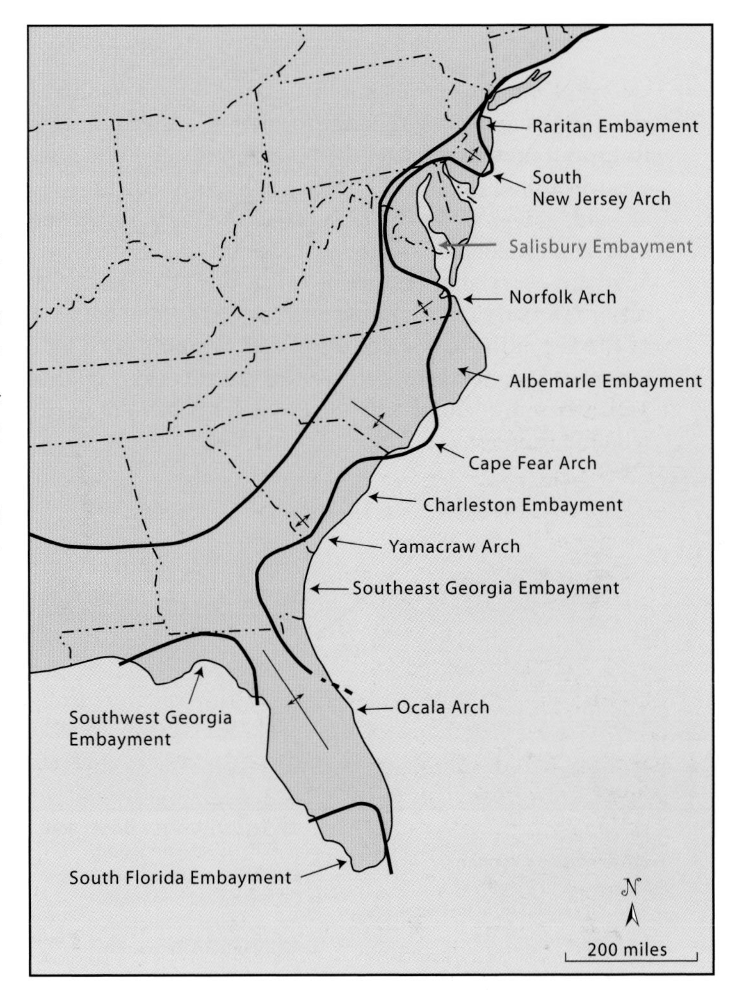

Embayments along the Atlantic coast in Tertiary time. Without the Coastal Plain deposits of the Salisbury Embayment, there would be a gulf to the Fall Line at Wilmington, Baltimore, and Washington. —Modified from Ward and Powars, 2004

during pulses of high sea level. The Paleocene Aquia and Brightseat Formations, the Eocene Pamunkey Group, and the Miocene Chesapeake Group (today exposed at Calvert Cliffs) were laid down on top of the Cretaceous sediments. There are no Oligocene deposits here because of a drop in sea level at that time, but about 35 million years ago during this epoch, a 2- to 3-mile-diameter meteorite or asteroid is thought to have impacted southern Chesapeake Bay, leaving a crater about 50 miles in diameter. Today the Salisbury Embayment formations occupy more than half of Southern Maryland's surface southeast of the Potomac Group. They also stretch beneath Delmarva in a wedge thousands of feet thick.

After the Chesapeake Group was deposited, sea level fell and there was a period of time when the marine-deposited Miocene sediments were exposed and eroded. During Pliocene time, which began only 5 million years ago, silt, sand, and gravel were deposited in estuarine, fluvial, and deltaic environments on the eroded Miocene surface as water levels fluctuated below the mouths of ancestral rivers. In Southern Maryland—from Anacostia in D.C. through southern Prince

Georges County, central Charles County, and the middle of St. Marys County—the most widespread western shore formation was deposited by the ancestral Potomac River as it meandered laterally in a southeasterly direction. Containing layers that were deposited in estuaries when sea levels flooded the land, this broad formation of sand, gravel, and pebbles is called the Tertiary Upland Gravel 4, or TUG_4, by Maryland geologists, and today it forms a cap, or undissected plateau, over central Southern Maryland. Because it is flat, the TUG_4 has been used for the routing of major railroads and highways such as US 301 and MD 5. Elevations well in excess of 100 feet at the top of this formation suggest a good bit of uplift has occurred in Southern Maryland since the end of the Pliocene—uplift due, perhaps, to tectonic pressure from the mid-Atlantic divergent plate boundary or to isostatic (buoyant) rebound from erosive removal of material.

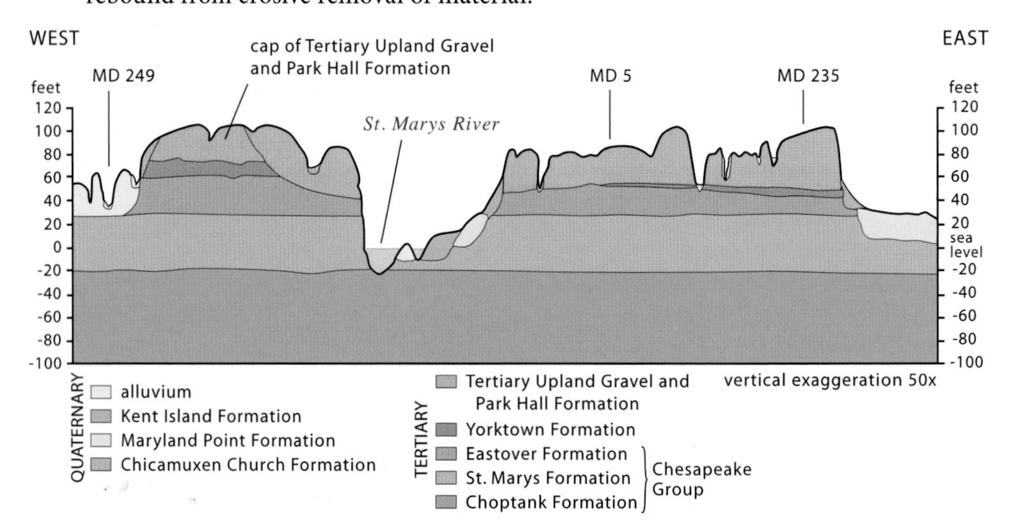

This vertically exaggerated profile shows the cap of Tertiary Upland Gravel in St. Marys County. —Modified from McCartan, 1989

Stream dissection into the cap and the underlying clays and sands increases as we move south and east, probably because the original upland gravel was thickest where the Potomac emerged from the Piedmont but then thinned out progressively farther downstream, making the resulting downstream areas more susceptible to erosion. Where streams have cut down through this cap, flowing west to the Potomac River and east to the Patuxent, the older, underlying formations are revealed. The regional uplift and the subsequent lowered sea levels during Pleistocene Ice Ages probably contributed to this stream downcutting.

Over on Delmarva, the ancestral Susquehanna and Delaware Rivers also deposited sand and gravel during Pliocene time. The oldest major surficial layer, the Beaverdam Formation, covers the spine of upper-central Delmarva and contains a lower stream-deposited unit and an upper estuarine unit. The two units may represent a complete marine cycle of regression and transgression. During regression, sea level dropped and streams deposited sediments across the low-lying lands. During transgression, when sea levels were higher, finer sediments fell out of suspension and covered Delmarva. Some geologists correlate the stream and estuarine TUG_4 with the Beaverdam, but others do not. Correlation means that formations in different locations are equivalent in geologic age and stratigraphic position.

PLEISTOCENE TIME

Although the glaciers of Pleistocene time did not cover Maryland or Delaware, they did occupy the headwaters of the ancestral Susquehanna and Delaware River drainage basins to the north. The retreating, melting glaciers produced large volumes of water that carried sediments eroded from the land by previously advancing glaciers. These sands and gravels (and some ice-rafted boulders) were washed down the ancestral Delaware and Susquehanna Rivers during a warm interglacial period about 500,000 years ago. The rivers, swollen with meltwater and sediment, spread in braided channels to lay down the Columbia Formation deposits on the Delmarva Coastal Plain. The land was not yet underwater because during global warming there is a time lag between increasing temperatures and significant rises in sea level. After this deposition, the Delaware and Susquehanna Rivers began to establish their modern channels. Today the gravelly Columbia Formation occupies much of central and southern New Castle County, Delaware; most of adjacent Cecil County, Maryland, east of the Elk River; and east-central Kent County, Delaware. The Cecil County map calls this gravel the Pensauken Formation. In northern Delmarva (southern New Castle County, Delaware), the meltwater flow eroded much of the Beaverdam Formation, but it remained over central Delmarva (Kent and Sussex Counties).

During interglacial periods, glaciers retreated and sea levels rose—a process occurring during our present global warming trend. During rises in sea level, the ancestral Delaware and Chesapeake Bays backflooded, turning low-lying land into shallow water where estuarine sediments were laid down. The waters invaded more than 5 miles into eastern Delaware

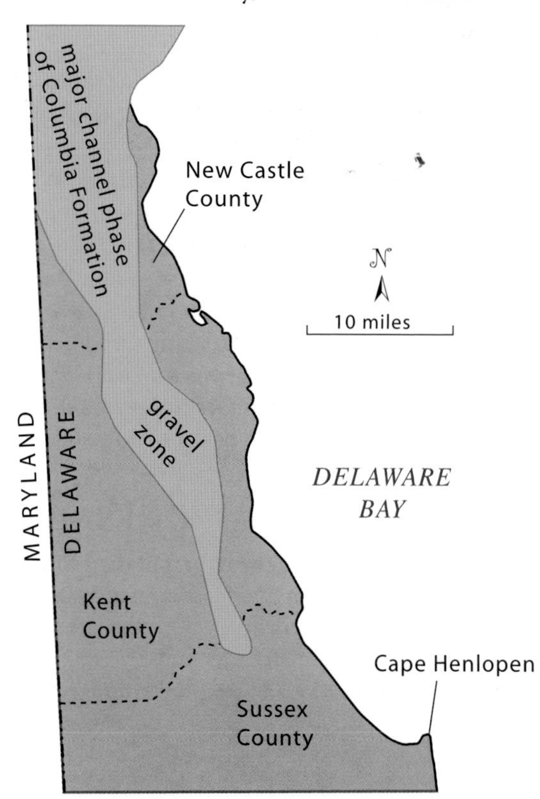

Gravelly outwash deposited by meltwater flowing south from melting Pleistocene glaciers to the north covered upper-central Delaware.
—Modified from Groot and Jordan, 1999

from the present Delaware Bay shoreline. They washed a comparable distance onto the Eastern Shore of Maryland and 1 to 2 miles onto the western shoreline of the present Chesapeake Bay. Sediments washed in from the nearby shores were reworked by wave action and deposited on the estuary bottoms. In Delaware the Lynch Heights Formation was deposited on top of the Columbia and Beaverdam Formations about 320,000 years ago. The Scotts Corners Formation was deposited on these formations about 120,000 years ago. In Maryland the Kent Island Formation was deposited about 40,000 to 30,000 years ago, and the Omar Formation is given different dates, depending on which map you see—in a range from 126,000 to 90,000 years, which would presumably correlate it with the Scotts Corners (but another source correlates part of the Omar with the Lynch Heights Formation).

In both states you can still see escarpments more than 100 miles long and 5 to 25 feet above sea level. The scarps were once cliffs on the shorelines of ancient estuaries, resembling the current Calvert Cliffs. At the toe of these scarps were deposited younger estuarine sediments. These deposits and scarps can be found facing not only present Chesapeake and Delaware Bays but also the previously flooded valleys of low-lying, estuarine rivers, such as the Choptank and Nanticoke. In Talbot County on Maryland's Eastern Shore, the Wicomico-Talbot Escarpment is clearly visible west of US 50.

Other Pleistocene formations, such as the upper part of the Omar and the Ironshire, both in eastern Delmarva just landward of the present-day lagoons of Chincoteague, Assawoman, and Rehoboth Bays, are thought to be beach, lagoonal, and back-barrier sediments deposited during high stands of the sea. These sediments are comparable to the sediments collecting today on the floors of the lagoonal bays separating Delmarva barrier islands from the mainland.

During glacial periods of the Pleistocene, winds winnowed fine-grained sediments from the arid, treeless coastal plain and reworked the remaining sand over large areas of the central Delmarva uplands. These wind-deposited sands are mapped as the Parsonburg Sand on Maryland maps, and in many areas they were shaped into parabolic dunes still visible today.

Carolina bays also developed on Delmarva. These circular to oval depressions are usually swampy. Their sand rims are thought to have formed in cold, windy climates or unforested terrain with fluctuations in groundwater levels.

The most recent deposits of the Coastal Plain are alluvium on river and stream floodplains. Swamp, marsh, and shoreline deposits were also laid down during the rise of sea level of the past 12,000 years.

CHESAPEAKE BAY

Chesapeake Bay is the largest of the 130 estuaries in the United States. An estuary is a semienclosed body of freshwater connected to the sea. Basically, it is a river valley that has been backflooded by the sea. The salty seawater is diluted by freshwater flowing in from streams and rivers draining the land—in about a fifty-fifty mix. Chesapeake Bay and its tidal tributaries (measured up to where they are 100 feet

TIME		MARYLAND'S WESTERN SHORE	MARYLAND'S EASTERN SHORE	DELAWARE
QUATERNARY		Holocene shore, marsh, swamp, and alluvial deposits are not included		
QUATERNARY	Pleistocnene	Kent Island (Talbot) Fm. Maryland Point Formation Ironshire Formation Omar Formation Chicamuxen Church Fm.	Parsonburg Sand (30,500 to 16,000) Kent Island Formation (40,000 to 30,000) Ironshire Formation (35,000) Sinepuxent Formation (120,000-80,000) Omar Formation (120,000 to 90,000)	dune deposits Cypress Swamp Formation Nanticoke dune deposits Carolina bays (100,000 to 70,000) Scotts Corners Formation (120,000) Omar Formation Lynch Heights Formation (320,000) Columbia Formation (500,000)
TERTIARY	Pliocene	Park Hall Formation Tertiary Upland Gravel 4 Tertiary Upland Gravel 3 Yorktown Formation	Pensauken Formation Walston Silt Beaverdam Sand Yorktown and Cohansey Formations	Beaverdam Formation (Carolina bays formed in these deposits)
TERTIARY	Miocene	Chesapeake Group Eastover Formation St. Marys Formation Choptank Formation Calvert Formation	Chesapeake Group "Manokin" Sand St. Marys Formation Choptank Formation Calvert Formation	Bethany Formation Chesapeake Group Cat Hill Formation St. Marys Formation Choptank Formation Calvert Formation
TERTIARY	Oligocene	no deposits in Maryland because sea level was lower in Oligocene time		unnamed glauconitic unit
TERTIARY	Eocene	Pamunkey Group Piney Point Formation Nanjemoy Formation Marlboro Clay	Piney Point Formation Nanjemoy Formation	Piney Point Formation Shark River Formation Manasquan Formation
TERTIARY	Paleocene	Aquia Formation Hornerstown Formation Brightseat Formation	Aquia Formation Hornerstown Formation Brightseat Formation	Vincentown Formation Hornerstown Formation
CRETACEOUS		Severn Formation Monmouth Formation Matawan Group Marshalltown Formation Englishtown Formation Merchantville Formation Magothy Formation Potomac Group Patapsco Formation Arundel Formation Patuxent Formation	Monmouth Formation Matawan Group Marshalltown Formation Englishtown Formation Merchantville Formation Magothy Formation Potomac Group	Navesink Formation Mount Laurel Formation Matawan Group Marshalltown Formation Englishtown Formation Merchantville Formation Magothy Formation Potomac Group (Marshalltown through Magothy: exposed in C&D Canal cut)

Stratigraphic column for the Coastal Plain. Correlations are based on dates from county geologic maps, and they involve many differences and contradictions, and much guesswork by the author. A blue box indicates the formation is only present in the subsurface. A pink box indicates that the epoch placement is uncertain. Spacing within an epoch is not scaled to time. Holocene shore, marsh, swamp, and alluvial deposits are not included. —Compiled from Owens and Denny, 1978, 1979, 1984, 1986a, 1986b; Glaser, 1998, 2003; Higgins and Conant, 1986; McCartan, 1989; Ramsey, 2003, 2005, 2007; Ramsey and Schenck, 1990; Andres and Ramsey, 1995; Andres and Klingbeil, 2006

wide) have more than 4,000 miles of shoreline in Maryland. Because it is about 50 percent freshwater, the bay is an accurate indicator of the degree of stream and river pollution, and as a result it was the first estuary in the nation to be targeted for restoration and cleanup—an effort that continues.

Chesapeake Bay is fed by many streams. Two tributaries, the Potomac and the Susquehanna—considered major rivers of the East—and many other smaller rivers that feed the bay have been flooded and widened at their mouths and lower reaches. They are also tidal, indicative of their estuarine connection to the ocean. For example, the mouth of the Potomac River is over 5 miles wide, and the Choptank River at Cambridge is about 2 miles wide. The Susquehanna contributes about 50 percent of the freshwater volume to the bay.

Most of Chesapeake Bay is shallow. The average depth is only 21 feet, and more than 80 percent of the bay is less than 33 feet deep. The deepest point is Bloody Point Hole at 174 feet below sea level, about 1 mile southwest of the southern tip of Kent Island. A bathymetric map of the bay and its tributaries shows that the deepest channels connect in what looks like a dendritic river system, with many small tributaries draining into one big central river.

At the height of the last Ice Age, about 18,000 years ago, the average global temperature was about 20 degrees colder than it is today. Seas were more than 200 feet lower than their current levels, and the shoreline of the East Coast was about 50 miles east of Ocean City, Maryland. The oceans and coasts looked so much different then because large volumes of the earth's water were frozen into expansive continental glaciers. There was no Chesapeake Bay.

The ancestral Susquehanna River ran between Maryland's western shore and Eastern Shore. The river's lower course, along with those of many rivers flowing east out of the Appalachians, is thought to have been determined by a large depression that ultimately became Chesapeake Bay. About 35 million years ago the impact of an asteroid 2 to 3 miles wide blew a 1-mile-deep, 50-mile-diameter crater (the sixth largest impact crater known on earth) in the Virginia portion of Delmarva, today lying beneath southern Chesapeake Bay and its mouth.

About 20,000 years ago global temperatures began to rise, and glaciers began to melt. About 8,000 years ago, rising seas began to fill a narrow bay in the lower reaches of the ancient Susquehanna. By 6,000 years ago, the sea and bay had risen to within

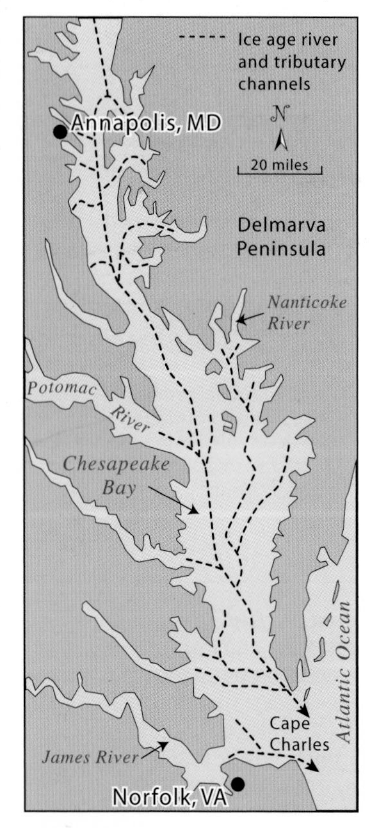

At the peak of the last glaciation, sea level was hundreds of feet lower than it is today, and the Chesapeake Bay area was drained by a branching river system. —Modified from U.S. Geological Survey, 1998

35 feet of today's level, and by about 3,000 years ago, the approximate boundaries of Chesapeake Bay had filled with water. A cross section of the bay preserves the deep contours of the ancestral Susquehanna River valley, even though it has been smoothed out by thousands of years of sedimentation. The flooded river gorge forms a naturally deep shipping channel.

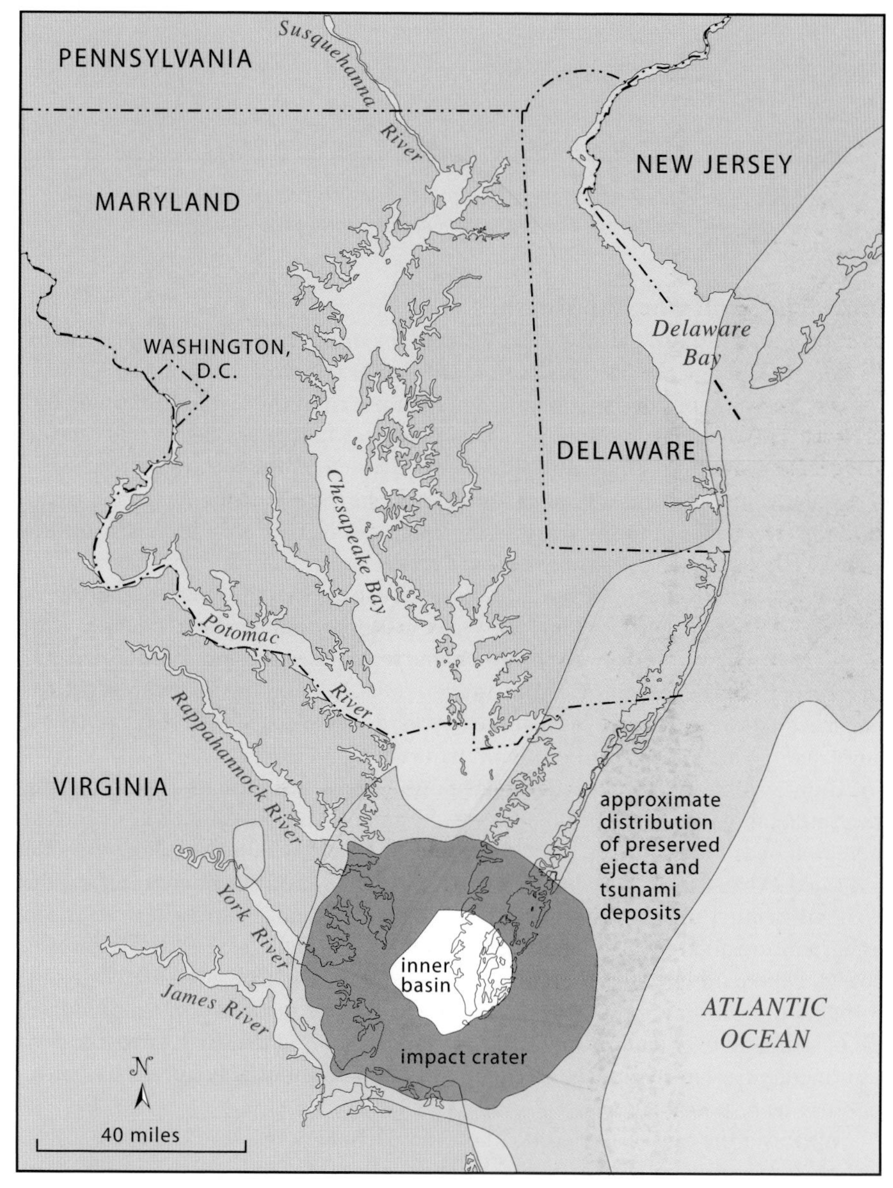

Evidence of a huge crater, blasted out by an asteroid impact about 35 million years ago, lies deep below Chesapeake Bay. —Modified from Powars and Bruce, 1999; Poag, 1997

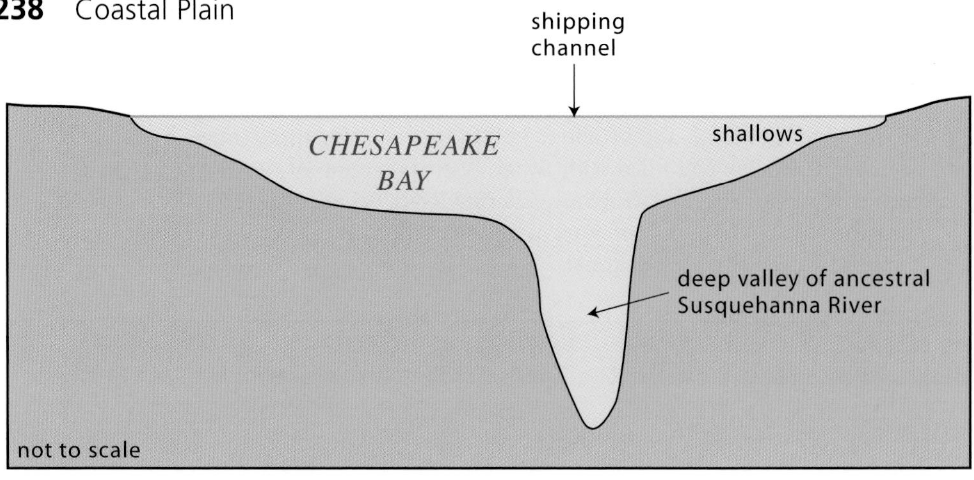

Chesapeake Bay is blessed (or cursed) with a natural shipping channel—the now-flooded, deeply cut valley of the ancestral Susquehanna River.

Subsidence and Coastal Marshes

A coastal marsh is also called a tidal wetland, or in the case of the marshes along the shore of Chesapeake Bay and its tidal tributaries, an estuarine wetland. Coastal marshes form where slow, sediment-laden rivers, such as the Little Blackwater and Blackwater Rivers, meet an ocean or bay. Because Chesapeake Bay is open to the ocean, incoming tidal currents collide with river currents, forming stretches of slackwater where clay and silt fall out of suspension. These areas of deposition occur from the river's mouth to the tidal reach, which can be tens of miles upstream. As sea level has risen, these areas have moved farther upstream.

Soft-stemmed emergent aquatic plants, such as cattails, begin to grow in shallows of a few inches to a few feet. Water salinity varies in these shallows with the tides and volume of river discharge. As the vegetation takes hold, it may further slow currents and cause more sedimentation. One study in the 1980s showed that marsh accretion rates along Chesapeake Bay were keeping pace with rising sea level. However, this may no longer be true in many areas.

Gain or loss of coastal marshes has both natural and human causes. During the American colonial period, widespread clearing and plowing of land caused increased sedimentation in the tidal areas, increased emergent plant growth, and rapid expansion of marshes. However, Delmarva land subsidence during the twentieth century may have been accelerated by widespread groundwater extraction for agriculture and for the burgeoning human population. On a much larger scale, Coastal Plain land subsidence over many thousands of years has contributed greatly to the natural formation of estuarine marshes. The accumulation of thousands of feet of sediment over tens of thousands of years has caused the earth's crust to sink into the mantle. Downslippage along buried faults in the deeply buried basement may also be responsible for the sinking of the land.

Whatever the causes, geologists know that Delmarva is subsiding, and because sea level is measured relative to land, Chesapeake Bay is rising faster than most other water bodies in the world. Rising water levels are both destructive and constructive.

Former farm fields and forests become marsh, while marshes near the water's edge become open water. In other words, the brackish marshes migrate landward with the rising water levels. Maryland geologist Benjamin L. Miller summarized this process well in 1926 in *The Physiography of Talbot County*: "Many square miles that had been land before this subsidence commenced are now beneath the waters of Chesapeake Bay and its estuaries and are now receiving deposits of mud and sand from the adjoining land."

During at least the last decade, the overall effect has been net loss. In 2005, Chesapeake Bay had about 283,950 acres of tidal wetlands, a decline of about 2,600 acres from ten years earlier. Another factor in rising bay levels is that much less water is being retained by upland wetlands within the bay drainage basin as a whole—meaning that more is running off. Today the bay watershed has about half as many acres of wetlands as it did during colonial times.

BARRIER BEACHES

The Atlantic Coast of the Delmarva Peninsula is not part of the mainland like beaches of the Pacific Coast, but rather a barrier island stretching 60 miles from a headland at Rehoboth Beach, Delaware, to near Chincoteague, Virginia, just south of the Maryland state line. Barrier islands are long, narrow ridges of sand that parallel the mainland coast but are separated from it by shallow lagoons or bays. In their natural state, these islands have relatively wide beaches backed by sand dunes. Most of the dunes are no more than 20 feet high.

Several hundred barrier islands line the Atlantic and Gulf Coasts from New York to Texas, all of them perched atop a thousands-of-feet-thick offshore wedge of deposits on the continental shelf. Barrier islands do not have bedrock. They are made solely of unconsolidated sand. The Delaware coast has more than 20 miles of alternating headlands (connected to the mainland) and barrier islands with back bays. The Maryland shoreline has a narrow, 30-mile-long barrier island.

The northern 10 miles of Maryland's barrier island is called Fenwick Island, but most vacationers know it as Ocean City, a heavily urbanized beach resort that with its condos, hotels, and various amusements is more concrete and asphalt than dunes and beach. At the island's southern end is Ocean City Inlet, a storm-punctured break in the island kept open by jetties to provide access to the ocean for recreational and commercial watercraft. The southern 20-plus miles of the island in Maryland is Assateague Island, a mostly natural barrier island that is a national and state park. The only man-made structures are campgrounds, a few park buildings, and a few miles of paved road. Assateague Island's broad, gently sloping beach is one of the finest on the East Coast because it is allowed to move naturally with the currents and storms of the ocean. Ocean City, anything but natural and constrained by economics not to move, pays dearly to retain its beach because densely packed oceanfront buildings have caused erosion to accelerate. We might say that the Ocean City beach is worth billions of dollars and costs billions of dollars; the beach of Assateague is free and priceless.

The sand for most of Delmarva's barrier islands was deposited during the last Ice Age. Up to about 15,000 years ago, the large volume of water frozen in thick, continental glaciers caused sea level to fall hundreds of feet. Shorelines were 40 to 90 miles seaward of present locations.

As the glaciers melted, sea levels rose steadily until about 3,000 to 4,000 years ago. Waves breaking offshore scoured sand from the seafloor and deposited it in long sandbars parallel to the mainland. These sandbars eventually became barrier islands. Marshy bays and lagoons separating the islands from the mainland were filled with saltwater by overwashing storm surges and tides moving through breaches and inlets.

Ocean levels have continued to rise slowly for the last 3,000 years. Although this rise has caused both mainland coasts and barrier islands to migrate inland, they have remained parallel to one another and maintained a fairly constant width of bay between them.

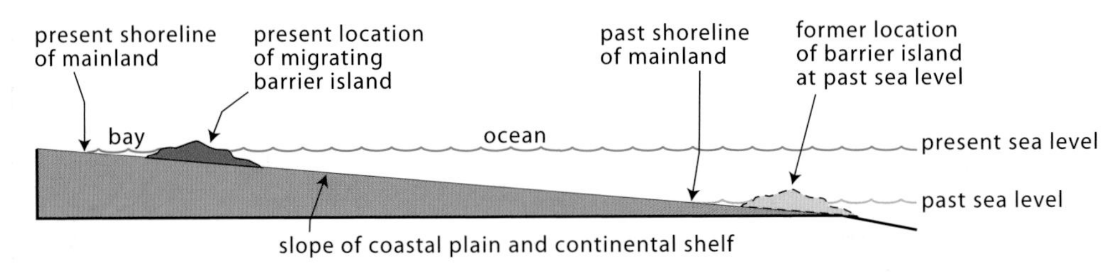

Barrier islands migrate up the coastal plain as sea level rises. —Modified from Titus and others, 1985

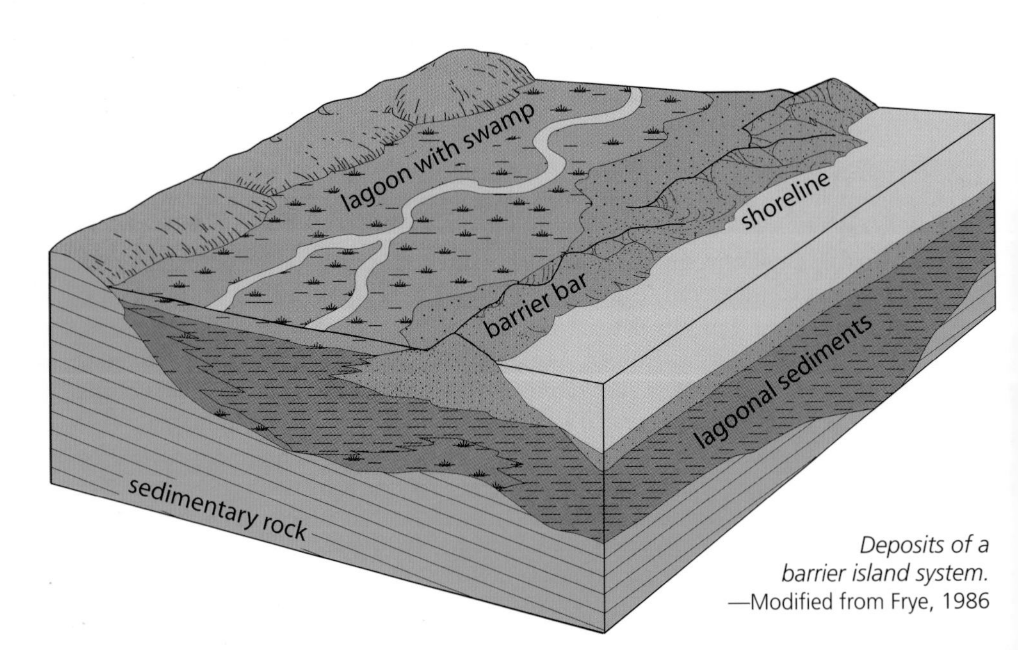

Deposits of a barrier island system. —Modified from Frye, 1986

In the ocean, the location of sediment deposition is largely determined by the size and weight of the sediments. Smaller, light sediments, such as silt and clay, remain in suspension and travel out to deeper waters. Sand, which is relatively large and heavy, is deposited near the shore. Because it is made of hard quartz, sand can withstand the repeated pounding of the waves and be moved along the coastline by longshore currents. These steady currents, which are caused by waves approaching the coastline at an angle, transport huge amounts of sand along the beach face. This part of the ocean, measured from the beach out to a depth of about 25 feet, is called the littoral zone. Thus, while individual quartz sand grains are highly resistant to erosion, masses of sand on the beach and offshore bottom are as active as the waves that push them. When a beach is "destroyed" by erosion, the

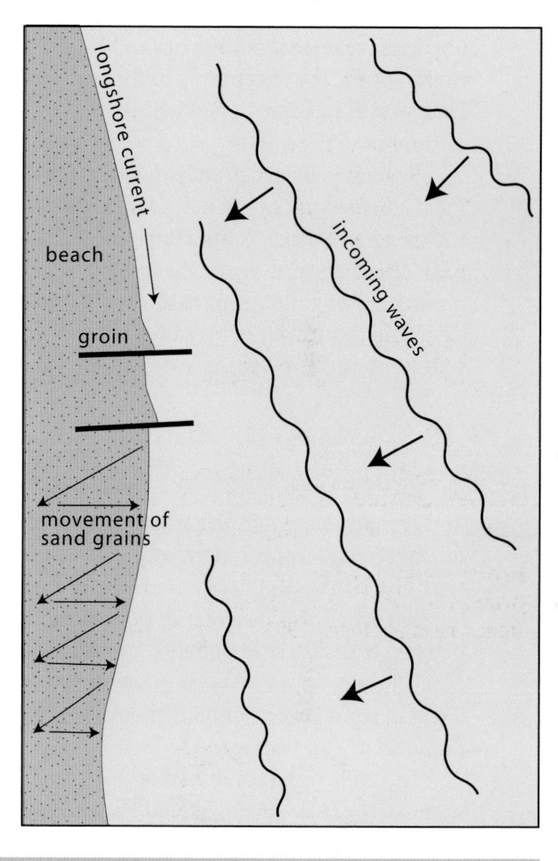

Predominant offshore winds or storm winds generate waves that approach the shore at an angle, creating a longshore current parallel to the beach. This current moves sand in a predominant direction along the beach. Groins can catch much of the sand and help prevent beach erosion in the immediate area. —Modified from Williams, Dodd, and Gohn, 1997

Sand at Ocean City, Maryland, is moved along the beach in a longshore current generated by the angled approach of waves.

sand is only being moved from one place to another, perhaps washed over the island into the back bay.

For the Maryland coast, the net transport of sand in the littoral zone is in a southerly direction and has been estimated at 150,000 to 175,000 cubic yards per year. Sand moves both north and south along the barrier beaches, depending on which direction the wind is blowing, but major winter storms blowing from the north move huge amounts of sand in just a few days, causing the net transport to be to the south. The exceptions to this rule are immediately south of large bays, such as Delaware Bay, so at Cape Henlopen and Rehoboth Beach, Delaware, net transport of sand is to the north.

Several geologic processes contribute to beach erosion and migration. The one most consistently in the news for the last couple of decades is rising sea level. Global warming and other factors have caused a rise of about 1 foot in sea level during the past one hundred years. Second, coastal storms and storm surges wash great quantities of sand from oceanfront beaches and deposit much of it on the bay side. The damming of streams and rivers also contributes to erosion by trapping sediment and preventing it from reaching the ocean to resupply the beaches and barrier islands.

THE FUTURE

Today many scientists are warning that global warming and sea level rise might be far more catastrophic than any volcano, earthquake, or tsunami. The devastation that Hurricane Katrina wreaked on New Orleans fully demonstrated what can happen when the hurricane, perhaps made stronger by unusually warm tropical oceans, transferred its energy to Gulf waters made higher by global warming. In 2003 Hurricane Isabel hit the Chesapeake Bay area and raised water to record levels, 9 feet in some places. On Smith Island in Chesapeake Bay, water entered houses that had never been flooded.

In a 2007 report prepared over six years by scientists from more than one hundred nations, the Intergovernmental Panel on Climate Change (IPCC) concluded that the centuries-long increases in human-generated greenhouse gases were "very likely" responsible for most of the 1.3-degree-Fahrenheit rise in average global temperature. In response to the report, the Maryland Department of Natural Resources (DNR) posted on its website an estimate of what the future might hold for the Maryland coastline. Chesapeake Bay has risen about 1 foot over the past one hundred years, nearly twice the global average reported by the IPCC. Natural land subsidence, currently estimated at a little over 1 millimeter per year, caused by the weight of ever-thickening Coastal Plain deposits, is responsible for the greater rise because sea level is always measured relative to land elevation. While the IPCC projects a global sea level rise of 7 to 23 inches by the end of this century, the Maryland DNR estimates that sea level will be 5 inches higher than that in the Mid-Atlantic region. These are conservative predictions. Maryland shores may see water rise as much as 2 or 3 feet by 2099, and another estimate goes to 4.5 feet.

Effects of climate change and rising waters are many. According to the same DNR website, Maryland is currently losing about 580 acres of land per year to shore erosion. Thousands of acres of wetlands have also gone under. Higher waters mean deeper and more extensive coastal flooding during major storms. Gradual inundation of low-elevation land has already occurred in Dorchester and Somerset Counties, Maryland. Changes in Chesapeake Bay salinity are also projected, but scientists aren't sure how it will change. Rising oceans may increase the salinity, or the projected increase in precipitation for the region may make it lower. Warmer bay waters will also mean a decrease in dissolved oxygen and an increase in biologically dead zones in bottom waters.

The earth has certainly experienced similar catastrophic changes before, but several billion people have never been around to observe which species and which groups will survive into the new world.

Road Guides to Southern Maryland 

Maryland 2
Annapolis—Solomons
44 MILES

It is difficult to see the area's geology from MD 2 because of development, but from it you can take short side trips. Much of the terrain surrounding MD 2 is not as flat as the terrain to the west of the Patuxent River in Charles County, where ancestral river-deposited gravels and sands of the Tertiary Upland Gravel absorb much of the precipitation that falls on them, reducing erosive runoff. Along MD 2 much of the Tertiary Upland Gravel has been eroded (probably because it was relatively thin to begin with, having been farther from the source of discharge than the land in Charles County), thus exposing the underlying marine-deposited sands and clays. Precipitation falling on these less permeable formations tends to run off the surface, forming gullies and stream systems. You can see and feel this uneven sculpting very well on MD 2 north of Prince Frederick.

Side Trip through Aquia and Nanjemoy Formations

Out of Annapolis, one alternate route south is to take Riva Road across the South River, turn southeast onto Beards Point Road, cross Central Avenue onto Brick Church Road, and continue until you reach MD 2. Parts of this route are sunken into the land. Although there are few places to stop, you can see the greensand and clay of the Aquia and the Nanjemoy Formations of the Pamunkey Group of Paleocene-Eocene time.

Smithsonian Environmental Research Center

From MD 2, turn east onto South River Clubhouse Road and drive through sandy clay roadcuts in the Miocene Calvert Formation. A 30- to 40-million-year period of

Unconsolidated clay and sand of the Pamunkey Group on the Coastal Plain near Annapolis.

nondeposition and erosion separate this formation from the Aquia and Nanjemoy Formations on the other side of MD 2. Cross MD 468 onto Contees Wharf Road (gravel) and drive to the Smithsonian Environmental Research Center. It is on the shores of the tidal Rhode River. This 2,800-acre area has trails through coastal woods and wetlands. It also has regular programs that include salt marsh hikes and canoe excursions along the shores of Rhode River.

Calvert Formation Roadcuts

From MD 2, head east on MD 256 (Deale Road) and turn south on Franklin Gibson Road to pass some excellent roadcuts in the Calvert Formation—excellent because on the highly developed western shore, there are precious few public-access locations where you can see this Miocene formation up close.

Upper Patuxent River

From where MD 2 and MD 4 join, take MD 4 north to Dunkirk. Turn west onto Ferry Landing Road and drive to the public hunting area at the end. Here you can walk about a half mile through the woods down to the Patuxent River. Near the river you descend into a gully that has cut through the overlying Calvert Formation down into the glauconitic Nanjemoy Formation. The river here is still tidal, 40 miles from its mouth.

Farther south on MD 2 at Huntingtown, you can follow the signs to Kings Landing Park and see the Patuxent River over 30 miles from its mouth. At the shoreline here you can see where the tidal waters have cut into the Quaternary floodplain to reveal gravelly layers.

Battle Creek Cypress Swamp

Battle Creek Cypress Swamp is a 100-acre sanctuary of one of the northernmost stands of bald cypress trees in North America. To reach the swamp, turn west onto MD 506 (Bowens-Sixes Road) about 2 miles south of Prince Frederick on MD 2. Then turn left on Grays Road. An elevated boardwalk winds through the wetland area. Stumps 10 feet in diameter have been found in underlying clay beds dating from 100,000 years ago, when cypress swamps were common here and mammoths roamed the area. Navigable until about one hundred years ago, this swamp is in the Calvert Formation.

Jefferson Patterson Park and Museum

The Jefferson Patterson Park and Museum is a great site for seeing the Patuxent River in a more natural state and learning about shore erosion control. Turn west onto Parran Road about 3 miles north of Lusby. Follow Parran Road to Mackall Road, and then follow Mackall Road until you see signs for the park. The visitor center has a map for trails to four sites on the Patuxent River where custom-fitted shoreline erosion-control solutions have been used.

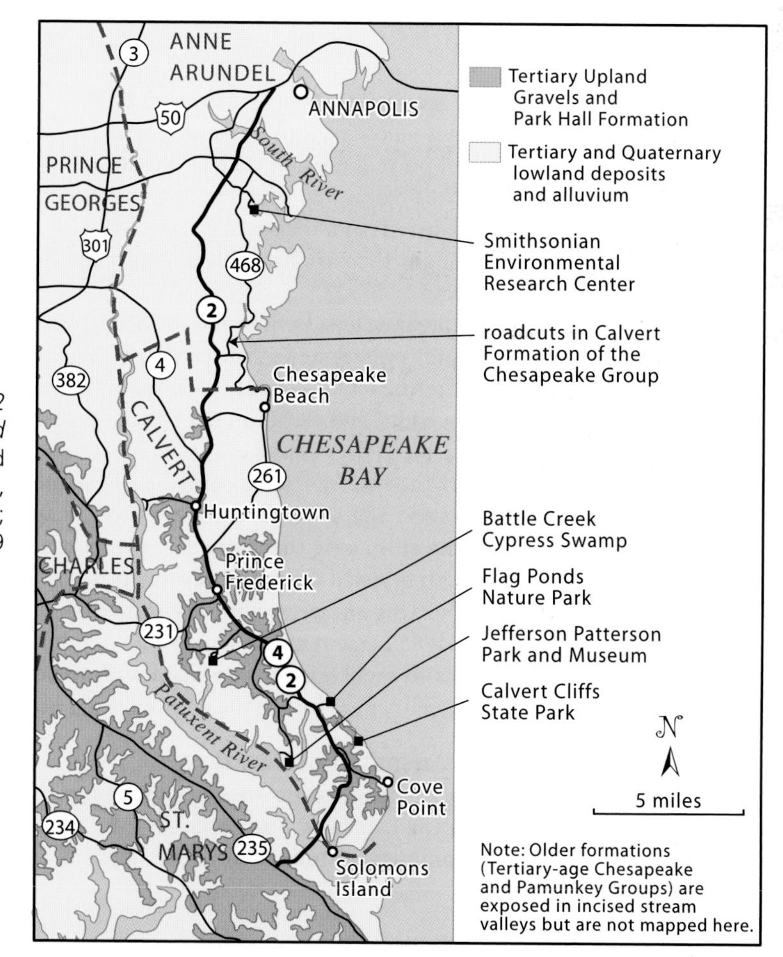

Geology along MD 2 between Annapolis and Solomons. —Modified from Cleaves, Edwards, and Glaser, 1968; McCartan, 1989

On wide, tidal open water, like the lower Patuxent River, waves from wind and motor boats can erode shorelines unless their energy is intercepted by erosion-control structures, such as these breakwaters.

Here you will not see the two traditional types of shore erosion control—the groin and the breakwater—both of which can ultimately do more harm than good. Although the groin can prevent erosion on its updrift side, it can increase erosion on its downdrift side. The breakwater, which reflects wave energy, actually increases erosion and scouring seaward from the barrier; this can cause the barrier to collapse if it is undermined enough. The park has instead constructed living shorelines and other innovative devices.

Site 1 uses planter breakwaters, built in the water near the shore and lined with beach fill and salt-tolerant vegetation in a T shape that bends and dissipates incoming waves. Site 2 has pocket beaches—small, concave inlets between breakwaters designed with a certain width and elevation to obstruct the movement of sand and preserve the beach area. Site 3 uses nearshore stone breakwaters, which are parallel to the shore, to protect the interspersed pocket beaches and valuable archaeological remains on the shore. Site 4, which can be visited by appointment only, has a line of stones (called a sill) along the shore that protects a marshy area behind it. All structures were sorely tested during Hurricane Isabel in September 2003, and all did their jobs of protecting and preserving the shoreline. The park and museum literature points out that 85 percent of Chesapeake Bay's tidal shoreline is privately owned, and that landowners will make a significant difference in determining how many wetlands and shorelines are lost in the coming years of rising sea levels.

Calvert Cliffs and Cuspate Forelands

In Calvert County, MD 2 travels over Miocene sediments deposited about 24 to 5 million years ago in the Salisbury Embayment. The Chesapeake Group, which represents the most complete section of Miocene deposits in the eastern United

States, is dramatically exposed in cliffs that tower nearly 100 feet over Chesapeake Bay. This group accumulated about 15 to 10 million years ago when sea levels were much higher. The sediments are derived from older Coastal Plain deposits and from the Appalachian and Piedmont highlands to the west, which were undergoing intense uplift at the time. Sand and clay were carried by rivers into the shallow marine setting of the Salisbury Embayment, an area that was rich in life. More than six hundred species of marine fossils have been found in the cliffs. The Calvert Marine Museum in Solomons has excellent exhibits featuring all species known to occur in the Calvert Cliffs.

Calvert Cliffs run for about 25 miles along the western shore of Chesapeake Bay between Chesapeake Beach and Calvert Cliffs State Park. Wave action at the base of the cliffs creates active, rapid collapse of the unconsolidated formations, leaving a cliff-face angle of about 70 degrees. The cliffs have little vegetation and, like some of the cliffs at Flag Ponds Nature Park, very narrow beaches at their toes. From bottom to top the cliffs include the diatomaceous clay and dark green to gray sandy clays and marl of the Calvert Formation (exposed only in cliffs in the northern half of Calvert County), brown to yellow sand and gray silt of the Choptank Formation, and dark gray sandy clay of the St. Marys Formation. These three formations of the Chesapeake Group, part of the seaward-dipping Coastal Plain wedge of sediments, lie above the Paleocene-Eocene Aquia and Nanjemoy Formations, visible to the west in Potomac River bluffs. The Chesapeake Group is topped by Tertiary Upland Gravel.

Locations for visiting the shores of the famous Calvert Cliffs in Calvert County are very limited because most of the county is developed and posted. You can view the cliffs from Flag Ponds Nature Park, Calvert Cliffs State Park, and Cove Point.

Flag Ponds Nature Park

Flag Ponds Nature Park is south of St. Leonard off MD 2. It has a path down a gully in the Calvert Cliffs, as well as swamps, freshwater marshes, a sandy beach, and good views of cliffs to the south. The beach is a coastal formation known as a cuspate foreland—often formed by bay or ocean currents carrying sand from opposite directions to a convergence point. Longshore currents are generated by

The sandy cuspate foreland at Flag Ponds Nature Park formed when longshore currents carrying sand in suspension slowed as they moved over an offshore platform of Miocene deposits and dropped their loads.

surface waves that approach the shoreline at an angle: the direction of those waves is determined by predominant winds. Therefore, the location of a cuspate foreland is determined in large part by the direction of the prevailing winds of a coastline. The cuspate projection at Flag Ponds Nature Park is thought to have formed when waves approaching from the north were broken up by an offshore platform of Miocene deposits. The slowing of the waves caused sand to fall out of suspension and collect on the bottom of the bay.

For the last 150 years, the Flag Ponds cuspate foreland has migrated south along the shore at an average rate of about 15 feet per year. With this wide beach protecting the toe of the cliffs from wave action, the cliffs eroded to stable slopes by landslides and slumping, rather than the rapid, more vertical collapse of cliffs that face the bay behind only a narrow sliver of beach. Thus, the slopes behind the relatively wide sandy foreland average about 30 degrees, compared to the 70-degree slopes of the cliffs at water's edge. Geologists have been surprised at the quickness, on the order of decades, at which a bluff can change from the 70- to the 30-degree slope after it becomes protected by a migrating foreland.

Wave action cuts away at the bottom of the unprotected cliffs at a relatively constant rate of 1 to 2 feet per year. Given this rate, geologists warn that nothing can protect a house located on the cliff edge, and alas, several expensive properties have become endangered. For example, in 2005 one house was 12 feet from the cliff edge, 35 feet closer than it had been two years earlier. Its fate is not known, although it can be reasonably predicted. Human devices that protect bluffs from undercutting wave erosion, such as groins or riprap, do not protect the top of the bluff from receding because other agents of erosion, such as freeze and thaw, groundwater seepage, and general weathering, continue to work from the top.

Park at the top of the cliffs and walk the half mile down to the beach. A fence at the southern edge prevents you from trespassing into the area of steep 70-degree cliffs. Many trails run through the upland, cliff, and pond areas, with interpretive signs providing valuable information along the way. Bring your binoculars for good views of the steep cliffs to the south of the fishing pier. Even better views can be had from the end of the cuspate foreland out in the bay. The beach is loaded with shells. If you know what you're looking for, you may even find Miocene fossils and shark teeth that have washed down from the cliffs. Collecting is permitted.

Calvert Cliffs State Park

The cliffs are best seen from boat, plane, or the 1,400-acre Calvert Cliffs State Park located just off MD 2. From the parking lot, you can take the 1.8-mile (one way) Red Trail for close-up views of the southern end of the cliffs. But be advised: the cliffs are closed to foot traffic and collecting. This closure was mandated because excavations by previous fossil hunters have accelerated erosion of the area and increased the danger of cliff-face collapse. Fossil collecting is permitted, however, on the 100-yard beach south of where the trail meets the bay. Collecting is most productive at low tide. From this beach you can see the liquid natural gas terminal out in the bay to the south of Cove Point.

In Calvert Cliffs State Park, the cliffs are actively eroding, collapsing, and retreating. They are closed to foot traffic and fossil collecting.

The walk from the parking lot to the beach is through woods and wetlands, giving you a good idea of how creeks erode the sand and clay of this area and carry the sediments downstream to form the marshy wetlands. Notice that the slopes to the south are vegetated and not very steep. Unlike the steep cliffs, this area has been protected for about 1,700 years by the southward-migrating Cove Point cuspate foreland, visible today almost 2 miles to the south. This foreland is larger and older than the foreland at Flag Ponds, but both migrate south as northerly winter winds erode sand in the north and deposit it on the downwind, downcurrent side of the foreland. Other trails through the park will give you a good feel for the steep slopes of this greatly dissected part of the Coastal Plain.

Since the 1970s the Calvert County coast has become the home of two giant energy facilities: the Calvert Cliffs Nuclear Power Plant (visible from Flag Ponds) and the Cove Point liquid natural gas terminal (visible from the Red Trail in Calvert Cliffs State Park). The nuclear power plant's two reactors generate about 20 percent of Maryland's electricity by using more than 40 tons of enriched uranium per year. Two by-products of this process are the release of more than 2 million gallons of warm water per minute into Chesapeake Bay and the storage of highly radioactive nuclear waste on the 2,000-acre site. The gas terminal receives cooled, liquefied natural gas pumped from tankers arriving from foreign countries. Pipes running under the bay carry the liquid from the pier out in the bay to large storage tanks on the 1,000-acre facility. As you might guess, access to either site is prohibited.

Although we often discuss geology in the past tense, we can look to the future in a 1970s U.S. Army Corps of Engineers' proposal concerning dredging of tidal lagoons behind barrier islands on the Atlantic Ocean. It concluded that sea level rise would inundate the entire Eastern Shore and that ocean surf would then erode the Calvert Cliffs. As more recent concerns about rising seas have become even graver, we might be seeing this erosion sooner than was originally projected, and we hope that nobody forgets about the radioactive waste perched within 1 mile of the cliff face. Remember that these Miocene sediments are unconsolidated—not solid rock.

Cove Point Lighthouse

To reach Cove Point Lighthouse, take Cove Point Road, MD 497, off MD 2 just south of Calvert Cliffs State Park. You will drive through a private community and pass a "No Outlet" sign on your way to the lighthouse, which has been located on the Cove Point cuspate foreland since 1828. You cannot go in the lighthouse, which is still in use, but from the viewing platform or the upper floors of the light keeper's house you can get good views of the Calvert Cliffs (and the liquid natural gas terminal) to the north. There is no access to the private beach.

While the Flag Ponds foreland to the north is thought to have been formed by an offshore shelf interrupting longshore drift of sandy sediment, formation of the Cove Point foreland may have been influenced more by the abrupt change of the Chesapeake Bay shoreline here, which shifts almost 90 degrees from southeasterly to northeasterly. Sand moving in the predominant longshore current piled up where the shoreline changed direction, resulting in a cuspate foreland jutting into the bay. Geologists have evidence that this triangular sandbar has been migrating in a southerly direction for at least 1,700 years. Large boulders on the shore act to deter any further migration, which might undermine the lighthouse structures.

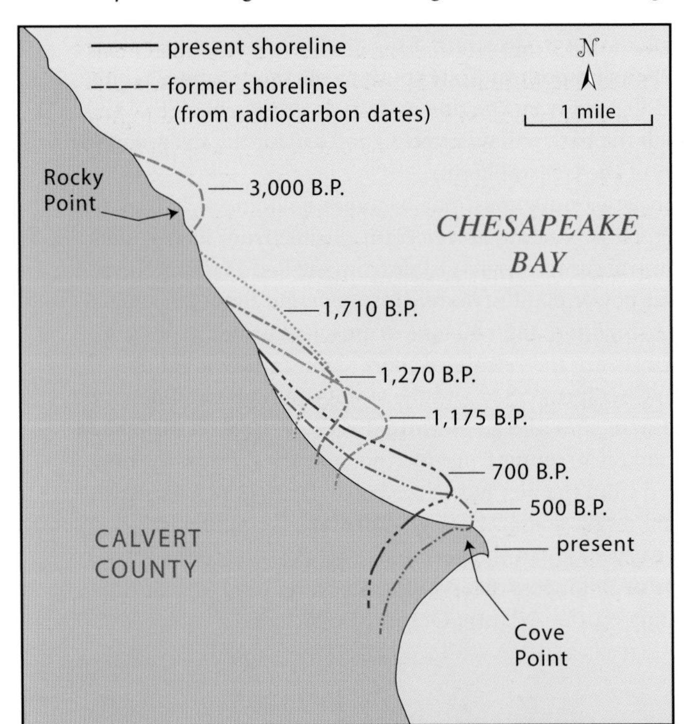

Longshore currents have moved the sands of Cove Point for many centuries. Former shorelines are dated in years before the present (B.P.) —Modified from Larsen and Clark, 2003

Solomons Island and the Lower Patuxent River

At the end of Solomons Island, which is also the end of MD 2, you can see the wide mouth of the Patuxent River. Note the boulders of gneiss that have been hauled in here to protect the bank from wave erosion. They probably come from the Baltimore area. In the town of Solomons you can walk on a boardwalk along the edge of the river. Like those of the Potomac and other rivers entering Chesapeake Bay, the lower reaches of the Patuxent have been flooded by rising sea level to form estuaries. These wide and shallow waters move up and down with the tides.

Maryland 5 and Maryland 235
Washington, D.C.—Point Lookout
54 MILES

MD 5 and MD 235 have been the straight-shot route from D.C. to "The Point" for many decades. The route follows the relatively smooth and flat top of the Tertiary Upland Gravel 4 (TUG_4), which forms the drainage divide for streams flowing northeast into the Patuxent River and streams flowing south or southwest into the Potomac River. These streams have cut valleys down through the Tertiary Upland Gravel into the underlying, older Tertiary sand and clay formations. Thus, on either side of MD 5 and MD 235 you will find hills and valleys—a stream-dissected terrain. While rolling landscape might be interesting to the eye, an even surface has always been easier for travel and road construction. Those who first worked out this route probably did not study geologic maps, but just followed the most even surface, the undissected spine of the Tertiary Upland Gravel cap.

The lack of dissection by erosion of this original plain of Tertiary Upland Gravel may be due to two factors. First, the formation is only a few million years old, relatively young in geologic time. Second, when precipitation falls onto a highly permeable surface like gravel and sand, the water drains internally; that is, it soaks in, as it does with a sponge, and little water is left on the surface for runoff. By contrast, clayey or silty soils do not readily allow for infiltration of water and thus generate runoff, which channels into streams that erode and dissect the terrain. From the banks and bluffs of the Patuxent and Potomac Rivers, which form the edges of this gravel formation, the streams began cutting their valleys down through the gravel to the underlying silt and clay formations and then upstream toward the drainage divide.

Why were the gravel and sand deposited here? Geologists think they were deposited in a kind of alluvial fan by an ancestral Potomac River carrying considerably more water and sediment than the present river. If you look at the course of the Potomac on a Maryland map, you can picture Charles and St. Marys Counties occupying a low, flat area where the ancestral river slowed and dropped its sediment load, as in a delta or an alluvial fan, after plunging rapidly down over the rocks of the Fall Line. Geologists think that Southern Maryland was later tectonically uplifted to establish the over-100-foot elevations of the cap today.

Rosaryville State Park

Rosaryville State Park, adjacent to US 301 and accessible on roads east of Clinton, is in the marine Calvert Formation of Tertiary age. At almost 200 feet above sea level, it is hilly and has an overlook. Over 10 miles of hiking and biking trails pass through forest and field.

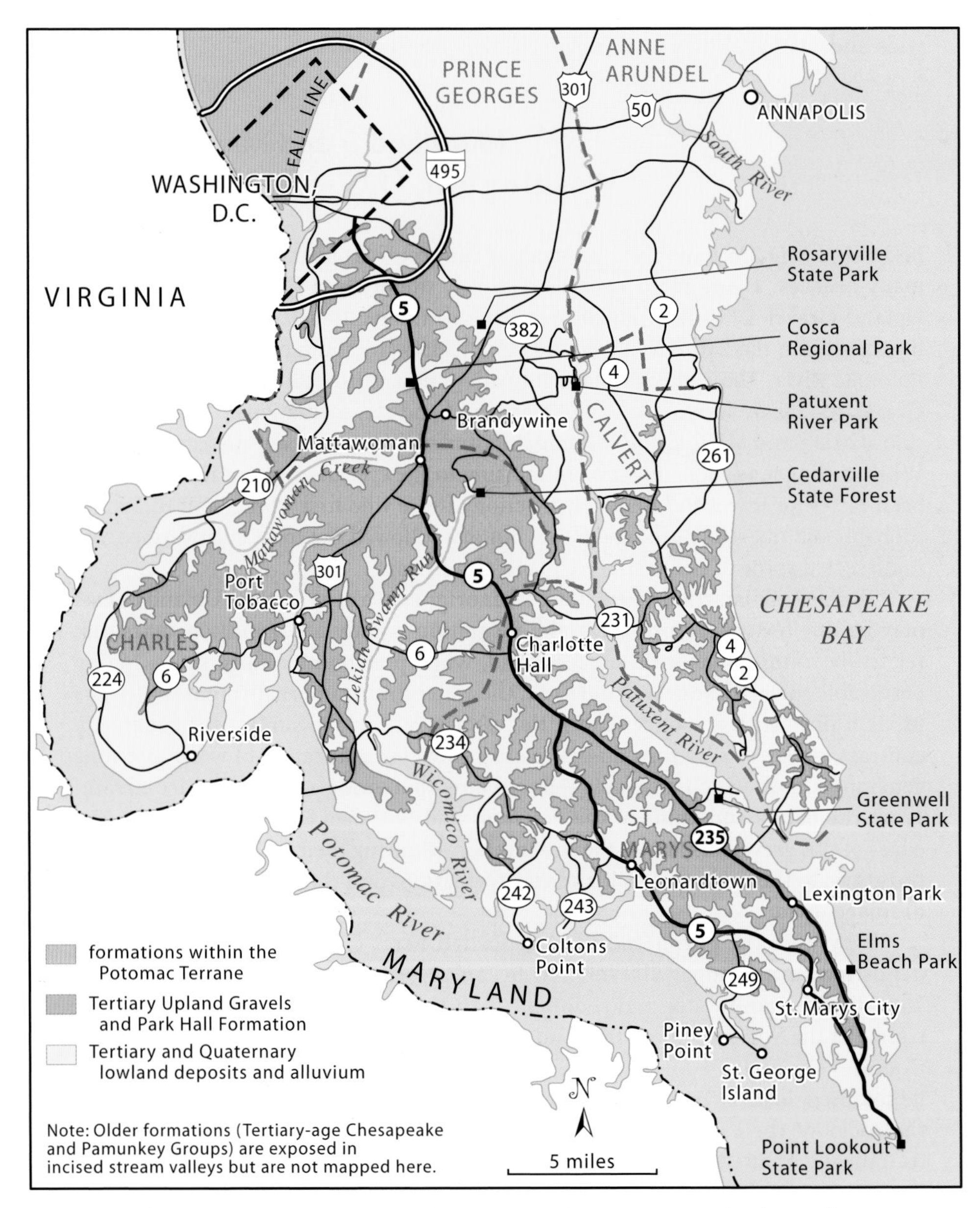

Geology along MD 5 and MD 235 between Washington D.C. and Point Lookout.
—Modified from Cleaves, Edwards, and Glaser, 1968; McCartan, 1989; Glaser, 2003

Cosca Regional Park

Cosca Regional Park has a unique nature center covered with lava rock. Five miles of trails in the park follow steep slopes and streams through woods in the Calvert Formation. Just below the nature center is a meandering stream that might remind you of places in western Maryland. To reach the park, turn west onto Surratts Road, south of MD 223, left (south) onto Brandywine at the light, and then right at the next light onto Thrift Road, and follow it to the park.

Patuxent River Park

For access to the many points of Patuxent River Park, take MD 382 (Croom Road) a few miles south of Upper Marlboro on US 301. You can also reach MD 382 by traveling north on US 301 from MD 5. This rolling drive passes through the wooded hills of the creek-dissected Calvert Formation—the same Tertiary sands and clays that form the Calvert Cliffs about 10 miles to the southeast. The creeks downcut to the east into the tidal Patuxent River, which is bordered by silted marshlands all the way upstream to the Patuxent National Wildlife Research Refuge northeast of D.C. These wetlands, of great value in helping control pollution and preserving wildlife, are protected by county and state natural areas.

Jug Bay and the Critical Area Tour are part of the larger Patuxent River Park. A special-use permit or reservation is required, as the area's 2,000 acres are designated as limited use. Hiking trails, such as the Black Walnut Nature Study Area, with

Valley-bottom floodplain formed by a meandering stream at Cosca Regional Park.

boardwalks and observation tower, provide access to both tidal and nontidal wetlands of the Patuxent. The difference between high tide and low tide in this area is about 2 feet. Many geologic formations in Maryland and Delaware were deposited in river and marsh environments like these, and here you can observe modern examples of these depositional environments.

The 4-mile-long Chesapeake Bay Critical Area Tour connects the Jug Bay area with the Merkle Wildlife Sanctuary. The drive is open to cars on Sundays only, but open daily for hiking and bicycling (except at Merkle from September through December). Interpretive signs along the route indicate nontidal wetlands, freshwater-tidal wetlands, and forest buffers. A 1,000-foot-long wooden bridge crosses the marsh, and a 40-foot observation tower gives wide views of the marshes and open river. To begin the tour, follow Croom Road 4 miles from US 301, turn left onto Croom Airport Road, and follow signs to the one-way entrance. Don't forget your permit from the Jug Bay Visitor Center.

The limited use policy is designed to protect the woodlands and wetlands of Patuxent River Park. The wetlands not only buffer tidal waters from development and the erosion and pollution that it can bring, but they also act as filters and retainers for water. Geologically, these wetlands are very important in the Coastal Plain because all of the formations here are unconsolidated and subject to rapid erosion without vegetative or wetland coverage.

Marshes like this one at Jug Bay on the Patuxent River occur where freshwater river currents meet incoming tidal flow. Marshes filter sediments and pollutants and provide habitat for many species of animals.

Add some Spanish moss, and Zekiah Swamp, Maryland's largest, could be mistaken for the Deep South.

Cedarville State Forest and Zekiah Swamp

On US 301/MD 5 just north of Mattawoman and the US 301/MD 5 split, you will see a sign for Cedarville State Forest at Cedarville Road. The headwaters of Zekiah Swamp, Maryland's largest freshwater swamp, are in this 3,500-acre forest. The swamp, up to 1 mile wide in places, runs south 20 miles to the Wicomico River. Prehistoric beaver ponds were probably initially responsible for slowing stream flow and creating the swamp.

There are almost 20 miles of trails in Cedarville State Forest, many of them with gravel alongside and underfoot. The gravel was not hauled in. This area is on the Tertiary Upland Gravel 4 (TUG_4), once known as the Brandywine surface—a mixture of pebbles of quartz, quartzite, and chert in a sandy matrix. The rounded surfaces and varied composition of the gravel indicate it was washed down by rivers from ancient inland highlands, forming a cap over earlier Tertiary deposits. The sandy gravel forms an excellent, spongelike aquifer for Zekiah Swamp to the south.

About 6 miles south of Mattawoman, MD 5 dips down from the upland gravel into the Tertiary Calvert Formation exposed by the downcutting of Zekiah Swamp Run. This same broad, sandy clay formation, a layer in the Coastal Plain wedge

This gravel is part of the Tertiary Upland Gravel; it was not trucked in for the trail surface.

beneath the Tertiary Upland Gravel, is found 20 miles to the east in Calvert Cliffs and in Potomac River bluffs 12 miles south near the US 301 bridge. Zekiah Swamp Run is a creek that has silted up to become a swamp, or wetland. Much of the siltation is due to human clearing and planting of the land.

St. Marys County

You can do a loop drive through St. Marys County. The following sites are organized along the loop clockwise, heading to Point Lookout via MD 235 and returning on MD 5. Though the Tertiary Upland Gravel underlies MD 5 and MD 235 from Charlotte Hall to Lexington Park, the Park Hall Formation underlies most of MD 5 from St. Marys City to Leonardtown, most of MD 235 between Lexington Park and Dameron, and much of the southern third of St. Marys County. Quarried heavily for gravel, this late Pliocene formation contains much sediment derived from the ancestral mountains to the west. It does not overlie the Tertiary Upland Gravel and appears to have been deposited in an area that had been moved downward by tectonic adjustments even more than other subsiding Coastal Plain areas. The coarsest (largest) pebbles and gravel in both the Tertiary Upland Gravel and the Park Hall Formation were deposited in the channels of fast-moving rivers during a period of lowering sea level.

The southern section of MD 5 goes up and down as it traverses dissected terrain, where tributaries of the Potomac have downcut into the Park Hall Formation. You may see silty, swampy streambeds, such as Glebe Run just east of Leonardtown.

Greenwell State Park

At Hollywood you can turn north on MD 245 from MD 235 and visit Greenwell State Park. A good trail follows the shore of the tidal Patuxent River, which here, about 6 miles from its mouth, is nearly 1 mile wide. Toward the upstream end of the shore trail, across a small cove, you can see river bluffs in the Choptank Formation. Notice on your drive down and back that you are in hilly, dissected terrain and no longer on the Tertiary Upland Gravel cap.

Elms Beach Park

About 5 miles south of Lexington Park on MD 235, watch on the east side for the sign to Elms Beach Park (south of the sign for the Environmental Center). At this rare public access to the western shore of Chesapeake Bay, you can see a natural beach, dunes, pond, and marsh, and views across the bay. The beach itself and shores to the south and north are on the Kent Island Formation, deposited in Quaternary time in the ancestral, high-standing Chesapeake Bay.

Point Lookout State Park

Most of the last 4 miles of the neck down to the point are in the Kent Island Formation, and some of the land is very low. The narrow causeway to the point has

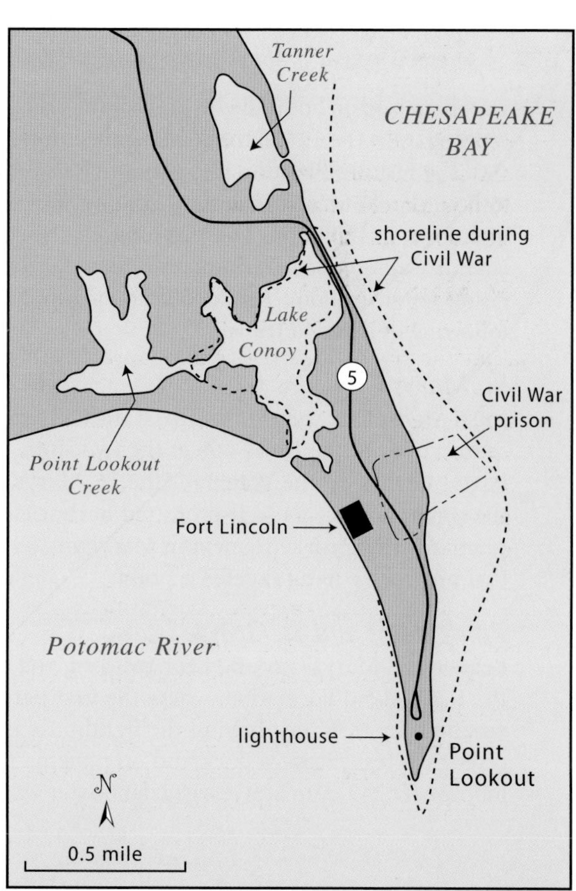

The prison at Point Lookout was used to hold Confederate soldiers during the Civil War. It was located on land that is now mostly submerged or eroded. —Modified from plaque at Point Lookout State Park

Boulders help prevent shoreline erosion at Point Lookout—for now. Note the change in water color and shade where the 5-mile-wide Potomac enters Chesapeake Bay (from the right of this photograph).

riprap stacked on both sides to prevent the rising bay from spilling over the road and eroding it. In the 1970s, the road perched undefended just above the waters.

The Fishing Pier provides good views of Chesapeake Bay, the shoreline down to Point Lookout, and the over-5-mile-wide mouth of the Potomac River. At road's end down at the point, you can look up the Potomac and see inlets, points, and islands—all composed of unconsolidated sediments subject to wave erosion and rising water. In Point Lookout State Park you can explore marshes and creeks by following designated trails.

St. Marys City

In St. Marys City you can walk past the Old State House to the St. Marys River and see a replica of the *Dove*, one of the two ships that landed in 1634 at St. Clements Island farther up the Potomac. The St. Marys River, with bluffs on both sides of the water, provided a well-protected harbor close to Chesapeake Bay for the first permanent English settlement in Maryland, St. Marys City, located on a river bluff that provided a natural defense point.

Piney Point and St. George Island

Between St. Marys City and Leonardtown, you can take MD 249 to Piney Point, on the Kent Island Formation, where the first permanent lighthouse on the Potomac was built in 1836. Just behind the lighthouse is a reminder of our dependence on geological resources—a dock and oil pipeline for a petroleum tank facility. If you follow MD 249 onto St. George Island, you will drive across a stretch of road that appears to be below the water line—yet another location showing signs of rising sea and estuary levels in Chesapeake Bay. Located at the mouth of the St. Marys River along the depositional inside bank of a bend in the Potomac River, this island probably formed as daily incoming tides slowed the currents and facilitated the deposition of sediments washing in from the St. Marys River.

Coltons Point and St. Clements Island

West of Leonardtown, take MD 242 south to Coltons Point, which sticks well out into the Potomac on the Kent Island Formation. From Coltons Point you can see St. Clements Island out in the river—site of the first English landing in Maryland. One member of the original 1634 landing party estimated the island at around 400 acres. Today it is only 60 acres because erosion and rising water have reduced its size. The island probably formed here because it is located on the depositional, or inside, bank of a long sweeping bend in the river. In addition, this area is supplied with sediments washing in from the Wicomico River and St. Clements Creek, with daily incoming tides slowing the currents and facilitating the deposition of sediments. Now a state park, the island can be visited by excursion boat from the Potomac River Museum at Coltons Point during summer months.

You do not need to drive or bike along the coast of Holland to appreciate the significance of the term Netherlands. *Try a trip along MD 249 (Piney Point Road) to St. George Island.*

Maryland 210, Maryland 224, and Maryland 6
Washington D.C.—Indian Head—Maryland Point—Port Tobacco
58 MILES

South Capitol Street out of D.C. becomes Indian Head Highway, MD 210, which provides access to many interesting sites along the wide, tidal Potomac River. MD 210 ends at Indian Head but you can continue south to Maryland Point on MD 224 and Port Tobacco on MD 6.

Fort Foote and Fort Washington

On MD 210 a few miles south of the beltway, you can take Old Fort Road to Fort Foote. This former Civil War fort was strategically perched above the Potomac River on the oldest of the Coastal Plain formations—the alluvial, or floodplain-deposited, Potomac Group of Cretaceous age.

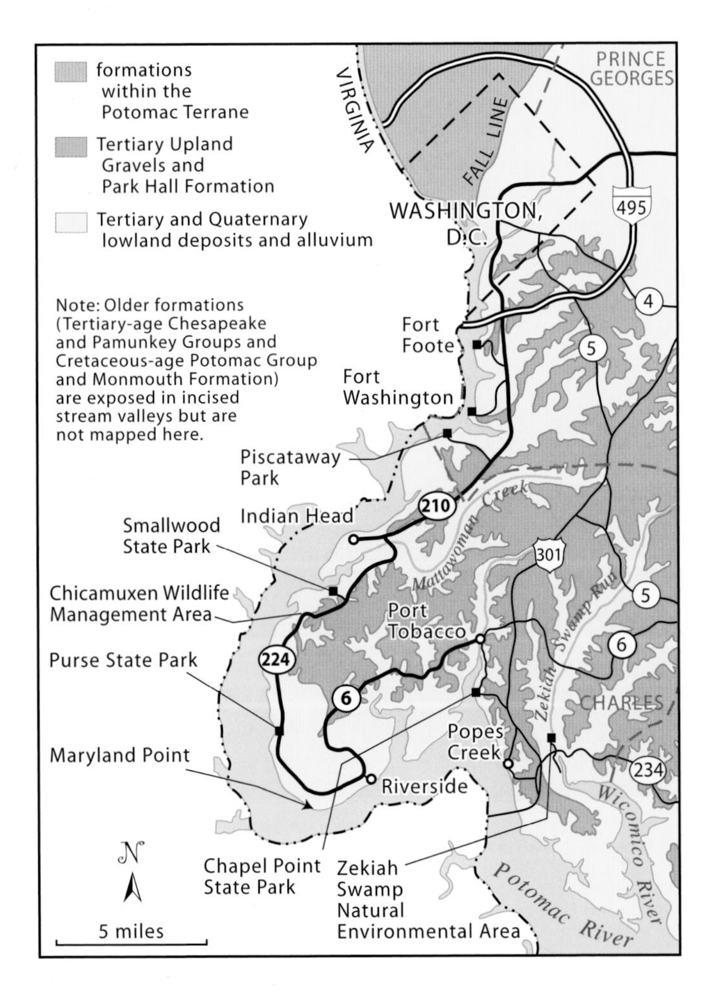

Geology along MD 210, MD 224, and MD 6 between Washington D.C. and Port Tobacco. —Modified from Cleaves, Edwards, and Glaser, 1968; Glaser, 2003; McCartan, 1989

An outstanding panoramic view of the Potomac is located just downstream at a right-angle bend in the river at Fort Washington, the first fort built after the Revolutionary War. From MD 210, take Fort Washington Road. You can immediately see why this fort was built. On a clear day you can see a dozen miles upstream to the city of Washington and downstream a half dozen miles. On this point above the wide Potomac River and Piscataway Creek, and just a couple of miles upstream from Mount Vernon in Virginia, Tertiary Aquia Formation sediments are topped by the Tertiary Upland Gravel. Here and beneath Fort Foote are two of the few locations in Maryland downstream from D.C. where you can visit bluffs above the river.

Piscataway Park and National Colonial Farm

Farther south on MD 210 at Accokeek, you can take Bryan Point Road to Piscataway Park/National Colonial Farm. Directly across the river from Mount Vernon, the park was established to preserve the view from Mount Vernon as it existed during George Washington's time. Before his time, the area was Moyaone, a Piscataway urban center that was the seat of the high chief since about 1400. Six trails take you across this large preserve. You can explore the flat floodplain of the Potomac River and a tidal wetland on a former river terrace from Quaternary time. You can also hike up into some wooded hills located in the same early Tertiary, sandy, marine-deposited Aquia Formation that makes up the bluffs across Piscataway Creek at Fort Washington—both uplands topped by the Tertiary Upland Gravel. The extensive gullying you see in the uplands was probably established during the Pleistocene Ice Ages when the smaller Potomac River ran at a much lower elevation, which facilitated the downcutting. Along the shoreline at the boat wharf, you can see gravels deposited during Quaternary time.

Smallwood State Park

About 5 miles south of Indian Head along MD 224 is a state park on land originally belonging to General Smallwood of Revolutionary War fame. Smallwood State Park is on Quaternary alluvial gravels near the mouth of the shallow but half-mile-wide

Along Mattawoman Creek in Smallwood State Park, you can see the instability of unconsolidated Cretaceous deposits.

Mattawoman Creek. It is hard to imagine that this large body of water is a creek, but the term would have been appropriate before postglacial sea level rise. If you camp overnight, consult a tide table and check the water levels below the footbridge connecting the boat ramps and campground. You will see that the tide does rise and fall significantly. If you can get out in a boat or canoe or walk the dock area a bit, you can see bluffs and bluff erosion on both sides of the creek. Some of this erosion is accelerated by the waves generated by high-speed watercraft—notice that the boat ramp has six lanes! Some geological processes are indeed influenced by humans.

Chicamuxen Wildlife Management Area

MD 224 travels over wooded, dissected terrain as it parallels the Potomac shoreline. About 3 miles south of Smallwood State Park, you can walk through some of this rolling woodland and down to marshlands along Chicamuxen Creek in Chicamuxen Wildlife Management Area, a hunting area during season.

Purse State Park

At Purse State Park, you can walk about a quarter-mile down to Wades Bay for rare public access to bluffs over the Potomac River. Here you can see direct evidence of an environment almost as old as the dinosaurs. At the river, which in this area is a 3-mile-wide tidal estuary, you will see a distinctive dark green layer packed with fossil shell fragments in the eroding bluffs of the Aquia Formation. During the first half of the twentieth century, Maryland geologists described the greensand and many fossils they found here—particularly a spiral, cone-shaped mollusk called

Greensand of the Aquia Formation at Purse State Park. The uncemented deposits slump and erode when undercut by wave action.

Turritella. Across the river in Virginia at Aquia Creek, geologists in the very early twentieth century listed the same fossils—particularly *Turritella*—and greensand in the reference section for the Aquia Formation. The characteristic green color is due to glauconite, an iron-rich mineral known to occur only in shallow marine-shelf environments where very few sediments are washing in. Thus, during late Paleocene time, sediments eroding from the land must have either slowed or been deposited elsewhere as the glauconites developed in the Salisbury Embayment.

Maryland Point

From Purse State Park, MD 224 heads south to Maryland Point, a wide neck in a large meander of the Potomac River. The road passes through the woods and swamps of a United States Naval Reservation along the way. If you turn south on MD 6 and drive to Riverside, you can park at the edge of the river, a 1.5-mile-wide tidal estuary at this point. You cannot walk to either side of the road or to the river because the land is private, but you can get a good view. This is about the only overlook of the river on Maryland Point because nearly all of the riverfront is either private or posted by the government.

Port Tobacco and Vicinity

Port Tobacco is 2 miles west of US 301 on MD 6. Park at the historic courthouse and walk to the rear of the building, where you will see a swamp. John Smith sailed here from Jamestown in the early 1600s and visited a Native American village. From 1770 to 1775, the port was one of the busiest in the colonies. It became landlocked,

This photo was taken from the MD 6 bridge over the Port Tobacco River. An important colonial port was located not far downstream from here but was closed more than two hundred years ago because of siltation, a process that continues today.

however, when widespread cutting of trees and clearing of land on plantations led to heavy soil erosion and the silting up of the Port Tobacco River channel. At the MD 6 bridge over the river, you can see evidence of ongoing soil runoff.

For a beautiful vista of the Potomac and the mouth of the Port Tobacco River, turn off MD 6 just east of Port Tobacco onto Chapel Point Road and drive to St. Ignatius Catholic Church, one of the oldest Catholic parishes in the eastern United States. An eighteenth-century traveler called this "a commanding eminence overlooking the Potomack." Chapel Point Road continues south and east to US 301. There is a good stopping area with a view of Zekiah Swamp about 1 mile east of US 301 on MD 234. With a drainage area of 105 square miles, it is Maryland's largest freshwater swamp. The swamp, a creek that has silted up since colonial times, has its headwaters in Cedarville State Forest and extends south for 20 miles, emptying into the Wicomico River. Colonial efforts to ditch and drain it for farming are still evident. Wetlands are very valuable for filtering water and for preventing flooding, but they are endangered worldwide. Fortunately, Zekiah Swamp is protected as a Maryland Natural Environmental Area.

You can take an interesting side loop on Popes Creek Road (the northern end is at Faulkner at US 301 and the southern end is off Edge Hill Road just south of the MD 234 junction). On the surface are the level ground and farms of the Tertiary Upland Gravel, but the road dips down through the Tertiary Calvert and Nanjemoy Formations, partially visible in vegetated roadcuts, as it goes into the incision made by Popes Creek. Bluffs on the Potomac near here are well documented by geologists, but today are privately owned. You can get a rough idea of the red clay-topped bluffs by stopping in the parking lot of the restaurant at the mouth of Popes Creek.

Road Guides to Maryland's Eastern Shore

US 50
Chesapeake Bay Bridge—Ocean City
106 MILES

US 50 runs from Ocean City on the Atlantic Ocean to San Francisco on the Pacific. For most Marylanders and many others, US 50 is the main route for traveling to the beaches of the barrier islands of the Delmarva Peninsula. There is much, however, between the bay and the ocean that gets missed. Because many miles of bay shoreline on Maryland's western shore or mainland are privately owned, the traveler has better opportunities for seeing and exploring the bay and its tributaries, peninsulas, and wetlands on the Eastern Shore, where there are many more public lands and access points. You can leave the heavy traffic of US 50 on many Maryland state routes and explore land and water in a relatively undisturbed condition.

Delmarva Peninsula makes up the eastern, geologically younger portion of the Coastal Plain physiographic province. Remember that the Coastal Plain runs from the Fall Line to the coast, including Chesapeake Bay, and far out on the ocean

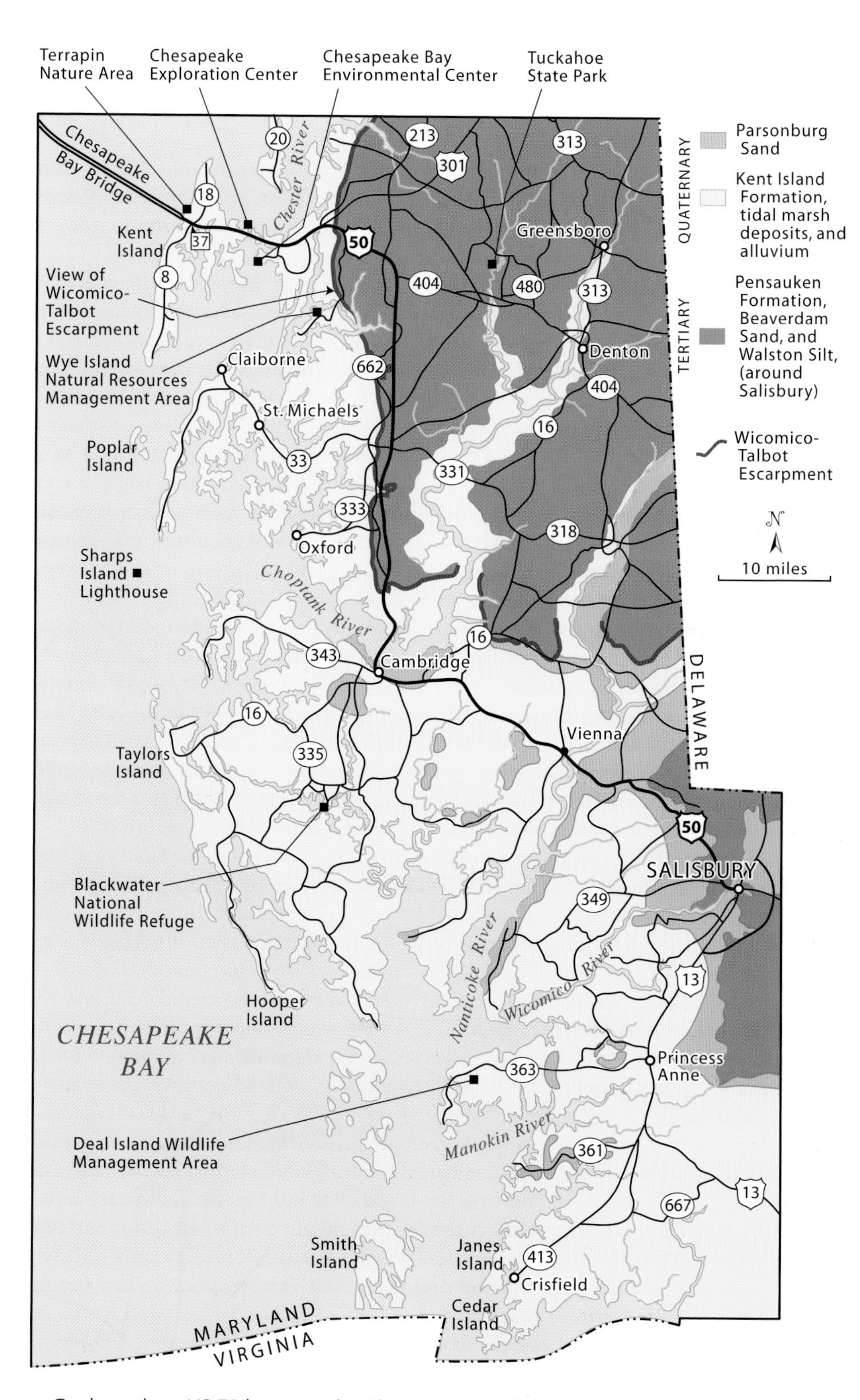

Geology along US 50 between the Chesapeake Bay Bridge and Salisbury. The Wicomico-Talbot Escarpment was the shoreline of Chesapeake Bay about 125,000 years ago. —Modified from Cleaves, Edwards, and Glaser, 1968; Owens and Denny, 1979, 1984, 1986a, 1986b; Glaser, 1998

floor. On Delmarva you will not see the exposed layers of solid rock that most people associate with geology because the Coastal Plain is a slightly tilted layer-cake of unconsolidated sediments: gravel, sand, silt, and clay. Eroded from the ancient Appalachians and Piedmont, these sediments were laid down in successive layers out on the bottoms of many different ancient seas that stood at heights much greater relative to the land than the sea stands today. These sediments did not lithify into sandstones or shales because the waters in which they were deposited lacked dissolved cements. Lying on the buried, solid basement rock, the wedge of sediment thickens from a few feet near the Fall Line to over 1 mile beneath Ocean City. Chesapeake Bay is but a shallow cavity in this huge wedge.

On Delmarva you will encounter few hills and valleys. Heading southeast on US 50 or northeast on US 301, you will see many fields that might make you think you are in the American Midwest—driving across farmland in, say, Illinois or Nebraska. On Delmarva you will see flat fields, woods, and wetlands. Agriculture is widespread, but there are a surprisingly large number of wooded areas. Many of these serve as windbreaks to help prevent wind erosion of soil—also a Midwestern problem.

The southernmost counties of the Delmarva peninsula—Maryland's Dorchester, Wicomico, Somerset, and Worchester, and Delaware's Sussex—all have primarily Quaternary deposits on top of the mile-thick wedge of deposits. Most of Dorchester, Wicomico, and Somerset southwest and south of US 50 are topped by the Kent Island Formation, deposited in the ancestral Chesapeake Bay, or recent tidal marsh deposits. Except for gentle slopes, dunes, or stream valleys, the land is flat, and the highest elevations do not reach even 100 feet. Many areas near Chesapeake Bay or the Atlantic Ocean are less than 10 feet, and they are losing elevation to rising sea levels.

Percentages of county land surfaces that are occupied by wetlands are high. Nearly all counties on the Delmarva Peninsula are over 10 percent wetlands, while Maryland's Dorchester and Somerset are about 40 percent, those two counties having over 1,000 miles of shoreline. The many different types of wetlands found on Delmarva include estuarine, tidal, and brackish-water marshes along the shores of Chesapeake Bay and its many tidal rivers and creeks; tidal salt marshes associated with the barrier islands and their lagoons; many freshwater wetlands such as forested swamps and bogs, and nontidal wetlands occupying floodplains or borders of streams and rivers; and the circular wetland depressions known as Carolina bays.

Along with abundant rainfall, geologic history has been the main factor in the wetlands becoming established here. Streams cut down through the landscape as they do everywhere, but because the land is flat and elevation above sea level is slight, they do not flow with fast currents. They do not cut deep channels and often have poorly defined outlets or no outlet. Also, in many low places clayey soils with low permeability and poor internal drainage result in water pooling into swampy bottoms. Low-lying areas near bay and sea have become backflooded by the sea, which has been rising for thousands of years since the last Pleistocene Ice Age.

As you drive, look about you and you will see these wetlands, even along highly developed US 50. Where you do not see swamp or marsh, you may see channels that

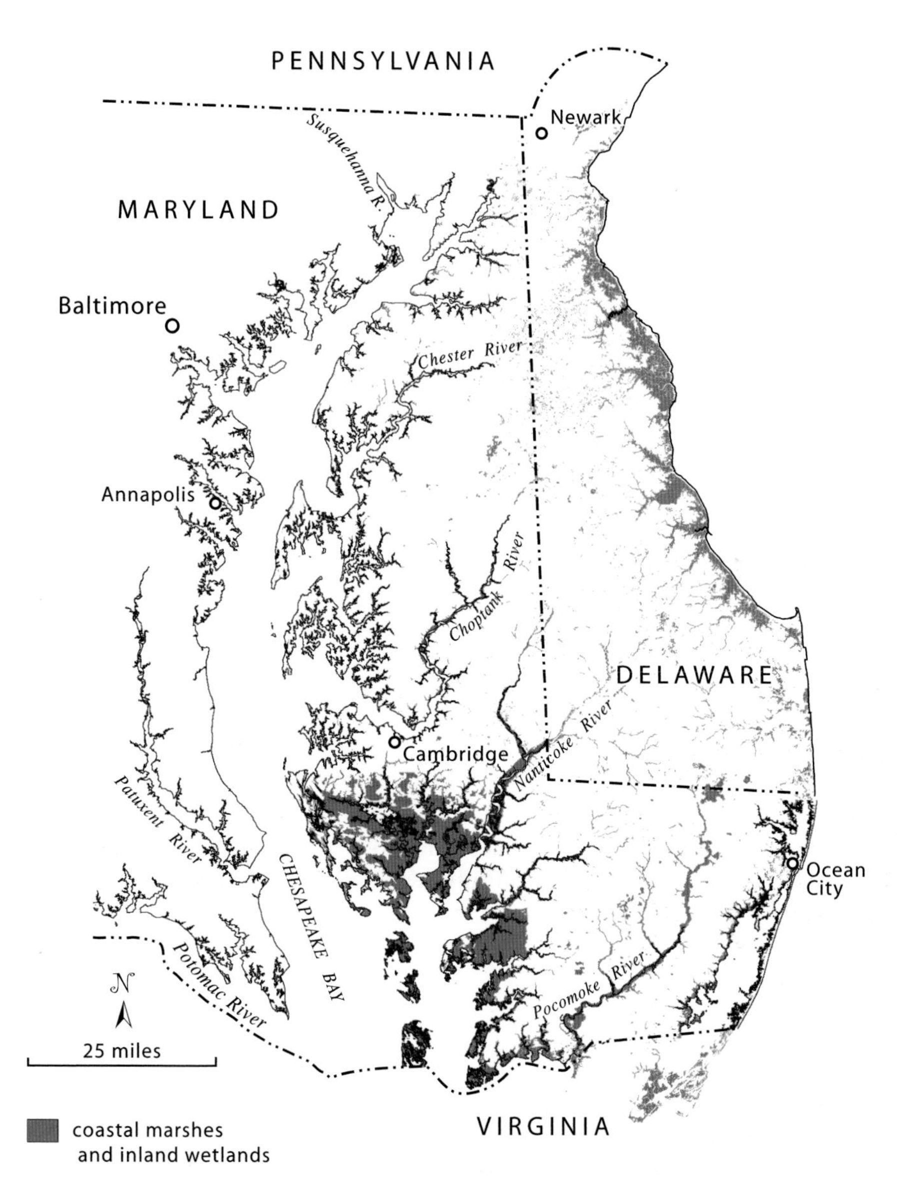

Wetlands of Delmarva. About 40 percent of the total area of Dorchester and Somerset Counties is tidal marshes. —Compiled from Delaware Geological Survey data, 2010

have been dug to drain fields for agriculture or ditches alongside the road. If you see water standing in these, you can usually assume that you are looking at the water table, the top of the groundwater, which is usually only a few feet below the land surface. Even though you may see large expanses of dry farmland on Delmarva, it is still a very watery area.

Chesapeake Bay Bridge

The Chesapeake Bay Bridge, its first span (the southern or eastbound one) built in 1952 across the narrowest part of Chesapeake Bay, connects the urbanized mainland Coastal Plain of Maryland—called the western shore—to Kent Island and the less populated Eastern Shore. The Bay Bridge is 4.5 miles long and rises very high, some say frighteningly high, above the water. The highest section of the bridge lies above the shipping channel, which connects Port of Baltimore to the Atlantic Ocean for large, sea-going vessels. The channel runs the length of the bay to its mouth at Portsmouth, Virginia. Was this channel dredged out so that ports could be established? No, just the reverse. Ports were established in colonial times because channels in the bay as well as in tributary rivers existed naturally. Well over 20,000 years ago, the ancestral Susquehanna and other rivers had carved valleys here into what was then land, not bay, during the Ice Ages when sea level was much lower. These flooded valleys now function as shipping lanes. See the Coastal Plain introduction for more discussion about Chesapeake Bay.

Kent Island

The east side of the Bay Bridge meets land at the western end of Kent Island. The first exit on the Eastern Shore, exit 37, gives access to points where you can get water-level views of Chesapeake Bay. To the north off MD 18, you can turn into Chesapeake Business Park and follow the signs to Terrapin Nature Area. There, a 0.4-mile trail leads to the bay. Be sure to turn left off the paved trail after about 100 yards and follow a gravel trail through a tidal marsh to the shore of the bay. From there you can see the Bay Bridge out in the bay. Two miles south of exit 37 on MD 8 is Matapeake State Park, where you can walk down to the bay and see the Bay Bridge from the south. Before the Bay Bridge was built, the Matapeake Ferry crossed to this point from Sandy Point.

Each location affords a good perspective for appreciating the size of this large estuary, which is 200 miles long, has a drainage basin spread across six states,

View to the south from Terrapin Nature Area of the Bay Bridge, which spans what was dry land a mere 10,000 years ago. Engineers located the bridge at one of the bay's narrowest widths—4 miles.

and has 5,600 miles of shoreline! On clear days can you see across the bay from Terrapin Nature Area to Baltimore and from Matapeake to Annapolis and the U.S. Naval Academy.

At the eastern end of Kent Island, a high bridge crosses the Narrows, a narrow channel that separates the island from the Eastern Shore.

Chesapeake Exploration Center

Just west of the Kent Narrows Bridge, you can get off on exit 41 and, at the Queen Annes County Visitor Center, visit the Chesapeake Exploration Center, an excellent exhibit on the human settlement of the Chesapeake Bay area and its geologic history. Three interesting displays here show: (1) Kent Island, Ocean City, and the southern Eastern Shore under the sea during a warmer climate 125,000 years ago; (2) the ancient Susquehanna River during the last Ice Age, draining land now covered by Chesapeake Bay; and (3) the accelerated erosion and siltation of natural harbors that resulted from extensive clearing and plowing of the land during colonial times.

From the visitor center parking lot, to the north you can see the mouth of the Chester River, a wide, tidal estuary itself. From here you can also explore the excellent Cross Island Trail Park. Get a map in the visitor center and walk, jog, or bike along the 6-mile paved trail for close-up views of marshes, woods, and tidal waters.

Chesapeake Bay Environmental Center

If you want to get a closer look at marshes from boardwalks and trails on narrow points of land, go a few miles east of the Narrows and get off exit 43B (or exit 44A if you're coming from the east) and visit Chesapeake Bay Environmental Center, formerly known as Horsehead Wetlands Center. By walking around here, you can get a good sense of how the rising sea level has flooded and is continuing to flood low-lying lands. In future years, if the present global warming continues to decrease the size of glaciers throughout the world, areas such as this will become bay bottom.

Wicomico-Talbot Escarpment

An escarpment, or scarp, is defined as a steep face that terminates high lands abruptly. The escarpment of the Eastern Shore, which runs for scores of miles to the west and east of US 50 in Queen Annes and Talbot Counties is no longer the steep face it once was, and the high land it terminates is only high comparatively speaking. Waves cut this escarpment when the level of Chesapeake Bay was up to 50 feet higher in Quaternary time. Wave action cut into the higher ground above the shore and created cliffs about 20 feet high in the unconsolidated material. Erosion of material from this higher ground by streams draining the area and by the wave action of the ancient bay resulted in deposition of gravel, sand, silt, and clay on the floors of these offshore waters.

The higher ground, once called by Maryland geologists the Wicomico Plain, is older than the deposits laid down off its shores—a seeming reversal of the law of superposition for sediments, which states that older deposits lie under younger ones. When the water levels dropped relative to the land, the younger, flat, former bay

From Wye Island Road off Carmichael Road approaching Wye Narrows bridge, you can see the Wicomico-Talbot Escarpment. The lower surface (foreground) was the bottom of Chesapeake Bay during interglacial times. The distant ridge was an escarpment, or cliff, similar to today's Calvert Cliffs.

floors were exposed. This low area at the toe of the escarpment was once known as the Talbot Plain. Today the lower plain is called the Kent Island Formation (Pleistocene) and the higher Tertiary surface, the Pensauken or Beaverdam Formation, depending on location.

The Kent Island Formation surface has elevations less than 20 feet, while the crest of the scarp is 50 to 60 feet. The scarp runs for almost 125 miles along the east side of Chesapeake Bay and is somewhat comparable to the present-day Calvert Cliffs in that it is wave-cut. The abruptness of the escarpment has been eroded over many years to a gradual slope in most places, but you can still detect the presence of an elevation difference. Just east of the US 50/US 301 split, you can feel the slope of the road as it crosses this ancient wave-cut shoreline from one plain to the other.

If you want a good view of the Wicomico-Talbot Escarpment where it has not been altered or obscured too much by human development, exit US 50 about 5 miles east of the US 50/US 301 split onto Carmichael Road. Drive about 6 miles to the first bridge, which crosses Wye Narrows, a wide tidal tributary of the Wye River. To the northeast, back where you have just driven, is a good view of the escarpment with the Wicomico surface above the Talbot surface. Here, with the present bay behind you, you can imagine it about 25 feet higher, its waves lapping against the upland surface, creating a cliff and washing sediments from the cliff onto the lowland surface where you are standing.

Side Trip to Tuckahoe State Park

To reach Tuckahoe State Park, turn east on MD 404. At Hillsboro you can turn north on MD 480, do an immediate left onto Eveland Road, and after a left turn onto Crouse Mill Road drive north to Adkins Arboretum or Tuckahoe State Park. The two areas connect, and each has trails along Tuckahoe Creek and its freshwater wetlands.

East of the creek on the state park trail leading south, you will see two distinct levels. The upper levels have been mapped as dunes, probably windblown sand deposits from a cold, Pleistocene glacial period. The Caroline County geologic map is not clear, but the lower level may be, in part, the Kent Island Formation, indicating that backflooding during interglacial times was so extensive that it reached this location over 60 miles upstream from the mouth of the Choptank River. The formation is definitely identified less than 10 miles downstream from here and an even greater distance inland on the main branch of the Choptank, up to Greensboro.

Just below the bridge and dam in the state park you can see a small bluff in the western bank. Here the creek has incised the top layer of Quaternary deposits down into the Chesapeake Group of Tertiary age, the same sediments found almost 50 miles west across Chesapeake Bay in the Calvert Cliffs. The Tuckahoe and many other Eastern Shore streams and rivers have incised their valleys into these formations.

Tuckahoe Creek at Tuckahoe State Park cuts down into the Calvert Formation of the Chesapeake Group, which forms the wooded slope across the creek.

Choptank River at Denton

MD 404 crosses the Choptank River at Denton. For a look at the river, follow Business 404 down to the old bridge over the Choptank River, or go to Martinak State Park off MD 404 just south of Denton. The Caroline County geologic map identifies here, in a band about 1 mile wide straddling the river, the Kent Island Formation of the ancestral, high-standing Chesapeake Bay. The upstream limit of the Kent Island Formation is at Greensboro. The sediment settled out of the bay water, which had inundated preexisting valleys.

Upstream on this river, via MD 313 from Denton, you can see where the Choptank has incised into the Choptank Formation, another formation of the Chesapeake Group. Just a little creek at Greensboro and 200 yards wide at Denton, this river becomes a 6-mile-wide estuary well over 60 miles downstream, below Cambridge.

Poplar Island and Other Vanishing Islands

If you want to explore one of the Chesapeake Bay peninsulas and see typical Eastern Shore landscape, woods, tidal creeks, and waterfront, along with two well-kept

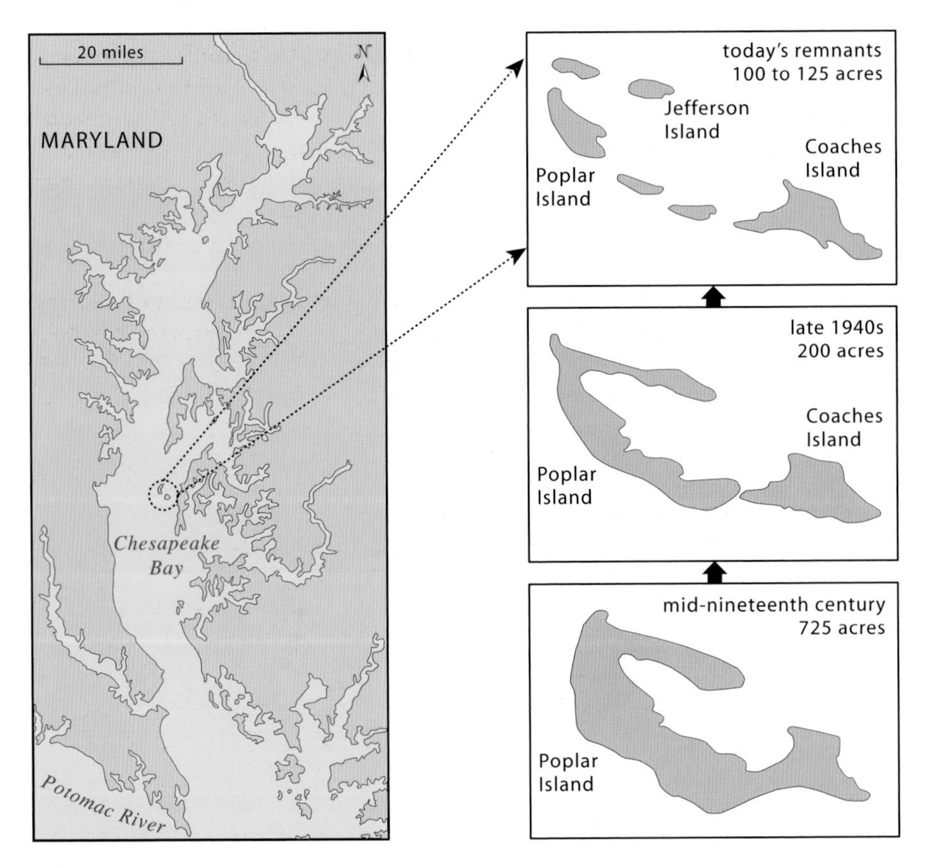

Rising sea level has reduced the size of Poplar Island. —Modified from Chesapeake Bay Program, 1994

towns with maritime history that goes back to the 1600s, visit St. Michaels and Oxford. From US 50 at Easton, you can get on MD 322, also called Easton Parkway, and access MD 33 to the former shipbuilding and present-day yachting center of St. Michaels and MD 333 to Oxford, a colonial port on the Choptank River that once rivaled Annapolis. MD 33 continues out to a small peninsula from which you can see islands that are slowly disappearing beneath the rising bay waters.

Past St. Michaels on MD 33 you can turn off to Claiborne and see Eastern Bay and, in the distance, Kent Island. In the late 1800s Claiborne was the ferry dock for the boat-and-rail line from Baltimore to Ocean City.

Traveling out the peninsula on MD 33, you can stop at Lowes Wharf and see several islands in the bay. On one of the farther ones you might be able to see barges and equipment bringing dredged material from the bottom of bay shipping channels to restore the greatly eroded Poplar Island to a size of 1,200 acres. In 1608 John Smith estimated this island at 2,000 acres, but by the late 1990s it had been reduced to small, low sandbars. Today the filling is a trial solution for two Chesapeake Bay problems: the disappearing shore and island acreage, and the disposal of the huge quantities of sediments that must be regularly dredged from shipping lanes.

MD 33 ends on Tilghman Island at a wide, asphalt parking area beside a naval installation. Stop here and look southwest into the bay for a leaning lighthouse. Today Sharps Island lighthouse is surrounded by water over 10 feet deep. In the 6-mile-wide mouth of the Choptank River, Sharps Island once was a 600-acre island (some say 900) with farms, a school, and a hotel. Rising sea level has submerged it. Another island across the Choptank mouth, James Island, has been reduced from 1,300 to less than 85 acres.

Choptank River

North of Cambridge is the Choptank River. The old Route 50 bridge is now a fishing pier that can be accessed from the north side of the river or from the south, including from the Dorchester County Visitor Center, which is in Cambridge off the first exit south of the river. The pier and bridge are over 2 miles long. The width of this river gives you a good measure of the extent of the post-Ice Age flooding that has raised ocean levels.

Blackwater National Wildlife Refuge

South of Cambridge, you can explore the bay side of lower Delmarva, a series of peninsulas separated by wide tidal rivers that do not have bridges due to the sparse human populations here. Dorchester and Somerset Counties have, respectively, about 500 miles and 600 miles of shoreline where conditions are favorable for the formation of tidal marshes. Almost half of Dorchester County is occupied by wetlands. Blackwater National Wildlife Refuge, along with adjacent Fishing Bay Wildlife Management Area, forms the largest area of coastal marsh in Maryland. Follow the signs for the refuge from Cambridge down MD 16 and MD 335. The refuge is over 20,000 acres of marshes and woods that are habitat and wintering areas for over 250 species of nesting and migrating birds, including bald eagles.

See it while you can, for in forty years Maryland's largest wetland could be underwater, submerged by growing Chesapeake Bay.

Blackwater was one of only nineteen sites identified as "globally significant" by an international group of scientists in 1971 and is only one of six "priority wetlands" identified by the North American Waterfowl Management Plan.

At Blackwater National Wildlife Refuge you can stop at the visitor center and observation platform on Key Wallace Drive for a wide view of the marsh or drive around Wildlife Drive for closer views. The drive has many places to pull off, as well as walking trails in marsh and woods. You can also see the area fairly well from Shorters Wharf Road, off Key Wallace, or from the bridge on MD 335.

Blackwater, which has been called the "Everglades of the North," contains a marsh and soil chemistry that is mostly brackish (mixed salt and freshwater), although some impoundment areas are freshwater. Although originally 28,000 acres, the refuge has lost about 8,000 acres to erosion and sea level rise. Vegetated brackish marshes have been flooded into more salty shallow ponds. Shallow-water marshes have deepened into open-water expanses. Low-lying areas with evergreen trees have been submerged into swamps. Overall, the refuge has been losing about 400 acres per year to drowning, and the entire area could possibly submerge by 2050. In addition, and most recently, the refuge is now threatened by increased runoff and contaminants from developments being built in the drainage basin to the north.

The refuge management is taking measures to help save its wetlands. Large volumes of dirt were recently moved to create impoundments that were then fitted with water gates for directing flow and controlling depths. A draft plan by the U.S. Army Corps of Engineers, which as of yet has not received a feasibility study, involves pumping millions of cubic yards of dredged bay sediment into the watery areas and restoring the marsh there. Each year the bay floor receives huge quantities of sediment from its many tributaries (400 pounds per second!) and in order for shipping channels to be kept open to Baltimore and the C&D Canal, the Corps of Engineers must dredge the bottom regularly. Disposal of this mud is a problem, but for Dorchester County, this bay sediment may be the only source for the large volumes needed to recreate the marshes in the flooded areas. Cost is a big factor because the dredged sediment would have to be barged from bay areas around and north of Baltimore and then pumped into the refuge, which is too shallow for barges to enter. Congress must also approve such a plan. The next twenty years will tell.

Why not let nature or humans take their course and continue the destruction of wetlands that has proceeded rapidly during the last 250 years? What's going on at Blackwater is indicative of larger patterns: less than half of the United States's early colonial wetland acreage exists today, and less than half of Maryland's remain, with the Eastern Shore sustaining the greatest losses in the state because so many wetlands have been drained or filled for agriculture. During the 1950s, 1960s, and 1970s, the United States lost almost a half million acres per year, and the Chesapeake region almost 3,000 acres per year.

Preserving wetlands of all types is important for many reasons. For Chesapeake Bay and coastal regions, a very important value of tidal or coastal marshes is as a buffer during major storms. Hurricanes in the early 2000s in the Gulf of Mexico have demonstrated the importance of having wetlands to store floodwaters temporarily and to act as bulkheads to slow and absorb some of the energy of fast-moving storm surges. With marshes as barriers, many lives, properties, and shorelines can be saved.

During normal runoff events, wetlands help to trap and filter sediments that can damage bottom-dwelling organisms and nutrients that can overfertilize subaquatic vegetation, which then grows rapidly and consumes oxygen in the bay. Farther upland, nontidal wetlands are important not only as filters but as areas of groundwater recharge. Last, but certainly not least, wetlands provide food and habitat for over 40 percent of the nation's endangered species, as well as many others.

Taylors Island and Hoopers Neck

If you follow MD 16 southwest from Cambridge across typical Eastern Shore farms and wetlands to its end at Taylors Island and then follow MD 335 (Hoopers Island Road) to its end, you might encounter the road at high tide or on a windy day when it is almost or completely underwater. Continuing with roadside geology on such a day is not recommended. With sea level continuing its relentless rise, we do not know how long it will be until places like Hoopers Neck are completely underwater. Other islands in Chesapeake Bay have already gone under.

If you follow MD 335 to its end on Middle Hooper Island, you will see more evidence of threats from rising waters. Along the bay, a long stack of huge boulders absorbs the energy of waves to help prevent erosion of the shoreline. Sometimes, low elevation roads here are flooded, and local residents drive them, it is said, by feel rather than sight. Waters from the low-end prediction of a 1-foot sea level rise by 2100 would flood the road that connects these islands to the mainland, while waters from the high-end prediction of about 3 feet would inundate the islands and their fifty or so properties. From Upper and Middle Hooper Islands you have good views of Chesapeake Bay—on a clear day all the way across, about 10 miles, to Calvert and St. Marys Counties.

On Middle Hooper Island discoloration on the boulders shows you the level of high tides—about the same elevation as the road on the other side of the boulders.

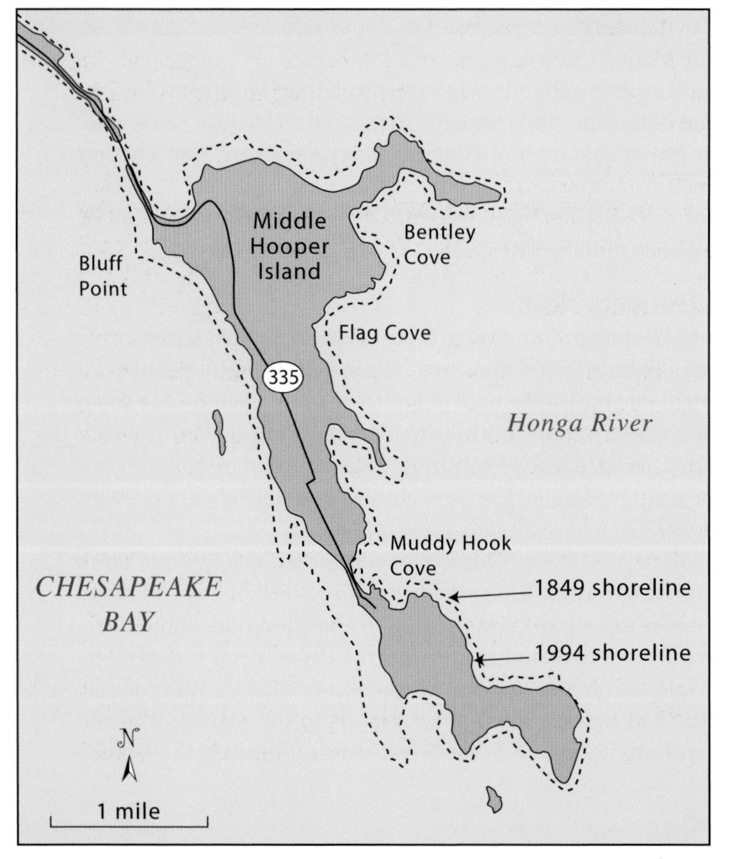

The area of Middle Hooper Island has diminished significantly during the last 150 years of rising water levels in Chesapeake Bay. —Modified from Maryland Geological Survey, Coastal and Estuarine Geology Program, 2001

White sands of the Parsonburg Sand off Decoursey Bridge Road just east of Transquaking River.

Parsonburg Sand Southwest of Vienna

On roads southwest of Vienna, watch for deposits—dunes or low ridges—of the Parsonburg Sand. Wind blew and deposited these sands during arid, cold, late Pleistocene glacial periods.

Nanticoke and Wicomico Rivers

Another peninsula of lower Delmarva, this one in Wicomico County, is bordered by the Nanticoke and Wicomico Rivers. Follow MD 349 west from either US 50 or US 13 in Salisbury, and once again you will go through flat fields and forests. MD 349 swings over to run along the Nanticoke just north of Bivalve and from there down to Waterview—mostly on Parsonburg Sand that tops the Kent Island Formation. You can stop at three different public areas for good views of the 2-mile-wide, tidal lower Nanticoke and, in the distance on the far bank, the wild coastal marsh of Fishing Bay Wildlife Management Area. Lying adjacent to Blackwater National Wildlife Refuge, Fishing Bay is another area susceptible to flooding of its marshes.

Side Trip to Deal Island

MD 363 off US 13 south of Salisbury at Princess Anne leads you down the Somerset County peninsula formed by the Wicomico and Manokin Rivers and crosses many miles of beautiful coastal tidal marsh that make up Deal Island Wildlife Management Area. Just northwest of Princess Anne and north of MD 363, Pine Pole Road and Ridge Road follow, more or less, a ring of five dunes made of Parsonburg Sand, but they are often difficult to discern on the landscape—*ridge* being a relative term in the flatlands here. A 1-mile section of MD 363 utilizes the Parsonburg Sand as a higher corridor through part of the Deal Island marsh. You can see standing water on both sides of the road as you cross the marsh. The Parsonburg has been frequently used for routing roads through marshy areas of the Eastern Shore because the sand's low elevations are higher than the near-zero elevations of the adjacent marshes.

After you cross the bridge to Deal Island, pull off immediately to the right on Parsonburg Sand for a view of Chesapeake Bay which, except for a few tiny island bumps, looks like the ocean. Here you are almost 20 miles from the opposite shore of the bay at Point Lookout, at the mouth of the Potomac River. If you follow MD 363 to its end at a dock, you can look south to the southernmost land areas in Maryland, Janes Island and Smith Island.

Ocean City

Ocean City, built on a barrier island of sand, marks the eastern end of US 50. Known as Fenwick Island, the sandy substrate hosts a resort town. The beach retreated an average of 1.9 feet per year from 1850 to 1980, but after the explosive growth of large buildings during the 1960s and 1970s, Ocean City and the state of Maryland felt the need to reduce this rate to zero. Walls of rock known as groins were built into the Ocean City beach and perpendicularly out into the ocean to trap sand moving to the south and retain it on the beaches in front of hotels and condos.

The many groins along the Ocean City beach are familiar structures to those who like to walk the beach. You can see sand piled up on the sides of the groins, usually the north side. The wind-generated waves from the ocean strike the shore at an angle and generate what is known as a longshore current out in the surf zone. As with any flow of water in nature, this current can transport sediment, in this case suspended sand. The general direction of movement is called longshore drift, and for the Delmarva coast it is from north to south, thus the sand collecting on the north sides of the groins. Ocean City and Assateague share a common supply

Stacked boulders called groins (hauled in from the Piedmont) project into the sea and are designed to prevent upscale, high-rent sand from being moved out of the resort area of Ocean City.

of sand, so what Ocean City keeps for its beachgoers and for structural stability is not available for Assateague Island.

Another structure, larger than any of the groins, was built in 1934–35 at the southern end of Ocean City. In August 1933 a hurricane blew a breach in Maryland's once-continuous barrier island and created what is known today as Ocean City Inlet. Three or four days of continuous, heavy rain caused flooding in the bays between the barrier island and the mainland, and the high water broke across the island toward the ocean, cutting the inlet and carrying the railroad bridge, fishing camps, and three blocks of town out to sea as debris. If you walk down to the end of the boardwalk or drive down to the Municipal Parking Lot, you can walk out on the long jetty that is the last sand-catching structure north of Assateague. To the

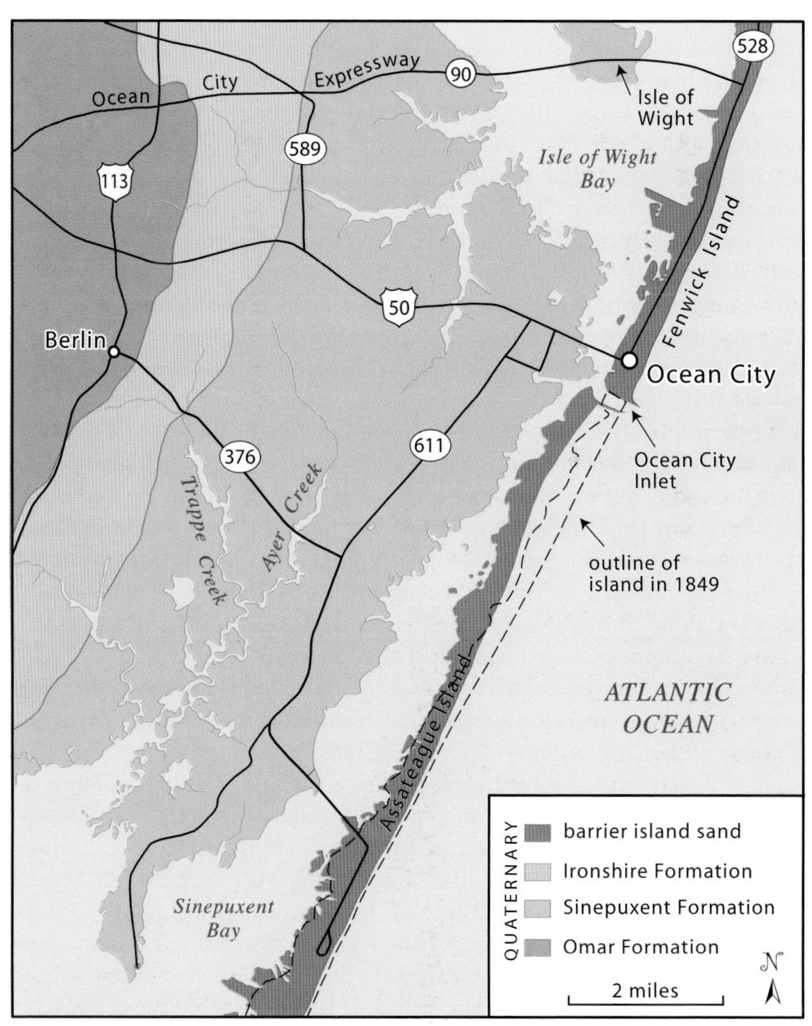

The northern end of Assateague barrier island, deprived of sand by Ocean City groins and undergoing natural migration, has moved landward of Fenwick Island, with which it was continuous until 1933. —Modified from Owens and Denny, 1978; Williams, Dodd, and Gohn, 1997

north you can see the triangular-shaped sand deposit that widens the beach here by several hundreds of feet. If not so trapped, this sand would naturally drift south, fill the inlet, and supply the northern end of Assateague Island. The inlet has been kept open since the 1930s because it provided easy access to the ocean from the bay and mainland for many types of watercraft—a great economic benefit to the area.

To the south of the jetty you can see the inlet and another jetty, but you no longer see Assateague Island directly to the south in its 1933 position. Unlike Ocean City, which was rebuilt after the storm, Assateague remained unpopulated and in a natural state of landward migration. With the inlet jetties trapping some of the annual longshore drift of more than 196,000 cubic yards of sand and diverting a large portion of it offshore, the northernmost 5 or 6 miles of Assateague became deprived of sand accelerating the landward migration.

If you look at a map, you can see that the northern part of the island has migrated much farther inland than the rest of Assateague—about a third of a mile more. Its rate of migration, averaging about 30 feet per year, has been about twenty times greater than that of the more southernly sections, which manage to receive sand from the longshore drift after its diversion around the inlet. To see the northern end of Assateague from Ocean City, you need to walk as far west as you can on top of the sea wall on the northern side of the inlet. You can also see it from the US 50 Harry W. Kelly Memorial Bridge or from streets near the water in West Ocean City.

The long jetty, or fishing pier, is a good reference point for another important event in the history of Assateague. In early March 1962, Ocean City was hit with what older residents remember as the worst storm, in this case a nor'easter—a storm out of the northeast that usually occurs during winter. The motel that you see adjacent to the inlet pier was flooded from the inlet to about 3 or 4 feet over the present parking lot. Down on Assateague, most of the island had been purchased by a corporation, and six thousand lots had been sold. Three days of tides and winds wiped out the development plans and moved the shoreline hundreds of feet. Afterward, the U.S. Army Corps of Engineers, sometimes considered the environmental bad guy, declared Assateague unsuitable for development. In 1965 Congress established Assateague Island National Seashore, which included a state park, and the island was saved from the fate of Fenwick Island.

Many seaside resort cities have spent millions of dollars to widen their beaches and construct dunes for storm protection. In the late 1980s the U.S. Army Corps of Engineers, the State of Maryland, Worcester County, and Ocean City began a multiphase beach replenishment project. Phase I included dredging more than 2 million cubic yards of sand from the offshore bottom about 1 to 2 miles out and pumping it through huge pipes onto the 8.3-mile-length of Ocean City shoreline from Third Street to the Delaware line. Initially this work resulted in a beach over 200 feet wide, a breadth designed to offer protection against 10-year frequency storms and provide plenty of room for the eight million annual visitors. Phase II was designed to provide protection against the one-hundred-year storm, a direct hit from a hurricane. It included installation of 1.6 miles of steel bulkhead and concrete cap to protect the boardwalk from Fourth Street to Twenty-seventh Street, 7 miles

At Ocean City, vegetated dunes and fencing help prevent natural migration (erosion) of the barrier island from beneath billions of dollars of real estate.

of beachgrass-vegetated dunes protected by sand fencing from Twenty-seventh to the Delaware line, and over two hundred beach-access crossovers. A third phase of periodic nourishment took into account the average annual beach erosion rate of about 2 feet per year by providing the addition of about a million cubic yards of sand every four years.

Overall and so far, the project has been successful. For example, two northeast storms in early 1998 caused no damage to any property and only slight damage to the beach and dunes. However, in other Atlantic beach cities, similar measures have not fared so well. Winter nor'easters have been known to wipe out over half of a restored beach only months after the replenishment. In addition, high artificial dunes and bulkheads can reflect the energy of high storm waves back onto the beach and actually increase erosion there.

Isle of Wight

If you want to see Ocean City from the context of our dynamic planet rather than from within the glitz and distraction of the resort city, at Sixty-second Street take the MD 90 bridge over to Isle of Wight Nature Park, or come in from the west on

Skyscraper resort properties sit out in the ocean atop the migration-prone sands of the barrier island—in harm's way of a nor'easter or hurricane. View from Isle of Wight.

MD 90 from US 50 or US 113. Walk out onto the pier. To the east you will see the city "floating," top-heavy as it were, out on the ocean. Here is a good place to consider how precarious is the location of this and other coastal cities.

Assateague Island National Seashore

The sand erosion processes at Assateague are very different from those at artificially buttressed Ocean City. If you walk the beach anywhere in Ocean City and then do the same on Assateague, you will immediately see the differences in beach width and profile. On Assateague there is no development to defend. Broad beaches and low dunes, kept in a natural state by law, absorb wave energy. Sand that is washed over the island during storm surges or that is blown over by winds is absorbed into the vegetated salt marshes on the bay side. The island may move or migrate landward, but its sand is not reflected back into the ocean. It maintains its natural contour and its broad beaches.

As discussed above, because the groins in Ocean City trapped much of the sand headed south with the longshore currents toward Assateague Island, the accelerated migration and erosion of the northern end of Assateague reached some 6 miles down the island by the 1990s. The lack of incoming sand began to affect the state park beach and dunes. Big storms during four different years in the 1990s removed practically all of the dunes that protected campground areas from the beach. Before these storms, you could not see the ocean from the campground because of the dunes. In the early 2000s you could stand in the campground and see the ocean. Assateague was thus placed in the position of its island neighbor to the north, Ocean City: to maintain its location, it would have to initiate a sand replenishment effort. Assateague's beach and dunes, like Ocean City's, were going to cost money to preserve.

Another joint effort of national, state, and local government agencies and the U.S. Army Corps of Engineers resulted in a plan to rebuild dunes within the state park and to provide sand to the beach system to restore the shoreline in the national park area to the north, toward Ocean City. By 2003, the dunes in the state park had been restored over a length of about 1.5 miles to a height of over 20 feet and a width of about 80 feet. Hundreds of thousands of dune grass plants and many miles of fencing were added to keep the dunes stabilized.

On Assateague the natural beach forms a wide, gradual slope. Fencing and planting of dune grass control storm overwash and wind erosion.

While humans cannot control the rising sea or the storms that rage across it, we can control one source of sand erosion—human foot traffic. The millions of visitors per year at Ocean City and Assateague need to stay off the dunes to avoid disturbing the vegetation. Use only the crossover access paths when you go to the beach.

To visit Assateague, follow MD 611 south from US 50 at West Ocean City, or if you are on US 113, take MD 376 from Berlin to MD 611. Just before the bridge to the island, you can stop at the Barrier Island Visitor Center and walk onto the bridge for an elevated view of the barrier island across Sinepuxent Bay.

From the MD 611 bridge, look north, west, and south at the mainland. From Sinepuxent Bay to a line about 3 or 4 miles inland lies the Sinepuxent Formation of Pleistocene age. With ages from 120,000 to 80,000 years before present, its sediments were deposited during a warm interglacial period when glaciers retreated and sea level rose. It was deposited in salt marshes, bays, and other backwaters near the ocean. It may correlate with the Scotts Corners Formation in Delaware. Its western boundary is marked by a bay-shoreline scarp that resembles the scarp between the Scotts Corners and Lynch Heights Formations in Delaware. The Sinepuxent Formation lies beneath Isle of Wight, West Ocean City, MD 611, and the town of Public Landing (to the south). Out in the water it lies beneath Assateague and Fenwick (Ocean City) barrier islands.

On Assateague Island you first come to the state park, where you can walk miles of undeveloped beach. If you head north onto the national seashore, on a clear day you can see in the distance the stabilized Ocean City sticking out in the ocean. It is clearly visible, of course, because the north end of Assateague has been migrating landward for over seventy years—almost a quarter of a mile along the island's northernmost mile. On the national seashore south of the state park, you can hike the Oceanside Backcountry Trail for a wilderness experience or explore the marshy areas of the bay side. If you visit during summer, remember that this is basically a desert with no freshwater but with plenty of mosquitoes.

Except for Ocean City Inlet, which is kept open artificially, there are no breaks in the barrier island. This has not always been the case, and a future storm may cut a new break. In 1920 a storm cut an inlet opposite Snug Harbor a few miles south of Ocean City. It closed naturally, leaving a marshy remnant of its tidal shoal, before the 1933 inlet was cut. A 1962 storm cut a shallow inlet opposite Ocean City Airport. Two storms in 1998 nearly cut through about 1 mile north of the MD 611 highway bridge. A 3- to 5-foot-deep and half-mile-wide section of sand was washed out, and the shoreline migrated west over 300 feet. This erosion was filled by the U.S. Army Corps of Engineers. These events are evidence of the fragility and transience of barrier islands.

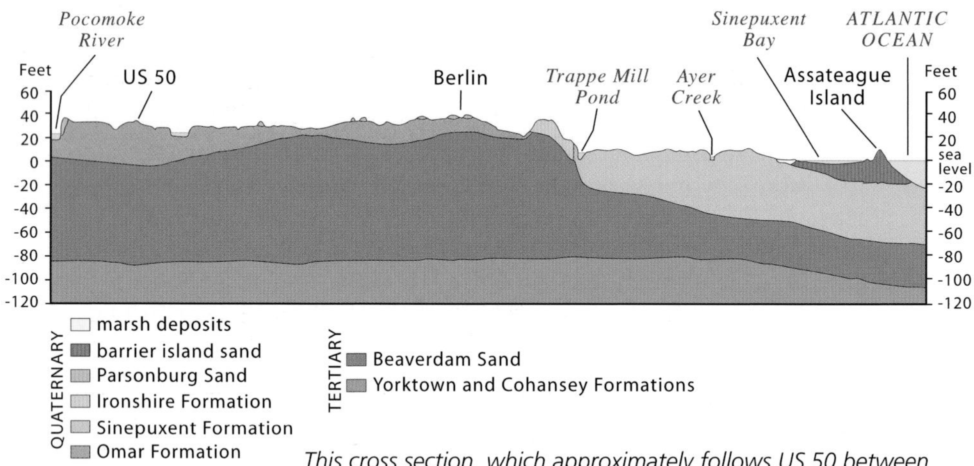

This cross section, which approximately follows US 50 between the Pocomoke River and Berlin and MD 376 between Berlin and Assateague Island, underscores the newness and transience of a barrier island of sand. —Modified from Owens and Denny, 1978

US 113
Berlin—Pocomoke City
29 MILES

US 113 crosses US 50 about 8 miles west of Ocean City, and it is a good road from which to see a variety of formations deposited during Pleistocene interglacial periods in different environments. During the times between worldwide glaciations, global warming caused glacial retreat, and rising seas flooded the lowest elevations of the ancestral Coastal Plain. The Omar Formation (120,000 to 90,000 years) and Ironshire Formation (about 35,000 years) were deposited as ancestral barrier islands and back bays, and the Kent Island Formation (40,000 to 30,000 years) was deposited when the ancestral Chesapeake Bay, a freshwater estuary, encroached on the land. The Parsonburg Sand (30,500 to 16,000 years) was deposited as dunes piled up by winds blowing across an unvegetated Coastal Plain during cold glacial times.

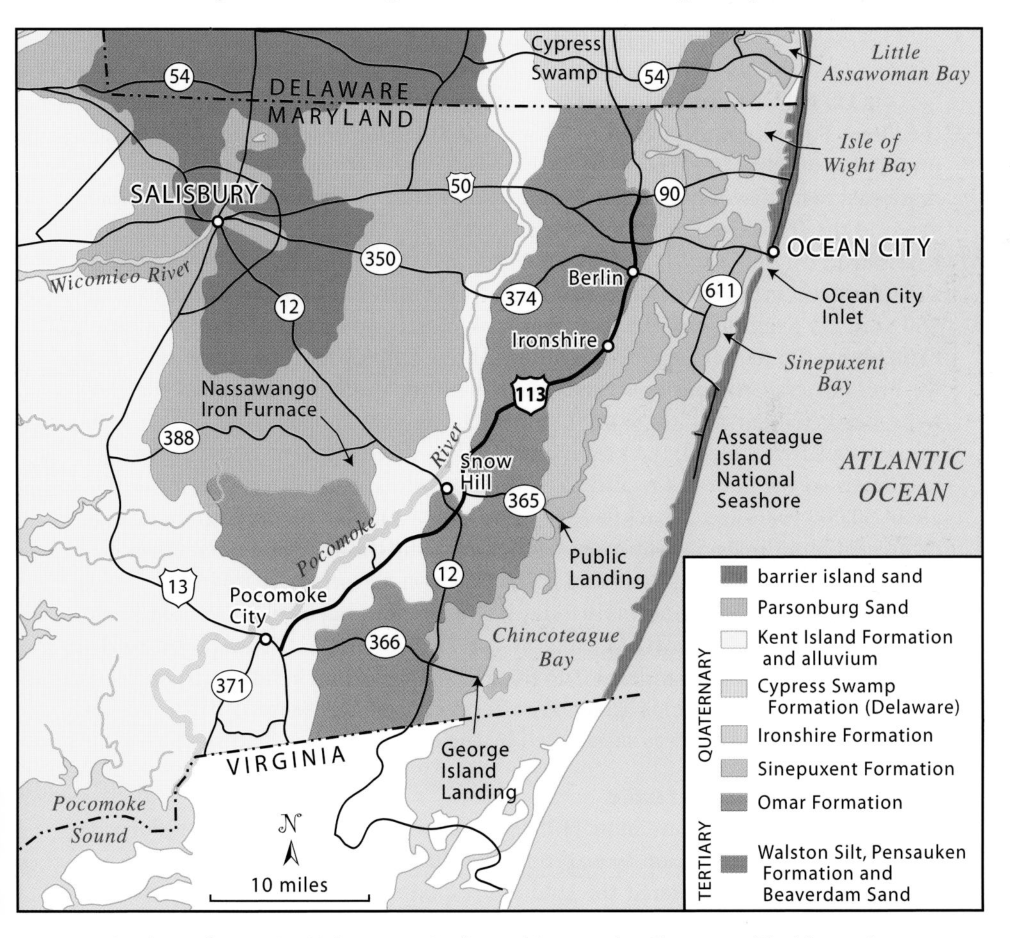

Geology along US 113 between Berlin and Pocomoke City. —Modified from Cleaves, Edwards, and Glaser, 1968; Owens and Denny, 1978, 1979, 1984

Porters Crossing Road climbs from the present floodplain of the Pocomoke River to an ancestral one made of Kent Island Formation and laid down during a much higher stand of the sea.

On US 113 between Berlin and Ironshire, watch for fields of dark silt and clay of the Omar Formation, deposited in a back bay or lagoon environment like those of present bays that lie between barrier islands and the mainland. At Ironshire, if you drive east just a few hundred yards on Mason Road, you will see the light-colored sands of the Ironshire Formation, which overlie the Omar Formation and are also back bay in origin.

About 1.5 miles south of Ironshire and west of US 113 on Downs Road about 200 yards is a subtle dune, and if you look closely you can see the Parsonburg Sand above the darker Omar Formation. About 4 miles north of Snow Hill, turn west on Porters Crossing Road and you will climb another low dune of the same wind-deposited Parsonburg Sand. If you continue west another mile and cross the Pocomoke River, you will turn right and climb up onto the Kent Island Formation, laid down on the floor of the ancestral, high-standing, freshwater Chesapeake Bay about 30,000 years ago. It backflooded up to here—30 miles up the Pocomoke from its present mouth—and many miles farther north up the ancestral river. On US 113 between Porters Crossing Road and Snow Hill, you can see more ridgelike dunes.

Established in 1686, Snow Hill today is a rural town, but until the Revolutionary War it was a thriving port. Located at the head of colonial navigation on the Pocomoke River about 25 miles inland from Chesapeake Bay, it did much commerce with England, but like Joppa Town, Port Tobacco, and Bladensburg of the western shore, its value as a port was supplanted by the deep harbor of Baltimore.

Nassawango Iron Furnace

Off MD 12 northwest from Snow Hill about 3 miles is Nassawango Iron Furnace. This swampy area may seem like an unlikely site for finding iron ore, but the vast pine forests that once covered the land are responsible for concentrating the iron. As naturally acidic rainwater lands on pine needles of an evergreen forest floor and soaks through, its acidity increases. Because the acidic groundwater is an excellent solvent, it readily dissolves iron that is present in the sandy soils. Where the groundwater emerges in low areas such as springs, creekbeds, or swamps, the contact with air causes the iron to oxidize into a form of hematite, an iron oxide mineral, called bog iron.

In the 1970s, prehistoric hearths and graves were found upstream a few miles from the iron furnace. The Native Americans processed the red ochre in the soil to paint their dead. White settlers first discovered the bog iron here in 1788, but the furnace did not begin operation until the early 1830s. It operated with periods of interruption until the late 1840s when higher grade ores were found in western Maryland. The furnace was charged or loaded with organic and geologic materials similar to those used in many other iron furnaces, and all were available nearby. Trees were clear-cut and smoldered by colliers into charcoal, which when fired would chemically remove the oxygen from the iron oxide ore, yielding pure, elemental iron. Also needed was a flux to remove the sand. While western Maryland furnaces used limestone for flux, the Nassawango furnace used oyster shells from nearby Chincoteague Bay. Both contain calcium carbonate, the compound that in the heat of the furnace combined with the silicon dioxide of the sand and floated off as a glassy material called slag. Water was also needed, and nearby Nassawango Creek was the source for channeled water used to power the giant bellows that pumped

In the early nineteenth century, a large industry existed at Nassawango Iron Furnace in the swamps of the Eastern Shore.

air into the furnace. A canal was built to provide a shipping connection to the deep Pocomoke River and Chesapeake Bay.

Each charge or loading of the furnace used about 500 pounds of sandy iron ore, 40 pounds of oyster shells, and 25 bushels of charcoal. In the extreme heat of the furnace, the chemical reactions occurred and the liquid slag floated to the top, where it was periodically drawn off, cooled into its glasslike form, and then discarded. The hot, heavier liquid iron flowed out the bottom of the furnace and was channeled into molds that resembled a sow nursing her piglets—hence the term *pig iron*.

The furnace produced over 20 tons of pig iron per week. Boats transported the iron out via the canal and returned with oyster shells. A company town of over three hundred people was located where the Furnace Town Living Heritage Museum stands today. The present structures are not re-creations but preserved nineteenth-century buildings moved here from other locations in Worchester County.

The original furnace, built on a cypress wood foundation to prevent it from sinking into the sandy soil, still stands. Near the furnace is the trailhead for a 1-mile walk into a wild, forested swampland along Nassawango Creek, preserved by the Nature Conservancy. You may still find orange bog iron seeps here.

George Island Landing and Public Landing

From George Island Landing and Public Landing (public docks) you can look to the east across the 5-mile width of Chincoteague Bay to Assateague Island. Here your view is of the island in its natural state—a longitudinal ridge or dune of sand sitting out in the ocean. Its sands are at meager elevations of 0 to 10 or 15 feet above sea level, and marsh deposits at the base of the barrier sands have been radiocarbon-dated at about 1,800 years.

MD 365 to Public Landing follows a nineteenth-century wagon road from Snow Hill to the dock for vessels that could pass through inlets that once existed in Assateague Island. About 8 miles northeast across Chincoteague Bay, the navigable North Beach Inlet existed in the midnineteenth century. The two Egging Beach Islands, on the bay side of Assateague but not visible from here, are remnants of the tidal shoal of this now-closed passage through Assateague.

Pocomoke River

About 3 miles southwest of Snow Hill on US 113, you can drive to Shad Landing on a meander of the deep Pocomoke River, the southernmost Maryland river of the Delmarva Peninsula. Winter may be preferable if you do not care for mosquitoes. The Pocomoke River originates in Cypress Swamp on the Delaware-Maryland border and is thought to be the deepest river of any of comparable width in America. The ancestral river must have incised deeply here as it drained to a much lower ocean during a Pleistocene ice age. We are not sure why the Pocomoke River is so deep for its width, but an asteroid did impact just south of here about 35 million years ago, creating a deep crater into which this river might have flowed. From Pocomoke City you can drive on MD 371 (Cedar Hall Road) to a boat ramp near its mouth.

Crisfield

From US 13 north of Pocomoke City, you can drive MD 413 or MD 667 south to Crisfield, on the most southern bay-side peninsula in Maryland. In the 1800s Crisfield was filled with hundreds of watermen who harvested huge quantities of oysters from Chesapeake Bay. Wetlands were lost here when oyster shells were used to fill marshes, and today the land from the city docks to the center of town is composed of billions of these crushed, tightly compacted shells—a kind of man-made limestone, which might be considered the only solid rock in lower Delmarva. Crisfield is still a good place to get seafood.

Nearby Janes Island, accessible only by boat, is over 3,000 acres of marsh and land with many trails and beaches. Smith Island, 13 miles across Tangier Sound from Crisfield, is Maryland's only permanently inhabited unbridged island. It is accessible only by daily and seasonal passenger ferries. Like other Chesapeake Bay islands, Janes and Smith Islands face rising water levels and loss of land area.

Maryland 213
US 301—Elkton
53 MILES

If you are heading north from the Bay Bridge, you can see some good terrain and bay views from several Maryland routes. MD 213 crosses the tidal Chester River at Chestertown and connects with MD 20, which goes out to Rock Hall and Eastern Neck National Wildlife Refuge. In the refuge, 6 miles of trails take you to great views of Chesapeake Bay or the miles-wide mouth of the Chester River. The Rock Hall Public Beach affords excellent bay views. The estuarine Kent Island Formation covers southern Kent County, including Eastern Neck and up the margins of the Chester River to US 301, and it covers northwestern Kent County west of a line roughly from Fairlee to Betterton.

Bluffs on Sassafras River

Betterton is on the southern shore of the Sassafras River. From the beach pavilion on a bluff, you can see across the wide mouth of the Sassafras River to the bluffs on its north bank, which have Cretaceous-age sediments at their base. From here you are

The south-facing bluffs of the Sassafras River include Cretaceous deposits overlain by Tertiary ones.

less than 20 miles from the Fall Line, where Cretaceous sediments butt up against the hard rocks of the Piedmont. On a clear day you can see up to Elk Neck, made of Cretaceous sediments, and Susquehanna Flats, the northern part of Chesapeake Bay below the mouth of the Susquehanna River. If the day is very clear, you can see Piedmont highlands across the bay to the west.

For a closer look at the beautiful Cretaceous and Tertiary Sassafras bluffs, turn off MD 298 to the Sassafras River Natural Resource Management Area. A 2-mile hike, preferably during winter (due to thick brush) and not during hunting season, will bring you to the river. Across the water to the north are the bluffs. The Cretaceous Monmouth Formation is overlain by the Tertiary Pensauken Formation.

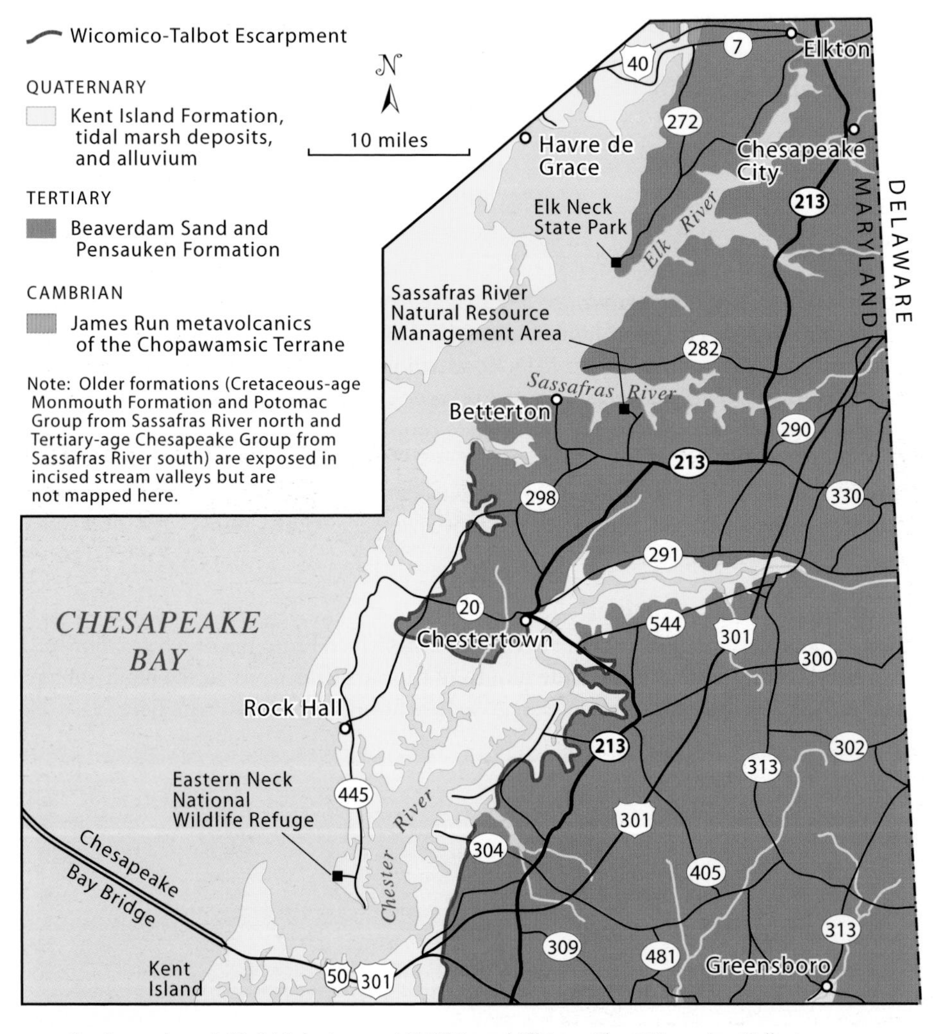

Geology along MD 213 between US 301 and Elkton. The Wicomico-Talbot Escarpment was the shoreline of Chesapeake Bay about 125,000 years ago. —Modified from Cleaves, Edwards, and Glaser, 1968; Owens and Denny, 1986b; Higgins and Conant, 1986

Chesapeake City Area

At the MD 213 bridge over the C&D Canal at Chesapeake City, you can risk a scary walk on the narrow bridge sidewalk or catch a quick look from your vehicle to the north on a clear day to see the gabbro outcrop of 340-foot-high Iron Hill projecting out of the Coastal Plain about 8 miles away. The Iron Hill Gabbro is an igneous pluton that intruded during or after the Taconic mountain building event. The gabbro is coated with iron oxides. The country rock into which it intruded was eroded to such a low elevation that Iron Hill became surrounded by Coastal Plain deposits. The museum near the C&D Canal bridge has a fossil collection.

Just north about 2 miles on MD 213 south and west of the Locust Point Road intersection is a Carolina bay, a circular depression with a sand rim. This type of wetland is thought to have formed from groundwater fluctuations during the cold, windy climate at the end of the last ice age.

Elk Neck State Park

Just west of Elkton you can veer south on Old Elk Neck Road, or farther west at North East head south on MD 272, and visit Elk Neck State Park, where the Elk River meets the upper Chesapeake Bay. From roughly north of I-95 at the Piedmont Fall Line down Elk Neck runs the unconsolidated Cretaceous Potomac Group, which lies with a thickness of over a thousand feet on the crystalline basement rock that gently slopes downward southeasterly from the Piedmont. You can see these unconsolidated sands, silts, and clays in bluffs above Chesapeake Bay in the state park. These sediments are thought to have accumulated through Cretaceous time when the Appalachians and Piedmont were uplifting and the Coastal Plain was

In Elk Neck State Park you can see up close some of the cliffs—Tertiary Pensauken Formation over Cretaceous Potomac Group—that border Chesapeake Bay. Because these sediments lacked the cements necessary to solidify them into rock, they have always been highly susceptible to wave erosion.

subsiding. They were laid down in nonmarine, fluvial environments—floodplains, fans, marshes—and they are the oldest exposed sediments of the Coastal Plain. Directly above the Potomac Group and exposed in the bluffs at the state park is the Pensauken Formation of Tertiary time. While the K-T boundary of dinosaur-extinction notoriety should be here, there is no mention of it by Maryland geologists. The evidential K-T iridium layer may have been eroded here prior to deposition of the Tertiary sediments.

Trails lead to Turkey Point and its lighthouse on hundred-foot bluffs, from which you can see to the east the broad, tidal Elk River and across it, the Eastern Shore of Cecil County. From the western bluffs of the park, you can look across the northernmost part of Chesapeake Bay to the mouth of the Susquehanna River. This shallow part of the bay, several square miles, is known as Susquehanna Flats because the bottom contains thick accumulations of sediments carried down by the Susquehanna River—billions of pounds per year.

You will encounter several hills on Elk Neck, most of them capped by Tertiary Upland Gravel, probably correlative with the TUG_4 of Southern Maryland. These are thought to have been deposited in alluvial fans by the ancestral Susquehanna River as it passed from the steeper gradient of the Piedmont to the flatter areas below the Fall Line—a process similar to that which produced Susquehanna Flats. Here, as stream velocity decreased, turbulence was no longer sufficient to transport the gravel. The gravels were laid down during Pliocene time when sea level was higher relative to the land than it is today.

Road Guides to Delaware's Coastal Plain

Delaware's Coastal Plain, extending south from the metamorphic rock of the Piedmont at the Fall Line, records Cretaceous, Tertiary, and Quaternary deposits. Cretaceous sands and silts of the Potomac and Matawan Groups run south past the Chesapeake & Delaware Canal (C&D Canal), hidden beneath younger deposits.

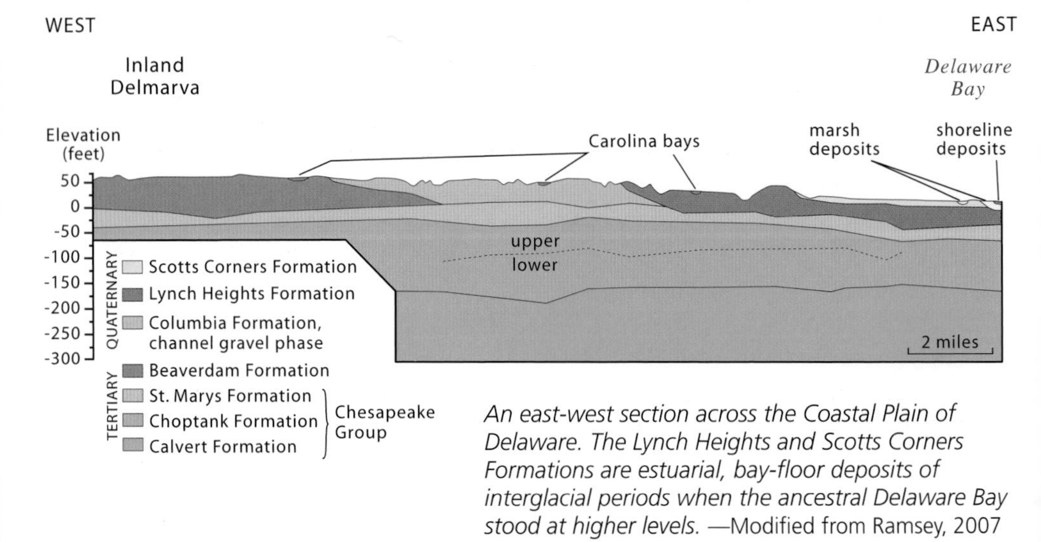

An east-west section across the Coastal Plain of Delaware. The Lynch Heights and Scotts Corners Formations are estuarial, bay-floor deposits of interglacial periods when the ancestral Delaware Bay stood at higher levels. —Modified from Ramsey, 2007

From there southward lie gravels and sands of Tertiary and Quaternary age. The inland core of the central Delmarva Peninsula is the Beaverdam Formation, deposited in river and estuary environments in the Pliocene Epoch of Tertiary time. The ancestral Delaware River deposited the glacial-outwash Columbia Formation in middle Pleistocene time over northern portions of the Beaverdam Formation. Reworking of the existing lower elevation Coastal Plain deposits by ocean waves occurred during Pleistocene interglacial periods of higher-than-present sea level, and these reworked formations have their own names.

All of these formations are unconsolidated, or loose material, a condition that accounts for many of inland Delmarva's geomorphic features, such as Carolina bays, inland dunes, swamps, ancient shoreline scarps, and relatively flat terrain.

US 301, US 13, and US 113
C&D Canal—Delmar
82 MILES

Chesapeake & Delaware Canal

The Chesapeake & Delaware Canal (C&D Canal), opened in 1829 after twenty-five years of excavation, is a shipping channel connecting Delaware Bay to Chesapeake Bay. It reduces the Baltimore-Philadelphia water route by 300 miles. At Summit Bridge and Deep Cut, accessible on dirt canal roads (open to the public) from roads exiting US 301, you could once find some of the best Cretaceous exposures in the Atlantic Coastal Plain. But today there are few unvegetated exposures, even during winter. Deep Cut, the excavation through the land of highest elevation, east of DE 71 on the north shore of the canal, might be your best bet to see shallow marine sands and silts of the Upper Cretaceous Merchantville, Englishtown, and Marshalltown Formations (all of the Matawan Group) deposited on top of the (not visible) thick Potomac Group. The Quaternary Columbia Formation tops the older formations. Three and a half million cubic yards of sediments were removed from the Deep Cut alone, and a tenth of this slid back into the canal before it was opened. It is no wonder that vegetation is allowed to cover the walls.

In nearby Lums Pond State Park is a Carolina bay, a swampy depression, right beside the Whale Wallow Nature Center. The formation of Carolina bays is not well understood, though the nickname "whale wallow" inspires an imaginative but incorrect explanation. See below for a geologic discussion of their formation. The pond was impounded in the early 1800s to supply water for the C&D Canal.

At the eastern end of the C&D Canal, a jetty projects into the Delaware River. Its purpose is to deflect sediment carried by the river and prevent the end of the canal from silting up.

The Levels

A few miles south of the C&D Canal on US 301 is the Levels, a flat expanse of the Columbia Formation undissected by streams. There are no exposures of this middle Pleistocene sand-and-gravel glacial outwash here, but you can see farmland

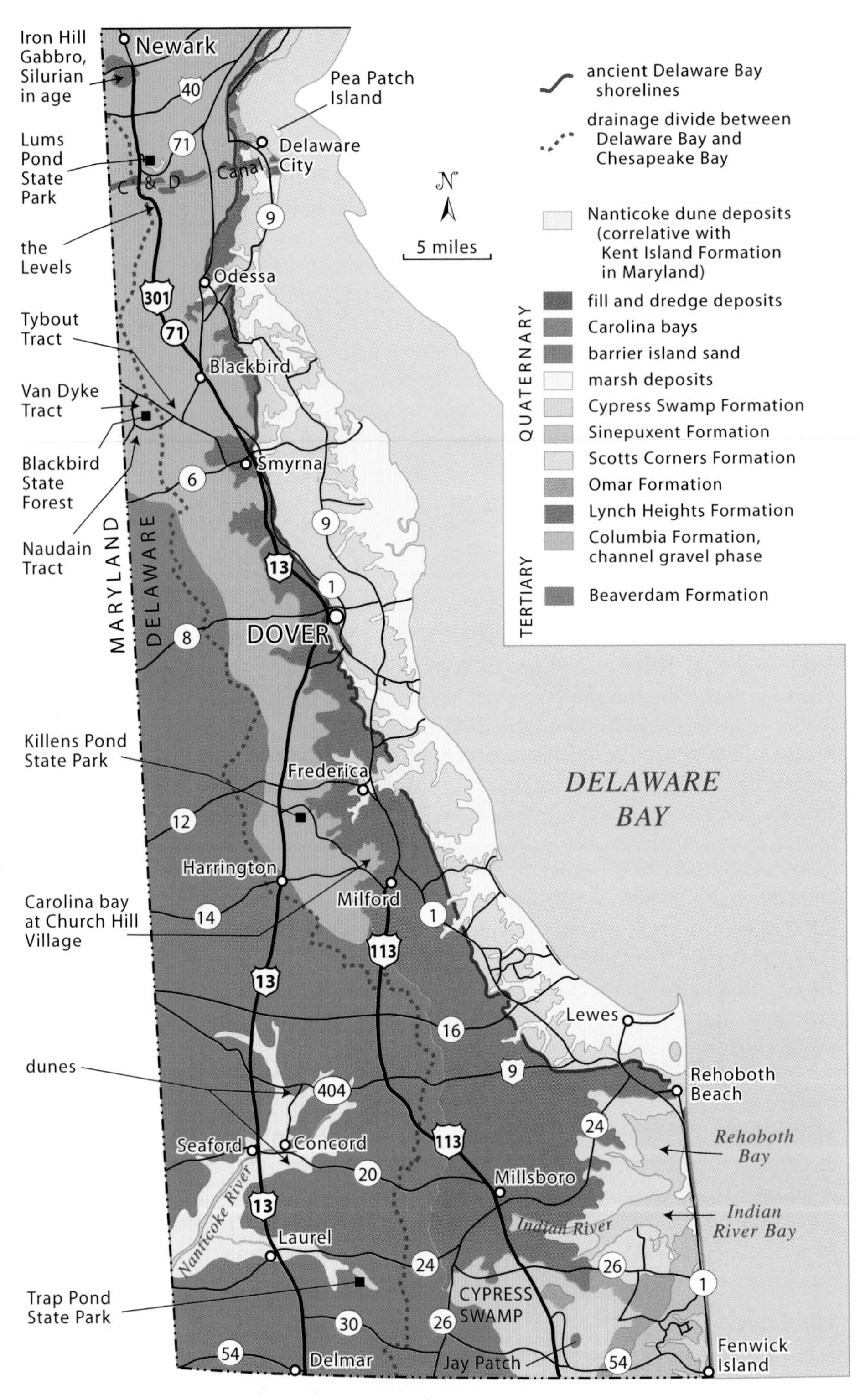

Geology of Delaware's Coastal Plain. Note the drainage divide between Chesapeake and Delaware Bays and also the shorelines formed during higher levels of Delaware Bay. —Modified from Ramsey and Schenck, 1990; Andres and Ramsey, 1995; Ramsey, 2003, 2005, 2007; Andres and Klingbeil, 2006

resembling the Midwest. This relatively high area, about a 3 square miles from just north of Mt. Pleasant to just south of Armstrong, straddles the drainage divide for the Delaware and Chesapeake Bays.

Carolina Bays at Blackbird State Forest and Vicinity

Carolina bays are circular to oval depressions, ranging from 0.1 to 3 acres, that contain swamp vegetation or ponds or both. They are often bordered by sandy rims a few feet to 20 feet above the depression. About five hundred thousand of these 100,000- to 70,000-year-old Carolina bays are oriented southeast to northwest along the eastern Coastal Plain from New Jersey to northern Florida.

In Maryland and Delaware these swampy pockets are also known as Delmarva bays, whale wallows (after a mythical origin), Maryland basins, potholes, sinkholes (a term customarily used for karst limestone depressions), roundponds, and loblollies. In Delaware, groups of Carolina bays can be found north of Smyrna in Blackbird State Forest and south of Dover near Prime Hook National Wildlife Refuge. In Maryland they are most abundant from Goldsboro in Caroline County to Millington Wildlife Management Area in Kent County.

What is the origin of Carolina bays? Nobody knows for certain, but theories abound: (1) a meteorite fragmented and exploded near the earth's surface, creating shock waves that blew out depressions in the sandy Coastal Plain soils; (2) groundwater dissolved layers below the surface, resulting in karstlike depressions; (3) artesian springs upwelled from groundwater pressurized by nearby uplands; (4) shallow water collected in dune fields or in wind-generated blowouts; and (5) underwater tidal eddies sculpted the depressions in lagoon bottoms during a higher

Carolina bays are difficult to photograph, but in Blackbird State Forest you can explore dozens of them and formulate your own theory about their origin.

stand of the sea. The Delaware Geological Survey favors wind as the major cause of Carolina bays. A report from the survey states that in Delaware, they appear to have formed when the climate was cold, vegetative cover reduced, and sea level lower than it is today. These conditions allowed wind to move considerable amounts of sand with few obstructions or impediments.

However they formed, many of the Carolina bays have been drained or filled for agriculture—an environment loss. As is the case for wetlands in general, these small swamp pockets retain water and help buffer the intensity of floods. They also provide habitat for many plants and animals, including vast swarms of mosquitoes.

Just off US 13 north of Smyrna, you can take the Blackbird Forest Road to the Tybout Tract of Blackbird State Forest and see Carolina bays across the road from the park office. If you don't mind a short walk into the forest, you can find good sites in Van Dyke Tract or Naudain Tract on Saw Mill Road off Van Dyke-Green Spring Road. A sign at the park office specifies that there are more than one hundred Carolina bays in the area.

Just over the line in Maryland, off MD 330, you can drive a few miles up Black Bottom Road and see a few Carolina bays. You can see more on the Big Stone Road, off MD 330 just west of the state line, and on Walnut Tree Road, which connects with MD 313 from Big Stone Road.

Dover to Delmar

US 13 between Dover and Harrington runs just east of the drainage divide of the Delaware and Chesapeake Bays. Numerous lakes and ponds are located along gentle, marshy streams draining the flatlands of the Columbia Formation. Just south of Harrington you will see a sign indicating that you have entered the Chesapeake Bay watershed. The drainage divide swings east here toward Ellendale Swamp. If you are driving southeast of Killens Pond on Sandbox Road, or east-west on DE 14 (Milford-Harrington Highway), look for the large Carolina bay at Church Hill Village.

Near Seaford on Coverdale Road you can see part of a dune of Parsonburg Sand, sculpted by wind between 30,500 and 16,000 years ago.

South and east of Harrington, from about DE 404 to Seaford, lies a series of dunes moved into position probably at about the same cold-climate period—30,500 to 16,000 years ago—as the Parsonburg Sand of Maryland in Pleistocene time. Sand previously deposited during a warm, interglacial period (40,000 to 30,000 years ago) in lagoons located where the Nanticoke River is today would have been worked into dunes during cold periods when vegetation was sparse. Because of modern development these are somewhat obscured, but if you turn east on DE 20 just south of the Nanticoke River, you will encounter a rare Delaware hill at Concord—part of a large dune. North on Concord Pond Road at the Concord Pond Fishing Area you can see a dune across the water.

One of these dunes can be spotted from DE 404 south of Bridgeville. Look for a slope just east of the Nanticoke River. One mile east of there, you can turn south on Coverdale Road (Rd. 525) and see a ridge to the west. It is a parabolic dune, a crescent form with ends pointing toward the northwest wind that formed it, but its shape is not discernible from the road

South of Laurel you can turn east from US 13 on DE 24 to Trap Pond State Park, where you can hike through wetlands and a bald cypress swamp.

To the east off US 113 you can turn east onto McCabe Road (Rd. 400) and drive north on Pepper Road to see the Carolina bay called Jay Patch. It is not much to see—just woods and swamp, but if you are on US 113, it's a quick look.

Cypress Swamp–Burnt Swamp

DE 54 west of US 113 crosses Cypress Swamp–Burnt Swamp, the historical headwaters for the Pocomoke River. Delaware geologists recently studied the geologically young sediments in the swamp and reconstructed environments that existed during and after the last Ice Age. Over the past 22,000 years, the area has gradually changed from a temperate estuary to cold-climate forests of evergreens and mosses and then to a warm, forested freshwater swamp. These changes reflect the global cooling and warming of glacial and interglacial periods during Pleistocene time.

Ditches and draining have lowered the water table of this swamp, reduced its size, and even redirected much of the water from the south-flowing Pocomoke River of Maryland to the northeast-flowing tributaries of the Indian River of Delaware. However, when you see water standing in a ditch, you can usually use it as an estimate of the water table, the surface of the groundwater.

Delaware 1
Dover—Cape Henlopen—Fenwick Island
60 MILES

Delaware 1 is a heavily traveled road, but from it you can see the characteristic fields, woods, marshes, and tidal rivers of Delaware. Between Dover Air Force Base and Little Heaven are roadside wetlands and a high bridge over the marshy St. Jones River.

Side Trip to Bowers Beach

Just south of Little Heaven, turn east onto Bowers Beach Road, which takes you to the town of Bowers Beach on the shores of Delaware Bay (see map on page 306). There is a beach access point near the Coast Guard station at the mouth of the Murderkill River. You can walk the sandy beach or drive around Bowers Beach and see houses built on pilings above the marsh. DE 1 crosses the marshes of Murderkill River at Frederica.

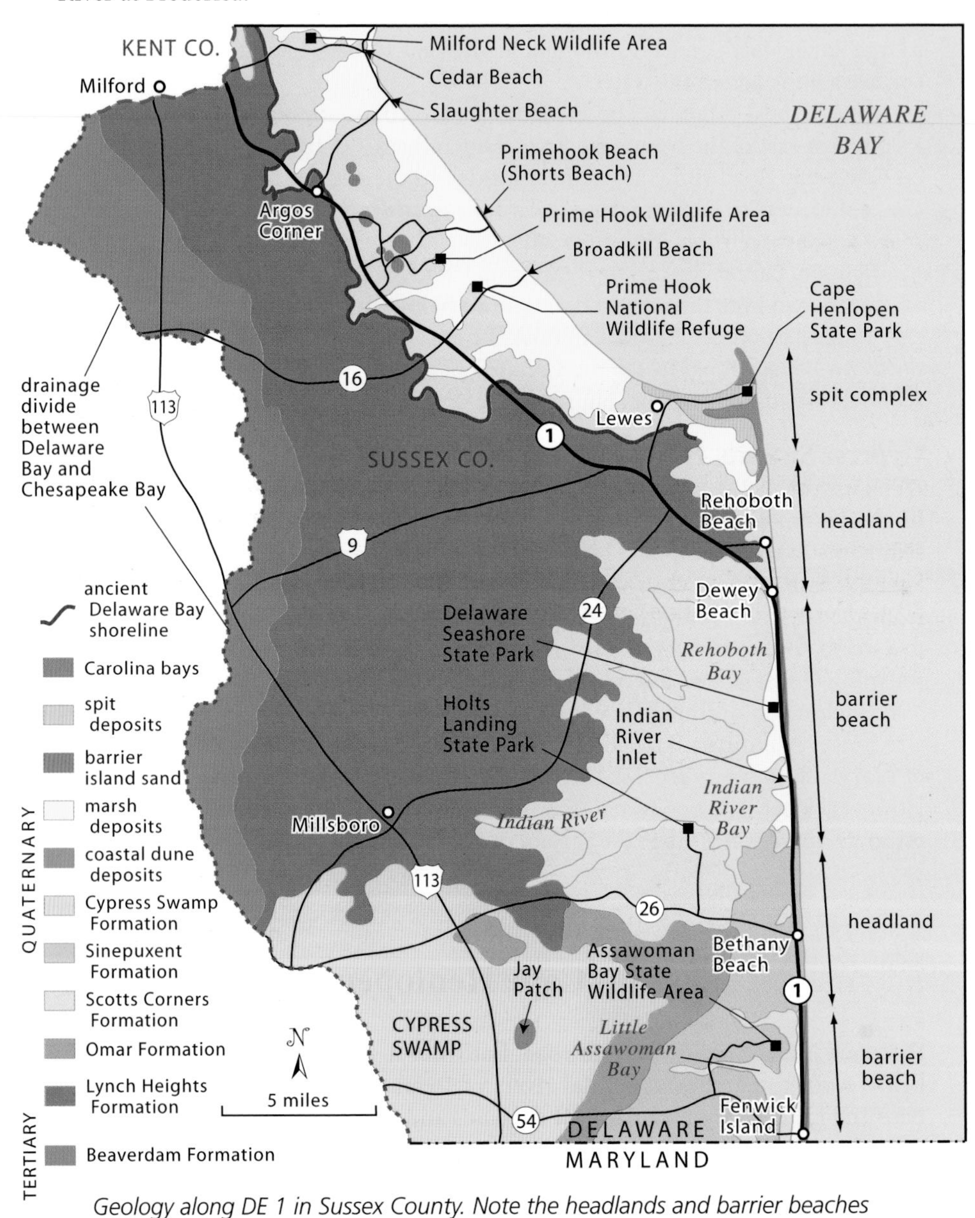

Geology along DE 1 in Sussex County. Note the headlands and barrier beaches between Cape Henlopen and Fenwick Island. —Modified from Ramsey and Schenck, 1990; Ramsey, 2003; Andres and Klingbeil, 2006

You can see from the ripples in the sand that erosion control barriers do not prevent water from reaching the sand, though they may prevent the sand from being carried away. Photograph was taken at the mouth of the Murderkill River on Delaware Bay.

Side Trip to Cedar Beach

Turn east onto Cedar Beach Road from DE 1 east of Milford and drive to the public access points on the bay. Several marshes can be seen along the way. Between DE 1 and the beach there is a scarp marking one of the Pleistocene shorelines when sea levels were high, but it is very hard to locate while driving. In the Frederica-Milford-to-Delaware Bay area Delaware geologists recognize five geomorphic surfaces. The segmented scarps represent Pleistocene shorelines and cliff erosion zones from the Columbia Formation of middle Pleistocene time through the Lynch Heights and Scotts Corners Formations of late Pleistocene time.

Side Trip on Prime Hook Road

South of Argos Corner, Prime Hook Road leads you to some of the best views of Carolina bays in Delaware and Maryland. For a discussion of the formation of these unusual features, see the section in **US 301, US 13/US 113, and C&D Canal—Delmar**.

The cluster of trees in this field marks a Carolina bay, formed about 100,000 to 70,000 years ago, perhaps when harsh winds blew out a depression in bare sand.

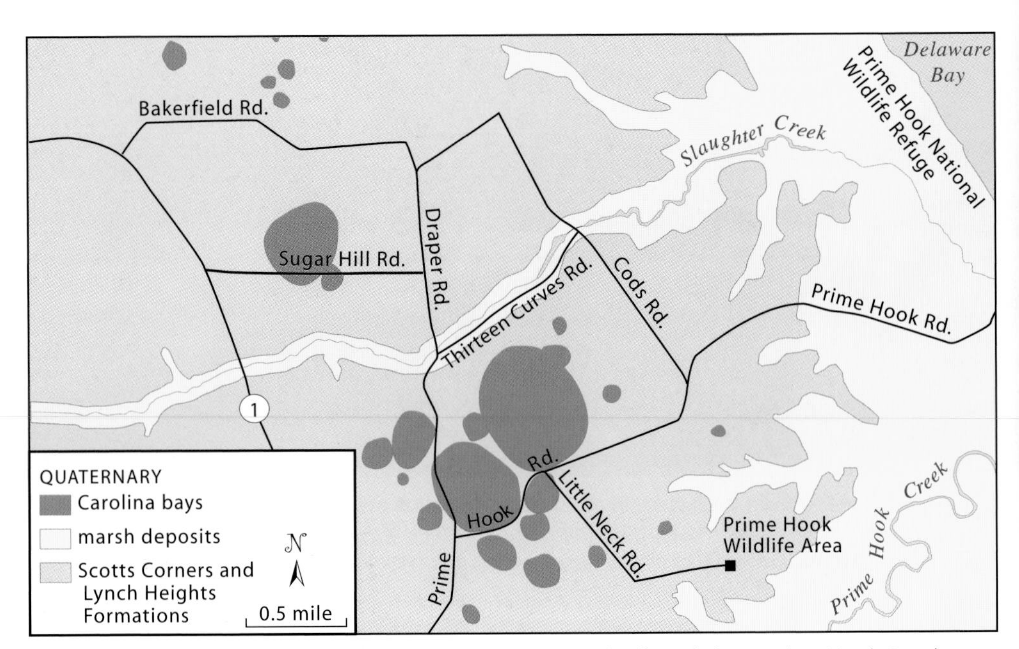

A large number of swampy, circular Carolina bays can be found along Prime Hook Road.
—Modified from Andres and Klingbeil, 2006

As you drive toward Prime Hook, look for low, freshwater-filled, circular or oval clusters of trees in the middle of farm fields. If you turn onto Little Neck Road and drive to Prime Hook Wildlife Area (not the same as Prime Hook National Wildlife Refuge), you will see more Carolina bays in the fields along the way. Little Neck Road ends at a coastal marsh.

From Prime Hook Road you can drive to Primehook Beach (or Shorts Beach) across the tidal marsh of Prime Hook National Wildlife Refuge. The beaches are private. On your return to DE 1, turn onto Cods Road and then onto Thirteen Curves Road, and you will follow the meandering, swampy Slaughter Creek. Turn right onto Draper Road and left onto Sugar Hill Road and look for more Carolina bays north of the road as you return to DE 1.

Cape Henlopen State Park

Jutting northeast into Delaware Bay between Lewes and Rehoboth Beach lie the marshes, dunes, beaches, and spit of Cape Henlopen State Park. At more than 5,000 acres, the park is one of the largest natural beach and dune areas remaining on the eastern seaboard. Miles of foot and bike trails provide access to the different depositional and erosional coastal features of the cape. The park is also home to the 80- to 90-foot-tall Great Dune (or Sand Hill), the highest point on the Atlantic Coast between Cape Cod, Massachusetts, and Nags Head, North Carolina.

The sands of Cape Henlopen (and those of all depositional, sandy coasts) have always been in flux as they respond to changing winds and waves. Longshore ocean currents have transported sand here from Pleistocene-sediment headlands located farther south. These sands have been moved north and deposited in northerly

Cape Henlopen is a depositional coast with dunes protected in their natural state. Notice the spit curving into Delaware Bay.

progressing spit systems over the past 2,000 years. A spit is a finger-shaped bank of sand deposited into open waters by a longshore current, and the present one is accumulating to the northwest at the mouth of Delaware Bay at a rate of about 100 feet per year. The remnants of older spits are still discernible in Cape Henlopen State Park to the south and east of the present spit. Because these spits curved inward on themselves, they created a tidal estuarine lagoon separated from Delaware Bay but open to the ocean. The brackish Gordons Pond and Lewes Creek Marsh (in a line between Lewes and Rehoboth Beach) are the site of this lagoon's position in 1400. Longshore drift and rising sea levels pushed the spits north and west, eventually causing the lagoon to silt in and become a marsh. The lagoon may have provided shellfish for Native Americans.

Sea level rise of about half a foot every 100 years for the past 2,000 years has contributed to this northwestern movement, but it has also caused the coastline to migrate landward about 3 feet per year for a couple of thousand years. Cape Henlopen Lighthouse was built in 1765. When completed, it was a quarter of a mile inland in the middle of a pine and cedar forest, but migrating sands caused it to fall into the ocean in 1926. The place where it once stood is now about 150 yards offshore, north of Herring Point.

The current spit is slowly moving to the northwest between the inner and outer breakwaters in Delaware Bay, seaward of Lewes harbor. Built in 1829 and 1890, these breakwaters, according to Kraft and John (1976) "significantly disturbed the sediment transport to the extent that the cuspate foreland accreted north into Delaware Bay, forming present-day Cape Henlopen." The northernmost area of Cape Henlopen State Park, the Point, is a partially vegetated dune field that is

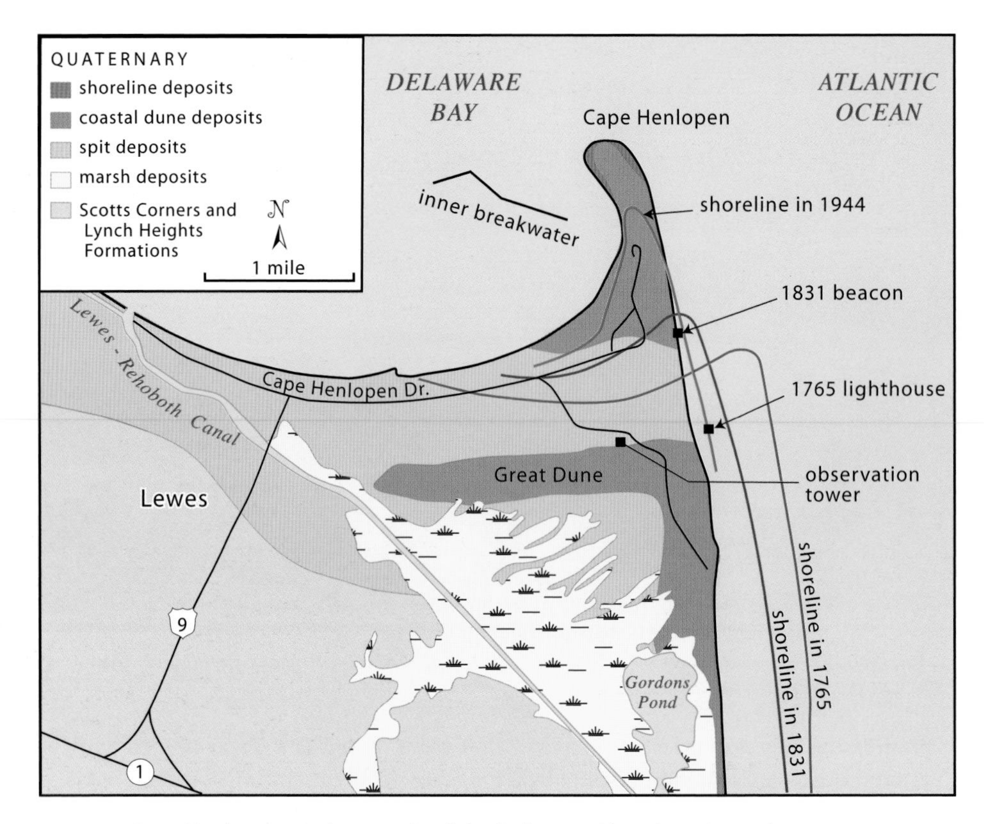

Cape Henlopen, at the mouth of the Delaware River, has always been on the move. —Modified from Kraft and Hiller, 1987; Ramsey, 2003; Andres and Klingbeil, 2006

closed to the public, but you can walk from the spit parking area down onto the beach facing Delaware Bay and the breakwaters and look northwest up the beach to see the curvature of the spit.

Natural beach dunes (as opposed to those constructed in Rehoboth and Ocean City), run parallel to the shoreline from the tip of the spit down to the northernmost developed part of Rehoboth Beach. As with all coastal dunes, these protect land areas behind them, but they are still subject to wind erosion and overwash sand transport during big storms, especially nor'easters. Big nor'easters in 1962 and 1998 washed sand over by tens of feet, depositing it on top of old recurved spits and in Lewes Creek Marsh. Sea waters were so high they covered the area north of the observation tower.

The truly unusual deposition of sand on Cape Henlopen is the Great Dune, which is over 2 miles long, 90 feet high, and, strangely, perpendicular to the coastline. Coastal geologists Kraft and Caulk (1972) think that the dune developed after the area was deforested in 1831 for construction of the inner breakwater—designed as a port of refuge for ships. Winds quickly eroded the bare sands and built them into a high dune, which has since migrated about 5 feet per year to the south, covering

The extensive, vegetated sands of the Great Dune might make you think you are at an inland desert. The tower was erected during World War II for observers to watch for German U-boats or other craft off the U.S. coast.

forest and historic recurved spits. The best place to see the Great Dune is from the top of the observation tower.

Rehoboth Beach

South of the spit complex of Cape Henlopen is the town of Rehoboth Beach, built not on a barrier island like Ocean City, Maryland, but on a headland where older, pre-Holocene deposits (older than 10,000 years) underlie the sands of the beach. Rehoboth Beach is like Ocean City, however, in that it has attempted to remain permanent while the sediments under it are subjected to wind and wave erosion. During the 1998 nor'easter, for example, an overwash just north of the town undermined the boardwalk. Like Ocean City, Rehoboth has spent millions of dollars on beach and dune replenishment. As of 2006, a multimillion-dollar project pumped sand from offshore to rebuild the beach and to create a vegetated, 6-foot-high dune that protects Rehoboth Beach and Dewey Beach.

Delaware Seashore State Park

A narrow barrier island runs about 7 miles from the headland of Dewey Beach south to the headland of Bethany Beach. Fenwick Island, a barrier island, begins a few miles south of that. Characteristic of barrier island geology, these islands

Longshore drift of sand would close Indian River Inlet without these jetties.

are separated from the mainland by lagoons: from north to south, Rehoboth Bay, Indian River Bay, and Little Assawoman Bay. With the exception of World War II watchtowers, Delaware Seashore State Park, like Assateague Island, is kept in a natural, undeveloped state. The beach and lagoons are accessible from many roadside parking areas along DE 1.

As with Ocean City Inlet, Indian River Inlet is kept open for fishing and recreational craft by two jetties that block longshore drift of sand. If you are careful of traffic, you can walk on the bridge for good views north and south along the barrier island. Notice that sand has been deposited hundreds of feet out into the ocean along the south jetty; this has not happened along the north jetty because the net longshore drift here is from south to north. Without the jetties, moving sands would eventually close the inlet.

Landward of Indian River Inlet you can observe a tidal delta, an area where sands have been washed in by tidal currents. The wide section of barrier island north of this inlet is composed of old tidal deltas associated with former naturally cut inlets.

When sea levels were higher, several geologic formations—today on the mainland—were deposited in bay, lagoon, and marsh environments similar to those found along the present Delaware Bay and along the Atlantic Coast. Should sea level drop, something not expected anytime soon, the present sediments on bay and marsh bottoms would become mainland formations.

Holts Landing State Park

About 3 miles west of Bethany Beach on DE 26, you can follow the signs north to Holts Landing State Park for views of the Delaware barrier island and Indian River Bay and Indian River Inlet. The Scotts Corners Formation sands in the park were deposited in marsh and esturine environments similar to the ones you can view today but during a Pleistocene interglacial period when sea level was higher.

Assawoman Bay State Wildlife Area

To reach Assawoman Bay State Wildlife Area from the town of Fenwick Island, follow DE 54 west, turn right (north) on County Road 381, and then turn right (east) on County Road 364A. If you are starting from Bethany Beach, follow DE 26 west, turn left (south) on Kent Avenue, left (south) onto County Road 363 and then left (east) onto County Road 364. Pick up a self-guided auto tour brochure at the entrance.

The marshy Assawoman Bay State Wildlife Area, created from nine small farms, lies at about 5 feet above sea level on the mainland side of Little Assawoman Bay. The ponds closest to the observation tower are impoundments for waterfowl management, but in the distance to the south you can see the high-rise buildings of northern Ocean City perched precariously on the barrier sands of Fenwick Island. Mulberry Landing gives you a view from water level of Fenwick Island, while Strawberry and Sassafras Landings have views of marshy, brackish, tidal Miller Creek of Little Assawoman Bay. At Stop 9, near the road to Sassafras Landing, is a Carolina bay.

Delaware 9
Delaware City—Dover
38 MILES

In a band about 5 miles wide, paralleling the Delaware River and Delaware Bay from Delaware City to Lewes, are the low-lying coastal marshes of the Delaware estuary. The underlying sands are among the youngest deposits in Delaware, less than 10,000 years old. Landward of the marshes at higher elevations are Pleistocene sands that were deposited when higher seas covered this area during warm, interglacial periods—the Lynch Heights Formation about 320,000 years ago, today at elevations of 30 to 50 feet, and the Scotts Corners Formation about 120,000 years ago, at elevations less than 25 feet. DE 9 is a lightly traveled, two-lane road, excellent for exploring coastal Delaware. South of Delaware City the road passes high over the C&D Canal, providing good views of some of the marshes you will cross for the next 8 miles.

Fort Delaware on Pea Patch Island

The current structure at Fort Delaware was built on Pea Patch Island just before the Civil War and was used as a prison for over thirty thousand Confederate soldiers. Pea Patch Island does not appear on colonial maps, but because it is in a bend of the Delaware River, where currents slow on the inside, there was probably a mud shoal here. One story about the origin of the island is that a ship ran aground on the shoal and spilled peas into the shallows. The peas then sprouted and grew, catching and adding mud deposits until they made an island. Geologically, this explanation could hold up, but it may not be accurate. The island could have accumulated by natural deposition. At any rate, the island first appears on a map in 1794, and by the Civil War it was about 75 acres. Later dredging and dumping of river channel sediments

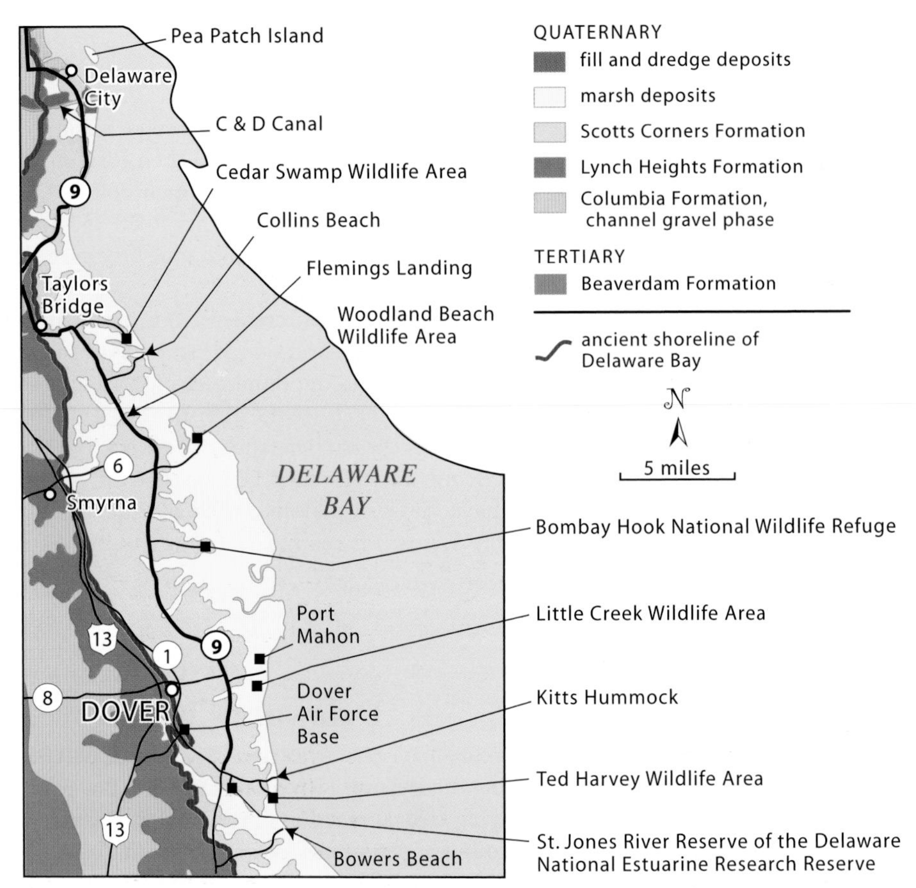

Geology along DE 9 between Delaware City and Dover.
—Modified from Ramsey, 2005, 2007

expanded the island to its present 300 acres. Today most of Pea Patch Island is a wetland nature preserve, although there are foot trails across some of it. Ferries from Delaware City give you excellent views of the tidal river and its shorelines.

Delaware Bay

If you look at topographic maps of eastern Delaware (U.S. Geological Survey quadrangles Taylor's Bridge, Smyrna, and Bombay Hook Island), you will see two distinctly different drainage systems for the streams of the low flatlands that border Delaware Bay. One system has meandering streams flowing from inland and emptying into Delaware Bay. The other system is dendritic, with tidal streams beginning close to the bay and spreading landward in the pattern of the limbs of a tree.

The meandering streams of Blackbird Creek, Smyrna River, Duck Creek, and Leipsic River have been in near-equilibrium with rising sea levels for hundreds of years; that is, the sediment carried and deposited near their mouths has maintained the land against the erosive power of rising estuarine waters. On the other hand, the

Because of heavy shipping traffic on the Delaware River, boulders defend Pea Patch Island from wave erosion.

broad flooded areas with the dendritically patterned waters reflect a system out of balance. They also tell the story of a coastline ravaged by storms and flooding.

Delaware Bay is a drowned, or flooded, river valley. With the ancient river channel more than 200 feet below sea level, Delaware River and Delaware Bay have a geologic history similar to Chesapeake Bay. The lands adjacent to both bays are unconsolidated sand, silt, and clay. These flatlands lie only a few feet above sea level, with the water table only a few feet below the surface. In the area's many wetlands, the water table is at the surface.

Because of the relatively protected ports afforded by the bays, Delaware, like Maryland, was one of the first areas on the East Coast settled by Europeans, and by the early 1700s many of the low-lying wetlands along Delaware Bay had been converted or, to use a modern term, *developed* into farmland. These early settlers built dikes along the shore of the bay in northern Delaware and drained the inland wetlands with ditches. With rising sea level documented back to at least the beginning of European settlement, by the late 1800s landowners observed that these drained farm fields lay 1 to 2 feet *below* Delaware Bay. The coastal Delaware fields were a sitting duck for a big storm. It came on October 23, 1878. Now considered to have been "only" a Category 1 hurricane with a storm surge that normally would have been about 5 feet, the surge catastrophically coincided with an incoming high tide in Delaware Bay. With the funnel-shaped bay magnifying the height toward its northern end, the surge swelled out of the bay in what was described by survivors as a tidal wave. Reports describe waters rising 4 to 5 feet in one hour and reaching 10 to 11 feet above mean low tide.

In addition to massive damage and destruction of nineteenth-century resorts along Delaware Bay, the shoreline was breached in several places. Hundreds of acres of "upland" farm fields, which were actually below sea level, were flooded. These breaches became tidal inlets of Delaware Bay, and the former fields became saltwater wetlands. Natural balance between land and sea was restored. Of course, farm acreage was greatly decreased.

The breaches were diked by 1880, washed out in 1882, and filled again in 1886. By the end of the 1880s, the fields had been reclaimed as dry land. In 1890 another storm surge broke through, but once again the breach was filled. An 1893 hurricane ruptured the dikes yet again, and in 1894 another storm washed over the remaining dikes. By the early 1900s the two tidal openings you can see today at Cedar Swamp Wildlife Area and Woodland Beach Wildlife Area had become established. These are two areas that exhibit the dendritic stream pattern, mentioned above.

Cedar Swamp Wildlife Area and Collins Beach

To reach Cedar Swamp Wildlife Area and Collins Beach, turn east on Thoroughfare Neck Road a few miles south of Taylors Bridge and drive across marshlands to isolated Collins Beach. Park in the lot that adjoins the boat ramp and look north and west to see hundreds of acres of salt marsh and open water. In the nineteenth century this area was farm fields and a cedar swamp—a freshwater wetland. The fields were taken back by the higher-elevation river and estuary when the shoreline dike was breached during the storms of the late 1800s. To the north, there was once a high shore bank and Collins Beach resort, but it was removed by the 1878 hurricane. The water by the boat ramp is the mouth of a dendritic tidal system that twice daily feeds in and out of the salt marshes.

Woodland Beach Wildlife Area

South of Flemings Landing, east of Smyrna, you can stop at the observation tower beside DE 9 at Woodland Beach Wildlife Area. Look east across Broadway Meadows, an area that was once farmed but today is marsh, and you'll see another dendritic

Delaware 9 crosses the Smyrna River at Flemings Landing, which is about 3 miles from the Delaware River, but not at sufficient elevation to escape the tide. Come during high tide and you may find water too deep to drive through.

tidal system with large areas of open water. In 1878 the coastal barrier sands east of here were blown out by a storm.

Bombay Hook National Wildlife Refuge

Bombay Hook National Wildlife Refuge is south of Woodland Beach and east of DE 9. Of the refuge's 16,000 acres, about 13,000 acres are tidal salt marsh. The storm of 1878 created the marshes when the storm surge left a vast lake in the area. Unlike Cedar Swamp and Woodland Beach, there is no tidal breach here.

The refuge supports thousands of migrating geese and ducks. You can take the 12-mile auto tour and gain access to five trails that have educational brochures. The Saltmarsh Boardwalk Trail brochure describes the high salt marsh under the boardwalk as an area that floods only during spring tides, about twice a month, or during big storms. In the distance, stretching across the meandering tidal stream (also called a "gut") all the way to Delaware Bay, lies a low marsh that floods and drains twice daily. The guts in this low marsh are winding tributaries that carry tidal flow in and out of the marsh from Delaware Bay. The salt marsh produces more vegetation per acre than a rain forest. It also filters sediments and contaminants and helps protect inland areas from storm-surge flooding. Salinity varies in the refuge's different marshes, but on average the brackish water is only about one-fourth as salty as the ocean. Heavy rains will lower salinity, while drought will raise it.

Little Creek Wildlife Area

East of Dover you can take Port Mahon Road to Delaware Bay across the low marshes of the Little Creek Wildlife Area. The road, which runs along the bay for awhile, may be blocked in places by high water or sand washed up by water. Either of these can quickly get your vehicle stuck. There are no houses or businesses out here, only a jet fuel terminal that supplies Dover Air Force Base. You can see the pipeline crossing the marsh.

Kitts Hummock Road

DE 9 ends south of Dover Air Force Base at Kitts Hummock Road, which leads to several interesting places. About three-quarters of a mile down the road is the St. Jones Reserve of the Delaware National Estuarine Research Reserve. The visitor center has interesting educational materials and exhibits. An excellent 2-mile out-and-back trail with a boardwalk crosses marshes and goes through a forest and fields. Seven-foot-tall big cordgrass towers above some parts of the trail. The difference between high and low tide in the gut just off the boardwalk is more than 5 feet. This low marsh is flooded at least once a day.

Farther east on Kitts Hummock Road, you can explore the Ted Harvey Wildlife Area, which has vast fields of open ground. The road leads to a boat ramp at the edge of a marsh near the bay. Kitts Hummock is at the end of Kitts Hummock Road on Delaware Bay. A hummock is a piece of ground elevated above a wetland, usually held in place by a root system. There is a narrow, sandy beach with public access along the bay.

•— Glossary —•

abandoned meander. The previous course of a curve or bend in a stream, left dry when the stream changed its course.
accretionary wedge/prism. A triangular mass of deformed sediment and sometimes oceanic basalt that forms at the edge of the overriding slab above a convergent-boundary oceanic trench.
Allegheny/Alleghany/Allegany. Alternate spellings, which often result in confusion: Allegheny Front, Allegheny Group, Allegheny Plateau; Alleghany (Alleghanian) mountain building; Allegany County (Maryland).
alluvial. Referring to deposits laid down by a stream or river—clay, silt, sand, or gravel.
alluvial fan. A relatively flat to gently dipping, fan-shaped wedge of unconsolidated sediments deposited by a stream, especially in semiarid regions, where it flows out of a narrow valley onto a wider, flatter area.
amphibolite. A dark-colored metamorphic rock consisting mainly of the minerals amphibole and plagioclase.
anticline. In layered rocks, a fold that is upward arching, with the oldest rocks in the center.
anticlinorium. A composite anticlinal structure of regional extent composed of several lesser folds.
aquifer. A body of saturated rock or sediment through which water can move readily.
axis (hinge line). The line about which a fold appears to be hinged or bent; the line of maximum curvature.
back bays. Shallow, tidal, brackish waters lying between barrier islands and the mainland.
backflooding. The creation of estruaries by the flooding of low-lying river valleys and their mouths by the ocean during periods of global warming, retreat of glaciers, and rising sea levels.
bar. Sand or other sediment above or slightly below water level in the form of a low ridge.
barrier island. A long, narrow, low-elevation, coastal, sandy ridge lying parallel to the shore but separated from it by a lagoon, marshes, or back bays.
basalt. A black or very dark gray volcanic rock that consists mainly of microscopic crystals of the minerals plagioclase feldspar, pyroxene, and perhaps olivine.
basement. The fundamental rocks of the continental crust, mainly granite, schist, and gneiss.
bedding. Layering in sedimentary rocks.
bedrock. Solid rock that is exposed in places or that underlies unconsolidated sediments and soil.
biotite. Dark mica, a platy mineral. It is a minor but common mineral in igneous and metamorphic rocks.
bituminous coal. Soft coal with a large content of volatiles, often banded with layers of plant material.
blue granite. A common but inaccurate name for the Brandywine Blue Gneiss, or Wilmington blue rock, of Delaware.
bog. A freshwater wetland in a cold region. Many (but not all) were formed over 10,000 years ago when the glaciers retreated.

boulder. A rounded stone larger than 10 inches in diameter.
boulder field/block field. A flat or gently sloping area covered with large angular blocks of rock derived from bedrock by glacial or periglacial freeze-and-thaw wedging.
brackish water. Water that is slightly salty (less than seawater), such as the waters of Chesapeake and Delaware Bays, where freshwater from rivers mixes with saltwater brought in by tidal currents.
breakwater. An offshore man-made structure that absorbs the energy of waves and allows for quiet water (and sometimes unwanted deposition) nearshore.
breccia. A sedimentary rock composed of angular fragments in a fine-grained matrix, the fragments having not moved far from their source area.
buffer. A band of grass, vegetation, or trees planted along the banks of streams to control erosion and help reduce turbidity in the stream.
bulkhead. A structure of wood, steel, or concrete designed to reflect the energy of storm waves in order to protect harbors or buildings in coastal areas.
calcareous. Rock consisting of more than 50 percent calcium carbonate.
calcite. Calcium carbonate, the most common nonsilicate mineral. Limestone and marble are made of calcite.
carbonaceous. Materials rich in carbon, such as coal.
carbonate rock. A sedimentary rock composed of carbonate minerals, such as calcite and dolomite.
Carolina bays. Shallow, oval-shaped basins or bowl-shaped depressions, usually swampy or wet, that occupy the Coastal Plain from Delaware to northern Florida.
cement. Solid material that precipitates from solution in the pore spaces of underwater sediments and binds together the grains into solid rock.
channel. The deeper part of a stream or river, often used for navigation.
chert. A sedimentary rock composed mainly of microscopic crystals of quartz.
clast. A grain or fragment of a rock produced by disintegration of a larger rock mass. Clastic rock is sedimentary rock composed of broken fragments derived from preexisting rocks.
clay. A sedimentary material composed of weathered minerals with grain sizes less than $1/256$ millimeter in diameter.
cleavage. A planar fabric in an unmetamorphosed or weakly metamorphosed, fine-grained rock.
coal. A sedimentary rock formed from the compaction of incompletely decayed plant material.
coarse-grained. A relative term used to describe the size of constituents in a rock. Said of igneous rocks with minerals larger than 0.2 inch in diameter. Said of sedimentary rocks with particles larger than 0.08 inch in diameter.
coastal marsh. A saltwater wetland in a cove or bay protected from ocean surf. Vegetation is herbaceous plants without woody stems.
coastal plain. A low, gently sloping region on the margin of an ocean.
cobble. A rounded particle 2.5 to 10 inches in diameter.
coke. The solid product of the incomplete combustion of coal, often used in iron furnaces.
compaction. Compression of sediment into a smaller volume by the increasing weight of sediments being deposited on top.
conglomerate. A coarse-grained sedimentary rock composed of pebbles, cobbles, or boulders set in a fine-grained matrix of silt or sand.
continent. A large landmass that is, or was, comparable in size to a modern continent. Laurentia, a Precambrian continent, evolved into the larger North American continent.
continental glacier. An ice sheet of regional or continental size, common during ice ages.
continental rise. The gently sloping edge of the continent that is adjacent to the deepest region of the ocean.
continental shelf. The gently dipping part of the continental landmass between the shoreline and the more steeply dipping continental slope.
continental slope. The most steeply sloped part of the continental margin, between the shelf and the rise.

convergent plate boundary. A boundary between two plates that are moving toward each other.
correlation. Determining historical time equivalency of physically separated rock units.
country rock. Any rock that was older than and intruded by a molten igneous body.
craton. A stable continental region that has not undergone mountain building activity since about 600 million years ago.
creep. Slow downslope movement of surface soil due to freeze and thaw of moist soil.
crossbed/crossbedding. A sedimentary bed, usually in sand or silt, that is at an angle to the main bedding.
crust. The upper surface of the lithosphere. Continental crust consists mainly of granite, gneiss, and schist; oceanic crust consists of basalt.
crystalline. Said of a rock formed of interlocking mineral crystals, usually igneous or metamorphic.
cuspate foreland. A sand deposit jutting into a bay, formed by longshore currents carrying sand from opposite directions to a point of convergence.
cutbank. The bank on the outside of a bend in a stream, where the faster moving current tends to erode or cut away at the existing bank.
Delmarva. The Coastal Plain area east of Chesapeake Bay, named for Delaware, Maryland, and Virginia.
delta. A nearly flat accumulation of clay, sand, and gravel deposited in a lake or ocean at the mouth of a river.
dendritic drainage. Drainage pattern of a river and its tributaries that resembles the branches of a tree or the veins in a leaf.
diabase. An igneous rock with the composition of basalt but which cooled far enough beneath the surface to have visible crystals.
diamictite. Poorly sorted conglomerate or breccia with a wide range of clasts and sizes. In Maryland the diamictite contains sediment debris from structurally higher thrust sheets and mélanges.
diatomaceous clay. A deposit composed of the fragmented silica-rich shells of the microscopic, single-celled plants known as diatoms.
differential weathering. Weathering that occurs at different rates due primarily to the chemical composition of the rock. Limestone, for example, weathers much more quickly than sandstone.
dike. A tabular structure of igneous rock that formed when molten magma filled a fracture in a solid rock. A dike is not parallel to the layering in the country rock.
dip. The sloping angle of a planar surface in rocks, such as a sedimentary bed or metamorphic foliation.
discharge. In a stream, the volume of water that flows past a given point in a unit of time, often expressed in cubic feet per second.
dissection (of terrain). The weathering and erosion of a level or continuous surface into stream valleys and gullies.
divergent plate boundary. A boundary between two plates that are moving away from each other.
dolomite. A calcium magnesium carbonate mineral, formed by alteration of calcium carbonate (limestone) by magnesium-rich solutions.
dolostone. A sedimentary rock composed primarily of calcium magnesium carbonate.
dome. A circular or elliptical uplift of rock, sometimes in the form of an anticline atop a mass of igneous rock. In the Piedmont, *domes* is a common term for the dozen or so large nappes in the province.
drainage basin. The land area drained by a stream and all of its tributaries.
drainage divide. A line of high elevations that outlines a drainage basin and demarcates the runoff flow into that basin from the flow into neighboring basins.
dune. A mound or ridge of windblown sand.
erosion. Movement or transport of weathered material. Agents include water, ice, wind, and gravity.

escarpment/scarp. A steep face that terminates relatively high lands along an edge. On Delmarva the escarpments mark wave-cut banks that were bluffs of ancient, high-standing seas.

estuary. Lower courses of rivers that have been backflooded into bays by sea levels that rose after the ending of the last Ice Age. Chesapeake Bay is the largest estuary in the United States.

extrusive igneous rocks. Rocks that solidify from magma on the surface of the earth.

falls. Name given to some streams and rivers that flow down across the eastward sloping Piedmont and the Fall Zone, such as Gunpowder Falls and Jones Falls.

Fall Zone/Fall Line. A line connecting locations on Piedmont rivers where navigators encountered the first major rapids—the boundary of Piedmont metamorphic rock and unconsolidated Coastal Plain. At this colonial limit of navigation were major ports: Washington, Baltimore, and Wilmington.

fault. A fracture or zone of fractures in the earth's crust along which blocks of rock on either side have shifted. A **normal fault** forms under extensional forces, and one side drops relative to the other side. A **reverse fault** forms under compressional forces, and one side is pushed up and over the other side. In a **strike-slip fault**, rock on one side moves sideways relative to rock on the other side.

feldspar. The most abundant rock-forming mineral group, making up 60 percent of the earth's crust and including calcium, sodium, or potassium with aluminum silicate. Includes plagioclase feldspars (albite and anorthite) and alkali feldspars (orthoclase and microcline).

felsic. An adjective used to describe an igneous rock composed of light-colored minerals, such as feldspar, quartz, and muscovite.

fine-grained. A relative term used to describe the size of constituents in a rock. Said of igneous rocks with minerals too small to see with the unaided eye. Said of sedimentary rocks with silt-size or smaller particles.

fissile (shale). Bedding that has thin layers less than 2 mm in thickness. Fissility is the tendency of some shales to break into thin flakes.

floodplain. The portion of a river valley adjacent to the river that is built of sediments deposited when the river overflows its banks during flooding.

fluvial. Refers to rivers or streams. For example, fluvial deposits are laid down by streams and rivers.

flux. A compound that helps fuse or separate metals; in an iron furnace, limestone (calcium carbonate) combines with impurities in ore to allow elemental iron to separate.

fold. A bend in a rock layer. Though hard and brittle today, the rock was once buried, under great pressure and at high temperatures, which made it plastic or bendable under tectonic forces of compression.

foliation. A textural term referring to planar, sheetlike arrangement of minerals or structures in metamorphic rock.

fore-arc basin. A depression with undeformed deposits that bury the top of the accretionary wedge adjacent to a convergent-boundary volcanic arc.

formation. A body of sedimentary, igneous, or metamorphic rock that can be recognized over a large area. It is the basic stratigraphic unit in geologic mapping. A formation may be part of a larger group and may be broken into members.

fossils. Traces or impressions of plants or animals preserved in rock. Any remains of once-living matter or its imprint preserved in rock.

freeze and thaw. A process of weathering during which liquid water flows into cracks and pores in rock during days when temperatures are above freezing, but at night the water freezes, expands its volume in doing so, and wedges the rock apart.

gabbro. A dark igneous rock consisting mainly of the minerals plagioclase and pyroxene in crystals large enough to see with a simple magnifier. Gabbro has the same composition as basalt but contains much larger mineral grains.

garnet. A family of silicate minerals with widely varying chemical compositions. Garnets occur in metamorphic and igneous rocks and are usually reddish.

gelifluction. Slow soil flow in saturated zones associated with thawing that does not penetrate the underground permafrost layer.
glacial outwash. Sediment deposited by the large quantities of meltwater emerging from the terminus of a glacier.
glacier. A large and long-lasting mass of ice on land, originating from the compacting of snow.
global cooling/global warming. Climatic changes during earth history that have occurred in cycles of various lengths—from tens of thousands to hundreds of years—during which average global temperature warms or cools by a few degrees or tenths of degrees Celsius. Today's trend is global warming, perhaps a natural cycle accelerated by human-generated greenhouse gases
gneiss. A coarse-grained metamorphic rock with a streaky foliation due to parallel alignment of minerals, usually in bands of light- and dark-colored minerals.
gorge. A steep-sided bedrock valley with a narrow bottom containing a stream or river.
graded bed. A sedimentary bed in which particle size progressively changes, usually from coarse at the base to fine at the top.
gradient. Similar to slope, the degree of inclination, or the rate of ascent or descent, as in stream gradient, measured in number of vertical feet in drop per mile of travel.
granite. An igneous rock composed mostly of the minerals orthoclase feldspar and quartz in grains large enough to see without using a magnifier. Most granites also contain mica or amphibole.
gravel. Rounded rock particles larger than 0.5 inch in diameter.
graywacke. A dark-colored sandstone with angular quartz and feldspar grains in a matrix of clay.
greensand. Bluish green sand deposit originating in shallow-shelf, marine environment and containing the mineral glauconite, often formed as pellets.
greenstone. A dark green, altered or metamorphosed basalt or gabbro. The green comes from the minerals chlorite, actinolite, or epidote.
groin. A structure or wall of boulders built out into the water (most notably, the ocean) at right angles to the shoreline to catch drifting sand and control beach erosion.
groundwater. Subsurface water occupying the saturated zone, contained in fractures and pores of rock and soil.
half graben. In normal block faulting, a downdropped block of crust that has tilted and slid downward on only one side in a vertical to nearly fault.
hanging wall. The mass of rock above a fault plane or above a coal seam in a mine.
headland. Any projection of the land into the sea, such as Cape Henlopen, Delaware.
hematite. An iron oxide; the principal ore of iron.
Iapetus Ocean. The ocean that existed in the general position of the Atlantic Ocean before the assembly of the continental masses that made up the Pangean supercontinent in Paleozoic time.
Ice Ages (Pleistocene). A span of geologic time, the last 1 to 3 million years, during which periods of extensive continental glaciation alternated with warmer interglacial periods of glacial retreat.
ice wedging. The physical wedging apart of rock by the expansion that occurs when water within the rock freezes. See **freeze and thaw**.
igneous rock. Rock that solidified from the cooling of molten magma.
impermeable. Having a texture that does not permit water to move through. Clay is often considered a relatively impermeable sediment.
incised meanders. Meanders that retain their S-shaped course as they cut downward, the land uplifts, or both—resulting in the stream lying below the landscape in a kind of gorge.
infiltration. The flow of water into the pores of soil or into the cracks and pores of rock, or downward movement of precipitation into soil and rock.
interbedding. Alternately layered as interbedded sandstone and shale.
interglacial period. A period of global warming between Ice Ages when glaciers retreat and sea level rises. Some scientists believe that we might now be in an interglacial period.

intrusive igneous rocks. Rocks that cool from magma beneath the surface of the earth. The body of rock is called an intrusion.

overwash. The flooding over of a barrier island by ocean storm waves, often accelerating the landward migration of the sandy barrier island.

isostatic rebound. The rise, or uplift, of a block of crust resting on the plastic mantle—a kind of buoyancy—to achieve equilibrium, similar to how a boat rises in water when unloaded.

jetty. A structure or rock wall built above water level to protect a harbor entrance or a barrier island inlet from sediment deposition and storm waves.

joint. A planar fracture or crack in a rock along which there has been no movement (no faulting).

karst. A rolling, pitted topography characterized by caverns, sinkholes, springs, and disappearing streams—caused by solution weathering of limestone bedrock.

lagoon. A semi-enclosed, quiet body of water between a barrier island and the mainland.

land subsidence. Lowering of the elevation of the land surface due to excessive groundwater withdrawal or to long-term accumulation of sediments.

lava. Molten rock erupted on the surface of the earth.

ledge maker. A highly resistant geologic formation or member, usually sandstone or quartzite, that forms elongate ridges above less weathering-resistant formations, such as limestone or shale.

limb. The side of a fold, or the portion shared by an anticline and the adjacent syncline.

limestone. A sedimentary rock composed of calcium carbonate.

lithification. Compaction and cementation of sediment into sedimentary rock.

lithosphere. The rigid outer shell or crust of the earth, comprised of numerous separate pieces or plates of different sizes.

longshore current. Current in the ocean breaker zone generated by the angular approach toward shore of wind-generated waves.

mafic. An adjective used to describe an igneous rock composed of dark minerals such as hornblende, biotite, and pyroxene.

magma. Molten rock within the earth.

mantle. The part of the earth between the interior core and the outer crust.

marble. Metamorphosed limestone.

marine. Pertaining to the sea.

marl. A calcareous clay.

marsh. A type of wetland that is frequently or continuously underwater; usually the water is shallow and the vegetation is grasses and reeds. Marshes may be freshwater, saltwater, or brackish.

massif. A large block of crust that has been faulted and separated by tectonic divergence from its original position.

massive. Said of a rock that is relatively homogenous; without layering.

matrix. The small grains in a rock that surround certain larger grains.

mélange. In the accretionary wedge associated with a subducting-plate ocean trench, a mass of chaotically mixed, brecciated blocks in a highly sheared matrix.

metamorphic rock. Rock derived from preexisting rock that has changed mineralogically, texturally, or both in response to changes in temperature and/or pressure, usually deep within the earth.

metamorphism. Recrystallization of an existing rock. Metamorphism typically occurs at high temperatures and often high pressures.

metasedimentary rock. The initial classification given in the field to a metamorphosed sedimentary rock.

meta-. A prefix attached to a rock name if the rock has been metamorphosed. For example, **metabasalt, metarhyolite, metagraywacke, metasiltstone.**

mica. A family of silicate minerals, including biotite and muscovite, that crystallize into thin flakes. Micas are common in many kinds of igneous and metamorphic rocks.

mountain building event. An event in which rocks are folded, thrust faulted, metamorphosed, and/or uplifted. Intrusive and extrusive igneous activity often accompanies it.

mouth. The outflow at the lowest end of a stream or river where it discharges into a larger body of water, such as a larger river, a bay, or the ocean.
mudstone. A sedimentary rock composed of mud.
muscovite. A common, colorless to light brown mineral of the mica group. It is present in many igneous, metamorphic, and sedimentary rocks.
nappe. An overturned or recumbent fold that has been moved significantly from its original position by overthrusting. These are commonly called domes in the eastern Piedmont.
nonmarine. Referring to deposits made in freshwater streams, rivers, and lakes.
nor'easter. A strong storm out of the northeast, usually during winter. Nor'easters often cause severe beach erosion and sometimes property damage along coastal Maryland and Delaware.
normal fault. A fault in which rocks on one side move down relative to rocks on the other side in response to extensional forces.
offshore wedge. A thousands-of-feet-thick, seaward-dipping, and seaward-thinning wedge of layered sedimentary deposits that have been eroded from the land for millions of years.
olivine. An iron and magnesium silicate mineral that typically forms glassy green crystals. A common mineral in gabbro, basalt, and peridotite.
orogeny. A major episode of convergent mountain building processes, including folding, faulting, metamorphism, volcanism, and igneous intrusion.
orthoclase. A potassium-rich alkali feldspar and a common rock-forming mineral. It forms at higher temperatures than microcline.
outcrop. A section of rock that is exposed on the land surface.
outwash. Sand and gravel deposited by meltwater from a receding glacier.
overthrust. A low-angle thrust fault with considerable displacement (measured in miles). An overthrust sheet is the body of rock that moved over the top.
overturned fold. A fold in which both limbs dip in the same direction and the axial plane is tilted beyond vertical.
oxidation. In chemistry, a reaction in which the loss of electrons occurs, as when an elemental metal has combined with atmospheric oxygen to form an oxide; many metal ores, such as hematite, are oxides.
Pangea. A supercontinent that assembled through plate convergences about 300 million years ago. It broke into the modern continents through divergences beginning about 200 million years ago.
peat. An unconsolidated deposit of semicoalified plant remains in a bog.
pebble. A rounded rock particle 0.5 to 2.5 inches in diameter.
pegmatite. An igneous rock, generally granitic, composed of extremely large crystals.
pelitic. Composed of clay-sized particles, such as shale or mudstone.
peninsula. An area of land almost completely surrounded by water except for a land bridge, or isthmus, connecting it to the mainland.
percolation. The flow of water through pores in soil or through cracks, cavities, or pores in rock.
permafrost. Ground that remains permanently frozen for many years.
permeability. The capacity of a rock to transmit a fluid, such as water or a liquid contaminant.
phyllite. A metamorphic rock intermediate in grade (and grain size) between slate and schist. Very fine-grained mica typically imparts a lustrous sheen.
pinnacle. A pointed, semi-pyramid-shaped limestone outcrop.
plagioclase. A feldspar mineral rich in sodium and calcium. One of the most common rock-forming minerals in igneous and metamorphic rocks.
plastic. Capable of being bent, molded, or deformed—descriptive of the slow flow within the viscous mantle.
plateau. An extensive elevated area that may be flat-topped or hilly, but that is significantly above surrounding areas, such as the Allegheny Plateau.
plate tectonics. A theory that the earth's surface is comprised of large crustal plates that move slowly and change in size, with intense geological activity at plate boundaries.
pluton. An large intrusion of igneous rock.

porosity. The percentage of a rock's volume that is open space—its capacity to hold water.

pothole. A cylindrical depression in rock eroded by the swirling, abrasive action of sand and gravel within a rapidly flowing stream.

precipitation. The discharge of water in liquid or solid state out of the atmosphere. Also, the precipitation of minerals from solution, which is the change of a compound from a dissolved state into solid form.

province. An area or region with a common geologic history, underlying structure, or composition.

pyroxene. An iron/magnesium-bearing silicate mineral, abundant in basaltic rock.

quartz. A mineral form of silica. Quartz is one of the most abundant and widely distributed minerals in rocks. It comes in a wide variety of forms, including clear crystals, sand grains, and chert.

quartzite. A metamorphic rock composed of mainly quartz and formed by the metamorphism of sandstone.

quartz vein. Veins in metamorphic rock, such as metabasalt, that form from precipitation of silica (quartz) from hot groundwater solutions that have been superheated by a magma body and circulated through cracks.

rain shadow. A relatively dry area on the lee side of a mountain resulting from prevailing winds moving upslope and dropping moisture on the windward side.

recurved spit. A spit having the end more or less strongly curved inward, as at Cape Henlopen, Delaware.

red bed. Sandstone, siltstone, or shale that is predominantly red due to hematite, an iron oxide.

rhyolite. A felsic volcanic rock; the extrusive equivalent of granite. It contains quartz and feldspar in a very fine-grained matrix.

rift. A long, narrow rupture in the earth's crust. A rift basin or rift valley is the trough formed by the rift.

ripple marks. Wavelike ridges formed on sediment surfaces produced by wind or water waves or currents.

riprap. Broken rock, usually boulders, laid or stacked to prevent soil erosion or to protect banks or bluffs from water waves or currents.

river terrace. A flat, steplike, remnant floodplain bordering a river that has been downcutting. These terraces may lie 50 to 100 feet or more above the current river and can be floored with gravels.

rock cycle. A concept that all rock is constantly recycled through weathering, erosion, lithification, and tectonic processes—and that it has been throughout geologic history.

runoff. Rainwater or meltwater running off the land surface and into streams, as opposed to that which soaks into pores of soil and rock.

salinity. A measure of the percentage of dissolved salts in a body of water. In Chesapeake Bay, for example, salinity varies with the amount of discharge of freshwater rivers into the bay.

salt marsh. A wetland in a cove or bay area that is protected from the surf. Made saline by the tides, the marsh supports herbaceous plants without woody stems. Also called a **tidal wetland** or **coastal marsh**.

sand. Weathered mineral grains, most commonly quartz, between $\frac{1}{16}$ millimeter and 2 millimeters in diameter.

sandbar. A ridge of sand built up to or near the water surface by river currents or coastal wave action.

sandstone. A sedimentary rock made primarily of sand.

schist. A metamorphic rock that is strongly foliated due to an abundance of platy minerals.

sedimentary rock. A rock formed from the compaction and cementation of loose sediment.

sedimentation/siltation. The filling of a navigable river channel or harbor with large quantities of water-born sediment, especially from muds running off from cleared or cultivated land. Many colonial shipping lanes and harbors were made too shallow by heavy erosion and deposition.

serpentine. A group of minerals derived from alteration of magnesium-rich silicate minerals and producing poor soils, as in the barrens of Soldiers Delight.
shale. A deposit of clay, silt, or mud solidified into more or less solid rock.
shale barren. Areas of western Maryland with sparse vegetation due to their location on lee-slope rain shadows and to the low permeability and porosity of shale and clay.
shear zone. The zone in which deformation occurs when two bodies of rock slide past each other.
silica. The compound **silicon dioxide**. The most common silica-containing mineral is quartz.
silicate. A large group of minerals whose main building block is one silicon atom surrounded by four oxygen atoms.
sill. An igneous intrusion that parallels the planar structure or bedding of the country rock.
silt. Weathered particles larger than clay but smaller than sand (between $\frac{1}{256}$ and $\frac{1}{16}$ millimeter in diameter).
siltstone. A sedimentary rock made primarily of silt.
sinkhole. A funnel-shaped depression in the land surface, generally in a limestone area, resulting from collapse over a solution cavern.
slag. Glassy material separated during the reduction of a metal from its ore, as in the early iron furnaces.
slate. Slightly metamorphosed shale or mudstone that breaks easily along parallel surfaces.
solution cavities/channels. Underground passageways that have been dissolved in limestone by naturally acidic groundwater.
source (of river). The most distant point from the mouth of a river or stream in its drainage basin from which water runs year-round or from which water could flow. Other definitions exist.
spit. A long, narrow, fingerlike ridge of sand extending into the water from the shore.
spring. A place where groundwater flows out of rock onto the land surface.
stone stream. A linear deposit of boulders up to a half mile long and 50 to 500 feet across that runs down the slope of a mountain. Usually headed by an escarpment, stone streams in Maryland were produced by periglacial solifluction during the Pleistocene Ice Ages.
storm surge. Wind-driven oceanic waves of larger than normal amplitude.
strata. Layers of sedimentary rock. A single layer is a stratum.
strike. The direction or trend of a feature, such as a dipping bed, where it intersects the horizontal.
strip/surface mining. Removal of vegetation and overburden of soil and rock to reach a vein of ore or coal.
subduction zone. A long, narrow zone where an oceanic plate descends into the mantle below a continental plate at a collision boundary.
submarine fan. Deposits under the sea below the mouth of a submarine canyon from which a sediment-laden water current discharges.
supercontinent. A clustering of all of the earth's continental masses into one major landmass; this has occurred at least three times in geologic history.
suspended sediment. Sediment in a stream that remains lifted or transported due to the turbulence and energy of the water. When the currents slow, suspended sediment falls to the bottom.
swamp. A wetland in which the dominant vegetation is woody plants. The bottom layer, usually saturated during the growing season, is typically peat.
syncline. In layered rocks, a downfolded trough with the youngest rocks in the center.
synclinorium. A composite synclinal fold of regional proportions with numerous smaller folds.
tailings. Residue left from a mining operation.
talus. An accumulation of rock fragments derived from and resting at the base of a cliff.
tectonics. A branch of geology dealing with the structure and forces within the outer parts of the earth, including continental plate movements.
tectonic window. An erosional hole through a thrust sheet that exposes the rock beneath the fault plane.

terrace. An erosional remnant of a former river valley, standing above the present river.
terrane. An assemblage of rocks that share a common origin and history.
thrust fault. A low-angle reverse fault. An **overthrust fault** is a thrust fault that transports large masses of rock many miles.
thrust sheet/thrust slice. A body of rock above a thrust fault.
tidal inlet. A narrow channel in a barrier island through which tidal currents flow. Ocean City Inlet is a tidal inlet.
tidal reach. The difference between high tide and low tide at a given point.
tide. The periodic rise and fall of ocean level resulting from the gravitational pull of the moon as it revolves around the earth.
tight fold. Wavelike structures with steeply dipping limbs in layered or banded rock.
transform boundary. A boundary where one tectonic plate slides laterally relative to the adjacent one.
trench, oceanic. A narrow, elongate depression that develops where the ocean floor begins its descent into a subduction zone at a convergent plate boundary.
tributary. A relatively small stream flowing into a larger one, adding water to it.
turbidite. A sediment or sedimentary rock deposited by a swift, bottom-flowing, gravity-propelled current laden with suspended sediment.
ultramafic. An adjective used to describe black to dark green rocks that are more mafic than basalt, consisting mainly of iron- and magnesium-rich minerals such as hypersthene, pyroxene, and olivine. They make up the oceanic crust and mantle.
unconformity. A break or gap in the geologic record where one rock unit is overlain by another that is not next in the stratigraphic succession.
unconsolidated. Referring to sediment grains that are loose, separate, or uncemented to one another.
undissected. A terrain that has not been cut by stream valleys.
uplift. Rise in elevation of a land area due to tectonic convergence or to isostatic rebound.
vein. A deposit of minerals that fills a fracture in rock.
volcanic arc. A chain of island volcanoes that formed above an ocean-floor subduction zone.
water gap. A pass in a mountain ridge through which a stream or river flows.
weathering. The physical disintegration and chemical decomposition of rock at the earth's surface.
wetland. An area in which most soil is saturated with water long enough during the growing season that there is no longer sufficient oxygen for land-based plants. Water may come from rain, runoff, groundwater, floods, or coastal waters. Types include marsh, swamp, and bog.
wind gap. A pass in a mountain ridge through which a stream or river flowed during a previous geologic time but whose course has since changed.

• — References — •

Many of the resources from governmental agencies are available online. Because the Internet and URLs are always in flux, their URLs are not included here. Generally, the easiest way to determine their online availability is to enter the title in an Internet search engine. To narrow down the results, you might also include the author's last name or other resource details in your search.

Abbe, C., Jr. 1899. A General Report on the Physiography of Maryland. In *Report on the Meteorology of Maryland*. Baltimore: Johns Hopkins Press.

Abbe, C. Jr. and others. 1900. Allegany County. Maryland Geological Survey. Baltimore: The Johns Hopkins Press.

Abbe, C. Jr. and others. 1902. *Garrett County*. Maryland Geological Survey. Baltimore: The Johns Hopkins Press.

Adkins, L. M. 2000. *50 Hikes in Maryland: Walks, Hikes, and Backpacks from the Allegheny Plateau to the Atlantic Ocean.* Woodstock, Vermont: Backcountry Guides.

Amsden, T. W., R. M. Overbeck, and R. O. R. Martin. 1954. *Geology and Water Resources of Garrett County.* Maryland Department of Geology, Mines, and Water Resources Bulletin 13.

Andres, A. S., and C. S. Howard. 2000. *The Cypress Swamp Formation, Delaware.* Delaware Geological Survey Report of Investigations No. 62.

Andres, A. S., and A. D. Klingbeil. 2006. *Thickness and Transmissivity of the Unconfined Aquifer of Eastern Sussex County, Delaware.* Delaware Geological Survey Report of Investigations No. 70.

Andres, A. S., and K. W. Ramsey. 1995. *Geologic Map of the Seaford Area, Delaware.* Delaware Geological Survey. Scale 1:24,000.

Andres, A. S., and K. W. Ramsey. 1996. *Geology of the Seaford Area, Delaware.* Delaware Geological Survey Report of Investigations No. 53.

Arnett, E., R. J. Brugger, and E. C. Papenfuse. 1999. *Maryland: A New Guide to the Old Line State.* Second edition. Baltimore: Johns Hopkins University Press.

Bernstein, L. R. 1980. *Minerals of the Washington, D.C., Area.* Maryland Geological Survey Educational Series No. 5.

Blackmer, G. C. 2004a. 102 Years of Wissahickon: An Introduction to the 69th Field Conference of Pennsylvania Geologists. In *Marginalia—Magmatic Arcs and Continental Margins in Delaware and Southeastern Pennsylvania: Guidebook*. West Chester, Pennsylvania.

Blackmer, G. C. 2004b. *Bedrock Geology of the Coatesville Quadrangle, Chester County, Pennsylvania.* Pennsylvania Geological Survey, 4th Ser., Atlas 189b, CD-ROM.

Blackmer, G. C. 2005. *Geologic Units of a Portion of the Wilmington 30- by 60-Minute Quadrangle.* Pennsylvania Bureau of Topographic and Geologic Survey Open-File Report OFBM 05-01.0.

Brezinski, D. K. 1992. *Lithostatigraphy of the Western Blue Ridge Cover Rocks in Maryland.* Maryland Geological Survey Report of Investigations No. 55.

Brezinski, D. K. 1993. *Geologic Map of the Smithsburg Quadrangle, Washington County, Maryland.* Maryland Geological Survery. Scale 1:24,000.
Brezinski, D. K. 1994. *Geology of the Sideling Hill Roadcut.* Maryland Geological Survey pamphlet.
Busch, R. A. (ed.) 2000. *Laboratory Manual in Physical Geology.* Fifth edition. Upper Saddle River, New Jersey: Prentice-Hall.
Cardwell, D. H., R. B. Erwin, and H. P. Woodward. 1968. *Geologic Map of West Virginia.* West Virginia Geological and Economic Survey. Scale 1:250,000.
Chesapeake Bay Program. 1994. *Chesapeake Bay: Introduction to an Ecosystem.* Ed. K. Reshetiloff. U.S. Environmental Protection Agency.
Chesapeake Bay Program. 2000. *The Impact of Susquehanna Sediments on the Chesapeake Bay*. Workshop Report.
Clark, W. B., and E. B. Mathews. 1906. *Report on the Physical Features of Maryland.* Baltimore: Johns Hopkins Press.
Cleaves, E. T., J. Edwards, Jr., and J. D. Glaser. 1968. *Geologic Map of Maryland.* Maryland Geological Survey.
Cloos, E. 1947. Oolite Deformation in the South Mountain Fold, Maryland. *Bulletin of the Geological Society of America* 58:849–918.
Cloos, E. 1950. *The Geology of the South Mountain Anticlinorium, Maryland.* Johns Hopkins University Studies in Geology, No. 16, Part 1. Baltimore: Johns Hopkins Press.
Cloos, E. 1958. *Structural Geology of South Mountain and Appalachians in Maryland.* Johns Hopkins University Studies in Geology, No. 17, Nos. 4 & 5. Baltimore: Johns Hopkins Press.
Cloos, E., T. W. Amsden, R. O. R. Martin, et al. 1951. *The Physical Features of Washington County.* Maryland Department of Geology, Mines, and Water Resources.
Cloos, E., G. W. Fisher, C. A. Hopson, and E. T. Cleaves. 1964. *The Geology of Howard and Montgomery Counties.* Maryland Geological Survey.
Conkwright, B. 2000. *MD Atlantic Coast Map 2.* Coastal and Estuarine Geology Program. Maryland Geological Survey.
Conkwright, B. 2008. *The Need for Sand in Ocean City, Maryland.* Coastal and Estuarine Geology Program. Maryland Geological Survey.
Crowley, W. P., J. Reinhardt, and E. T. Cleaves. 1976. *Geologic Map of Baltimore County and City.* Maryland Geological Survey. Scale 1:62,500.
Darling, J. M., and T. H. Slaughter. 1962. *The Water Resources of Allegany and Washington Counties.* Maryland Department of Geology, Mines, and Water Resources.
Delaware Geological Survey. 2007a. *Bringhurst Gabbro: A GeoAdventure in the Delaware Piedmont.*
Delaware Geological Survey. 2007b. *Exploring the Wilmington Complex (The Wilmington Blue Rocks): A GeoAdventure in the Delaware Piedmont.*
Delaware Geological Survey. 2007c. *National Hydrography Dataset. Data MIL.*
Diecchio, R., and R. Gottfried. 2004. Regional Tectonic History of Northern Virginia. In *Geology of the National Capital Region: Field Trip Guidebook*. U.S. Geological Survey Circular 1264.
Dryden, L., and R. M. Overbeck. 1948. *The Physical Features of Charles County.* Maryland Department of Geology, Mines, and Water Resources.
Duigon, M. T. 2001. *Karst Hydrogeology of the Hagerstown Valley, Maryland.* Maryland Geological Survey Report of Investigations No. 73.
Edmunds, W. E., and E. F. Koppe. 1968. *Coal in Pennsylvania.* Pennsylvania Bureau of Topographic and Geologic Survey Educational Series No. 7.
Edwards, J., Jr. 1978. *Geologic Map of Washington County.* Maryland Geological Survey. Scale 1:62,500.
Edwards, J., Jr. 1987. Baltimore Gneiss and the Glenarm Supergroup, Northeast of Baltimore, Maryland. In *Centennial Field Guide Volume 5: Northeastern Section of the Geological Society of America*. Ed. D. C. Roy. Boulder, Colorado: Geological Society of America.
Edwards, J., Jr. 1993. *Geologic Map of Howard County.* Maryland Geological Survey. Scale 1:62,500.
Ehlen, J., and R. C. Whisonant. 2008. Military Geology of Antietam Battlefield, Maryland, USA: Geology, Terrain, and Casualties. *Geology Today* 24(1):20–27.

Ellicott, A. 1792. *Plan of the City of Washington.* Engraved by Thackara and Vallance.
Elliott, W. D., and R. J. Montgomery. 1981. *A Survey of the Macroinvertebrates of the Antietam Creek.* Hagerstown, Maryland: Hagerstown Junior College.
Fauth, J. L. 1977. *Geologic Map of the Catoctin Furnace and Blue Ridge Summit Quadrangles, Maryland.* Maryland Geological Survey. Scale 1:24,000.
Fauth, J. L., and D. K. Brezinski. 1994. *Geologic Map of the Middletown Quadrangle, Frederick and Washington Counties, Maryland.* Maryland Geological Survey. Scale 1:24,000.
Fichter, L. S., and D. J. Poche. 1979. Sedimentary Paleoenvironments and the Paleozoic Evolution of the Appalachian Geosyncline in the Northern Shenandoah Valley. In *Guidebook for Field Trips in Virginia*. Ed. J. D. Exline. Richmond: Commonwealth of Virginia.
Fisher, A. 1993. *Country Walks Near Baltimore.* Third edition. Baltimore: Rambler Books.
Fisher, A. 1996. *Country Walks Near Washington.* Baltimore: Rambler Books.
Fitzpatrick, M. F. 1987. Active and Abandoned Incised Meanders of the Potomac River, South of Little Orleans, Maryland. In *Centennial Field Guide Volume 5: Northeastern Section of the Geological Society of America*. Ed. D. C. Roy. Boulder, Colorado: Geological Society of America.
Frye, K. 1986. *Roadside Geology of Virginia.* Missoula, Montana: Mountain Press Publishing Company.
Fuller, K. B., and P. S. Frank, Jr. 1974. *The Cranesville Pine Swamp.* Natural Resources Institute of the University of Maryland Educational Series No. 105.
Galgano, F. A., Jr. 2008. Shoreline Behavior along the Atlantic Coast of Delaware. *Middle States Geographer* 41:74–81.
Gates, A. E., P. D. Muller, and D. W. Valentino. 1991. Terranes and Tectonics of the Maryland and Southeast Pennsylvania Piedmont. In *Geologic Evolution of the Eastern United States: Field Trip Guidebook NE-SE GSA 1991*. Ed. A. Schultz and E. Compton-Gooding. Martinsville, Virginia: Virginia Museum of Natural History.
Geyer, A. R., and J. P. Wilshusen. 1982. *Engineering Characteristics of the Rocks of Pennsylvania.* Pennsylvania Geological Survey Environmental Geology Report 1.
Glaser, J. D. 1979. *Collecting Fossils in Maryland.* Maryland Geological Survey Educational Series No. 4.
Glaser, J. D. 1987. The Silurian Section at Roundtop Hill near Hancock Maryland. In *Centennial Field Guide Volume 5: Northeastern Section of the Geological Society of America*. Ed. D. C. Roy. Boulder, Colorado: Geological Society of America.
Glaser, J. D. 1994a. *Geologic Map of the Bellegrove Quadrangle, Allegany and Washington Counties, Maryland*. Maryland Geological Survey. Scale 1:24,000.
Glaser, J. D. 1994b. *Geologic Map of the Paw Paw Quadrangle, Allegany County, Maryland.* Maryland Geological Survey. Scale 1:24,000.
Glaser, J. D. 1998. *Geologic Map of Caroline County.* Maryland Geological Survey. Scale 1:62,500.
Glaser, J. D. 2003. *Geologic Map of Prince George's County, Maryland.* Maryland Geological Survey. Scale 1:62,500.
Godfrey, A. E. 1975. *Chemical and Physical Erosion in the South Mountain Anticlinorium, Maryland.* Maryland Geological Survey Information Circular 19.
Groot, J. J., and R. R. Jordan. 1999. *The Pliocene and Quaternary Deposits of Delaware: Palynology, Ages, and Paleoenvironments.* Delaware Geological Survey Report of Investigations No. 58.
Hack, J. T., C. C. Nikiforoff, and R. M. Overbeck. 1950. *Guidebook III: The Coastal Plain Geology of Southern Maryland.* Johns Hopkins University Studies in Geology No. 16, Part 3. Baltimore: Johns Hopkins Press.
Hahn, T. F. 1995. *Towpath Guide to the Chesapeake & Ohio Canal.* Shepherdstown, West Virginia: American Canal and Transportation Center.
Halka, J. P., and J. M. Hill. 1991. Bottom Sediments of the Chesapeake Bay: Physical and Geochemical Characteristics. In *Geologic Evolution of the Eastern United States: Field Trip Guidebook NE-SE GSA 1991*. Ed. A. Schultz and E. Compton-Gooding. Martinsville, Virginia: Virginia Museum of Natural History.

Hanner, C., S. Davis, and J. Brewer. 2006. *Formation and General Geology of the Mid Atlantic Coastal Plain.* U.S. Department of Agriculture, Natural Resources Conservation Service.
Hansen, H. J., and J. Edwards, Jr. 1986. *The Lithology and Distribution of Pre-Cretaceous Basement Rocks Beneath the Maryland Coastal Plain.* Maryland Geological Survey Report of Investigations No. 44.
Hatcher, R. D., Jr., W. A. Thomas, P. A. Geiser, A. W. Snoke, S. Mosher, and D. V. Wiltschko. 1989. Alleghanian Orogen. In *The Appalachian-Ouachita Orogen in the United States.* Ed. R. D. Hatcher, W. A. Thomas, and G. W. Viele. Boulder, Colorado: Geological Society of America.
Hennessee, L., and J. P. Halka. 2004. *Hurricane Isabel and Shore Erosion in Chesapeake Bay, Maryland.* Maryland Geological Survey Coastal and Estuarine Geology Program.
Herzog, M., C. Larsen, and M. McRae. 2002. *Slope Evolution at Calvert Cliffs, Maryland: Measuring the Change from Eroding Bluffs to Stable Slopes.* U.S. Geological Survey Open-File Report 02-332.
Higgins, M. W., and L. B. Conant. 1986. *Geologic Map of Cecil County.* Maryland Geological Survey. Scale 1:62,500.
Higgins, M. W., and L. B. Conant. 1990. *The Geology of Cecil County, Maryland.* Maryland Geological Survey Bulletin 37.
Hoskins, D. M. 1987. The Susquehanna River Water Gaps near Harrisburg, Pennsylvania. In *Centennial Field Guide Volume 5: Northeastern Section of the Geological Society of America*. Ed. D. C. Roy. Boulder, Colorado: Geological Society of America.
Kious, W. J., and R. I. Tilling. 1996. *This Dynamic Earth: The Story of Plate Tectonics.*
Kraft, J. C., and R. L. Caulk. 1972. The Evolution of Lewes Harbor. *Transactions of the Delaware Academy of Science* 2:79–125.
Kraft, J. C., and A. V. Hiller. 1987. Cape Henlopen Spit: A Late Holocene Sand Accretion Complex at the Mouth of Delaware Bay, Delaware. In *Centennial Field Guide Volume 5: Northeastern Section of the Geological Society of America*. Ed. D. C. Roy. Boulder, Colorado: Geological Society of America.
Kraft, J. C., and C. J. John. 1976. *The Geological Structure of the Shorelines of Delaware.* Newark: College of Marine Studies, University of Delaware.
Kunk, M., R. Wintsch, S. Southworth, B. Mulvey, C. Naeser, and N. Naeser. 2004. Multiple Paleozoic Metamorphic Histories, Fabrics, and Faulting in the Westminster and Potomac Terranes, Central Appalachian Piedmont, Northern Virginia and Southern Maryland. In *Geology of the National Capital Region: Field Trip Guidebook*. U.S. Geological Survey Circular 1264.
Larsen, C., and I. Clark. 2003. *So You Want to Stop Bluff Erosion? You'd Better Plan Ahead.* Field Trip, April 12. Assateague Shelf and Shore Workshop. U.S. Geological Survey Open File Report 03-243.
Little, H. P. 1917. *Anne Arundel County.* Maryland Geological Survey. Baltimore: Johns Hopkins Press.
Lutz, L. 2004. Nassawango's Furnace—and Forest—Rising from the Ruins. *Chesapeake Bay Journal*, June.
Mack, F. K. 1966. *Ground Water in Prince Georges County.* Maryland Geological Survey Bulletin 29.
Maryland Department of Natural Resources. 2007. *DNR Answers Questions about Sea Level Rise in Response to IPCC Report.*
Maryland Geological Survey. 1929. *Baltimore County.* Baltimore: Johns Hopkins Press.
Maryland Geological Survey Coastal and Estuarine Geology Program. 2001. *Shoreline Change Map Data. Honga quadrangle.*
Mather, L. B., Jr. 1937. *A Report on the Geology of the Patapsco State Park of Maryland.* Baltimore: Natural History Society of Maryland Proceeding No. 5.
McCartan, L. 1989. *Geologic Map of St. Mary's County.* Maryland Geological Survey. Scale 1:62,500.
Means, J. 1995. *Maryland's Catoctin Mountain Parks.* Blacksburg, Virginia: McDonald & Woodward Publishing Company.
Meyer, G., and R. M. Beall. 1958. *The Water Resources of Carroll and Frederick Counties.* Maryland Geological Survey Bulletin 22.

Middlekauff, B. D. 1987. *Relict Periglacial Morphosequences in the Northern Blue Ridge.* Ph.D. Dissertation, Michigan State University, Lansing.

Miller, B. L., H. W. Bennett, W. E. Tharp, et al. 1926a. *Queen Anne's County.* Maryland Geological Survey. Baltimore: Johns Hopkins Press.

Miller, B. L., H. W. Bennett, W. E. Tharp, et al. 1926b. *Talbot County.* Maryland Geological Survey. Baltimore: Johns Hopkins Press.

Miller, B. L., J. A. Bonsteel, R. Nunn, N. C. Grover, L. A. Bauer, and F. W. Beasley. 1926. *Kent County.* Maryland Geological Survey. Baltimore: Johns Hopkins Press.

National Park Service. 2008. *Chesapeake & Ohio Canal National Historic Park Map.*

Neal, W. J., O. H. Pilkey, and J. T. Kelley. 2007. *Atlantic Coast Beaches: A Guide to Ripples, Dunes, and Other Natural Features of the Seashore.* Missoula, Montana: Mountain Press Publishing Company.

Nutter, L. J. 1973. *Hydrogeology of the Carbonate Rocks, Frederick and Hagerstown Valleys, Maryland.* Maryland Geological Survey Report of Investigations No. 19.

Owens, J. P., and C. S. Denny. 1978. *Geologic Map of Worcester County.* Maryland Geological Survey. Scale 1:62,500.

Owens, J. P., and C. S. Denny. 1979. *Geologic Map of Wicomico County.* Maryland Geological Survey. Scale 1:62,500.

Owens, J. P., and C. S. Denny. 1984. *Geologic Map of Somerset County.* Maryland Geological Survey. Scale 1:62,500.

Owens, J. P., and C. S. Denny. 1986a. *Geologic Map of Dorchester County.* Maryland Geological Survey. Scale 1:62,500.

Owens, J. P., and C. S. Denny. 1986b. *Geologic Map of Talbot County.* Maryland Geological Survey. Scale 1:62,500.

Pelton, T. 2005. Blackwater Preserve's Natural Balancing Act. *Baltimore Sun*, October 2.

Pickett, T. E. 1987. Upper Cretaceous and Quaternary Stratigraphy of the Chesapeake and Delaware Canal. In *Centennial Field Guide Volume 5: Northeastern Section of the Geological Society of America*. Ed. D. C. Roy. Boulder, Colorado: Geological Society of America.

Plank, M. O., and W. S. Schenck. 1998. *Delaware Piedmont Geology, Including a Guide to the Rocks of Red Clay Valley.* Delaware Geological Survey Special Publication No. 20.

Plank, M. O., W. S. Schenck, and L. Srogi. 2000. *Bedrock Geology of the Piedmont of Delaware and Adjacent Pennsylvania.* Delaware Geological Survey Report of Investigations No. 59.

Poag, C. W. 1997. *The Chesapeake Bay Bolide: Modern Consequences of an Ancient Cataclysm.* U.S. Geological Survey.

Powars, D. S., and T. S. Bruce. 1999. *The Effects of the Chesapeake Bay Impact Crater on the Geological Framework and Correlation of Hydrogeologic Units of the Lower York-James Peninsula.* Professional Paper 1612.

Prothero, D. R., and F. Schwab. 1996. *Sedimentary Geology: An Introduction to Sedimentary Rocks and Stratigraphy.* New York: W. H. Freeman and Company.

Ramsey, K. W. 1997. *Geology of the Milford and Mispillion River Quadrangles.* Delaware Geological Survey Report of Investigations No. 55.

Ramsey, K. W. 1998. Post-Storm Observations. In *Summary Report of the Coastal Storms of January 27–29 and February 4–6, 1998, Delaware and Maryland*. Delaware Geological Survey Open File Report No. 40.

Ramsey, K. W. 2003. *Geologic Map of the Lewes and Cape Henlopen Quadrangles, Delaware.* Delaware Geological Survey. Scale 1:24,000.

Ramsey, K. W. 2005. *Geologic Map of New Castle County, Delaware.* Delaware Geological Survey. Scale 1:100,000.

Ramsey, K. W. 2007. *Geologic Map of Kent County, Delaware.* Delaware Geological Survey. Scale 1:100,000.

Ramsey, K. W., and M. J. Reilly. 2002. *The Hurricane of October 21–24, 1878.* Delaware Geological Survey Special Publication No. 22.

Ramsey, K. W., and W. S. Schenck. 1990. *Geologic Map of Southern Delaware.* Delaware Geological Survey. Scale 1:100,000.

Ramsey, K. W., W. S. Schenck, and L. T. Wang. 2000a. *Physiographic Regions of the Delaware Atlantic Coast.* Delaware Geological Survey.

Ramsey, K. W., W. S. Schenck, and L. T. Wang. 2000b. *Selected Geomorphic Features of Delaware.* Delaware Geological Survey.
Ramsey, K. W., and L. T. Wang. 2001. *Historical Coastline Changes of Cape Henlopen, Delaware.* Delaware Geological Survey Special Publication No. 26.
Reinhardt, J. 1974. *Geologic Map of the Frederick Valley, Maryland.* Maryland Geological Survey. Scale 1:62,500.
Ritter, D. F., R. C. Kochel, and J. R. Miller. 1995. *Process Geomorphology.* Third edition. Dubuque, Iowa: Wm. C. Brown Publishers.
Sando, W. J. 1957. *Beekmantown Group (Lower Ordovician) of Maryland.* Geological Society of America Memoir 68. Richmond, Virginia: William Byrd Press.
Schenck, W. S., M. O. Plank, and L. Srogi. 2000. *Bedrock Geologic Map of the Piedmont of Delaware and Adjacent Pennsylvania.* Delaware Geological Survey. Scale 1:36,000.
Schmidt, M. F., Jr. 1993. *Maryland's Geology.* Centreville, Maryland: Tidewater Publishers.
Schwarz, K. A. 1991. Sideling Hill Road Cut and Visitors Center: An Educational Opportunity Combining Outcrop and Classroom. In *Geologic Evolution of the Eastern United States: Field Trip Guidebook NE-SE GSA 1991*. Ed. A. Schultz and E. Compton-Gooding. Martinsville, Virginia: Virginia Museum of Natural History.
Shirk, W. R. 1980. *A Guide to the Geology of Southcentral Pennsylvania.* Chambersburg, Pennsylvania: Robson & Kaye.
Shosteck, R. 1968. *Potomac Trail Book.* Washington, D.C.: Potomac Books.
Southwick, D. L., and J. P. Owens. 1968. *Geologic Map of Harford County.* Maryland Geological Survey.
Southwick, D. L., J. P. Owens, and J. Edwards, Jr. 1969. *The Geology of Harford County, Maryland.* Maryland Geological Survey.
Southworth, C. S. 1996. *The Martic Fault in Maryland and Its Tectonic Setting in the Central Appalachians.* In *Studies in Maryland Geology*. Ed. D. K. Brezinski and J. P. Reger. Maryland Geological Survey Special Publication No. 3.
Southworth, C. S., and D. K. Brezinski. 1996. *Geology of the Harpers Ferry Quadrangle, Virginia, Maryland, and West Virginia.* U.S. Geological Survey Bulletin 2123.
Southworth, C. S., and D. K. Brezinski. 2003. *Geologic Map of the Buckeystown Quadrangle, Frederick and Montgomery Counties, Maryland, and Loudoun County, Virginia.*
Southworth, C. S., D. K. Brezinski, A. A. Drake, Jr., et al. 2002. *Digital Geologic Map and Database of the Frederick 30 x 60 Minute Quadrangle, Maryland, Virginia, and West Virginia.* U.S. Geological Survey Open-File Report 02-437.
Southworth, C. S., D. K. Brezinski, A. A. Drake, Jr., et al. 2007. *Geologic Map of the Frederick 30′ x 60′ Quadrangle, Maryland, Virginia, and West Virginia.* U.S. Geological Survey Scientific Investigations Map 2889. Scale 1:100,000.
Southworth, C. S., D. K. Brezinski, R. C. Orndorff, P. G. Chirico, and K. M. Lagueux. 2001. *Geology of the Chesapeake and Ohio Canal National Historical Park and Potomac River Corridor, District of Columbia, Maryland, West Virginia, and Virginia.* U.S. Geological Survey Open-File Report 01-188.
Southworth, C. S., A. A. Drake, Jr., D. K. Brezinski, et al. 2006. Central Appalachian Piedmont and Blue Ridge Tectonic Transect, Potomac River Corridor. In *Excursions in Geology and History: Field Trips in the Middle Atlantic States*. Geological Society of America Field Guide 8.
Spoljaric, N. 1967. *Pleistocene Channels of New Castle County, Delaware.* Delaware Geological Survey Report of Investigations No. 10.
Spoljaric, N. 1972. *Geology of the Fall Zone in Delaware.* Delaware Geological Survey Report of Investigations No. 19.
Srogi, L. 2007. Contrasting metamorphic histories within the Wissahickon Formation: Evidence for subdivision and tectonic implications Poster presentation presented at the 42nd Annual Meeting (12–14 March 2007) of the Northeastern Geological Society of America, Durham, New Hampshire.
Stanley, S. M. 1993. *Exploring Earth and Life Through Time.* New York: W. H. Freeman and Company.

Stegmaier, H. I. 1976. *Allegany County: A History.* Parsons, West Virginia: McClain Printing Co.
Stose, A. J., and G. W. Stose. 1946. *The Physical Features of Carroll County and Frederick County.* Maryland Department of Geology, Mines, and Water Resources. Baltimore: Johns Hopkins Press.
Strahler, A. H., and A. N. Strahler. 1992. *Modern Physical Geography.* Fourth edition. New York: John Wiley & Sons.
Tenbus, F. J. 2003. *Lithologic Coring in the Lower Anacostia Tidal Watershed, Washington, D.C., July 2002.* U.S. Geological Survey Open-File Report 03-318.
Tiner, R. W., and D. G. Burke. 1995. *Wetlands of Maryland.* Hadley, Massachusetts: U.S. Fish and Wildlife Service Ecological Services.
Titus, J. G., S. P. Leatherman, C. H. Everts, D. L. Kriebel, and R. G. Dean. 1985. *Potential Impacts of Sea Level Rise on the Beach at Ocean City, Maryland.* United States Environmental Protection Agency. EPA 230-10-85-013.
Travers, P. J. 1990. *The Patapsco: Baltimore's River of History.* Centreville, Maryland: Tidewater Publishers.
U.S. Department of the Interior. 1970. *The River and the Rocks: The Geologic Story of Great Falls and the Potomac River Gorge.* Washington, D.C.: U.S. Government Printing Office.
U.S. Environmental Protection Agency. 2008. *Coastal Zones and Sea Level Rise.*
U.S. Geological Survey. 1998. *The Chesapeake Bay: Geologic Product of Rising Sea Level.* Fact Sheet 102-98.
Van der Pluijm, B. A., and S. Marshak. 1997. *Earth Structure: An Introduction to Structural Geology and Tectonics.* Dubuque, Iowa: WCB/McGraw-Hill.
Van Diver, B. B. 1990. *Roadside Geology of Pennsylvania.* Missoula, Montana: Mountain Press Publishing Company.
Vogt, P. R., and R. Eshelman. 1987. Maryland's Cliffs of Calvert: A Fossiliferous Record of Mid-Miocene Inner Shelf and Coastal Environments. In *Centennial Field Guide Volume 5: Northeastern Section of the Geological Society of America*. Ed. D. C. Roy. Boulder, Colorado: Geological Society of America.
Vokes, H. E. 1961. *Geography and Geology of Maryland.* Baltimore: Waverly Press.
Wagner, M. E., L. Srogi, and E. Brick. 1987. Banded Gneiss and Bringhurst Gabbro of the Wilmington Complex in Bringhurst Woods Park, Northern Delaware. In *Centennial Field Guide Volume 5: Northeastern Section of the Geological Society of America*. Ed. D. C. Roy. Boulder, Colorado: Geological Society of America.
Wagner, M. E., L. Srogi, C. G. Wiswall, and J. Alcock. 1991. Taconic Collision in the Delaware-Pennsylvania Piedmont and Implications for Subsequent Geologic History. In *Geologic Evolution of the Eastern United States: Field Trip Guidebook NE-SE GSA 1991*. Ed. A. Schultz and E. Compton-Gooding. Martinsville, Virginia: Virginia Museum of Natural History.
Ward, L., and D. Powars. 2004. Tertiary Lithology and Paleontology, Chesapeake Bay Region. In *Geology of the National Capital Region: Field Trip Guidebook*. U.S. Geological Survey Circular 1264.
Ware, D. M. 1990. *Green Glades and Sooty Gob Piles: The Maryland Coal Region's Industrial and Architectural Past.* Crownsville, Maryland: Maryland Historical & Cultural Publications.
Wells, D. V., E. L. Hennessee, and J. M. Hill. 2003. *Shoreline Erosion as a Source of Sediments and Nutrients, Middle Coastal Bays, Maryland.* Coastal and Estuarine Geology File Report No. 03-07.
Whitaker, J. C. 1955. Geology of Catoctin Mountain, Maryland and Virginia. *Bulletin of the Geological Society of America* 66(4).
Wicander, R., and J. S. Monroe. 2004. *Historical Geology: Evolution of Earth and Life through Time.* Belmont, California: Brooks/Cole-Thomson Learning.
Williams, S. J., K. Dodd, and K. K. Gohn. 1997. *Coasts in Crisis.* U.S. Geological Survey Circular 1075.

• — Index — •

Page numbers in italics refer to photographs.

• — About the Author and Illustrators — •

John Means taught geology, physical science, and English at Hagerstown Community College for more than thirty years before retiring in 2005. He is the author of *Maryland's Catoctin Mountain Parks* and enjoys hiking, camping, and canoeing with his family.

Suzannah Moran, a geography professor at Hagerstown Community College, teamed up with her husband, **Matthew Moran**, a visual arts teacher at Seneca Valley High School, to create the maps.